Impact and Role of Digital Technologies in Adolescent Lives

Shaveta Malik
Terna Engineering College, India

Rohit Bansal
Department of Management Studies, Vaish College of Engineering, Rohtak, India

Amit Kumar Tyagi
Vellore Institute of Technology, Chennai, India

A volume in the Advances in Human and Social Aspects of Technology (AHSAT) Book Series

IGI Global
PUBLISHER of TIMELY KNOWLEDGE

Published in the United States of America by
IGI Global
Information Science Reference (an imprint of IGI Global)
701 E. Chocolate Avenue
Hershey PA, USA 17033
Tel: 717-533-8845
Fax: 717-533-8661
E-mail: cust@igi-global.com
Web site: http://www.igi-global.com

Library of Congress Cataloging-in-Publication Data

Names: Malik, Shaveta, 1987- editor. | Bansal, Rohit, 1982- editor. | Tyagi, Amit Kumar, editor.
Title: Impact and role of digital technologies in adolescent lives / Shaveta Malik, Rohit Bansal, and Amit Kumar Tyagi, editor.
Description: Hershey, PA : Information Science Reference, [2021] | Includes bibliographical references and index. | Summary: "This book provides insight on the impact and role of digital technologies on adolescent lives, and how they navigate developmental tasks and issues utilizing these technologies and how these technologies contribute to addiction, information processing for fake news, hinder or help identity development and formation, and can contribute to aggressive behaviors"-- Provided by publisher.
Identifiers: LCCN 2021018243 (print) | LCCN 2021018244 (ebook) | ISBN 9781799883180 (hardcover) | ISBN 9781799883197 (paperback) | ISBN 9781799883203 (ebook)
Subjects: LCSH: Internet and teenagers. | Information technology--Social aspects. | Digital media--Social aspects. | Mass media and teenagers.
Classification: LCC HQ799.2.I5 I47 2021 (print) | LCC HQ799.2.I5 (ebook) | DDC 302.23/1--dc23
LC record available at https://lccn.loc.gov/2021018243
LC ebook record available at https://lccn.loc.gov/2021018244

This book is published in the IGI Global book series Advances in Human and Social Aspects of Technology (AHSAT) (ISSN: 2328-1316; eISSN: 2328-1324)

British Cataloguing in Publication Data
A Cataloguing in Publication record for this book is available from the British Library.

Advances in Human and Social Aspects of Technology (AHSAT) Book Series

Mehdi Khosrow-Pour, D.B.A.
Information Resources Management Association, USA

ISSN:2328-1316
EISSN:2328-1324

Mission

In recent years, the societal impact of technology has been noted as we become increasingly more connected and are presented with more digital tools and devices. With the popularity of digital devices such as cell phones and tablets, it is crucial to consider the implications of our digital dependence and the presence of technology in our everyday lives.

The **Advances in Human and Social Aspects of Technology (AHSAT) Book Series** seeks to explore the ways in which society and human beings have been affected by technology and how the technological revolution has changed the way we conduct our lives as well as our behavior. The AHSAT book series aims to publish the most cutting-edge research on human behavior and interaction with technology and the ways in which the digital age is changing society.

Coverage

- End-User Computing
- Digital Identity
- ICTs and human empowerment
- Human-Computer Interaction
- Technology and Social Change
- ICTs and social change
- Activism and ICTs
- Technoself
- Public Access to ICTs
- Human Rights and Digitization

IGI Global is currently accepting manuscripts for publication within this series. To submit a proposal for a volume in this series, please contact our Acquisition Editors at Acquisitions@igi-global.com or visit: http://www.igi-global.com/publish/.

The Advances in Human and Social Aspects of Technology (AHSAT) Book Series (ISSN 2328-1316) is published by IGI Global, 701 E. Chocolate Avenue, Hershey, PA 17033-1240, USA, www.igi-global.com. This series is composed of titles available for purchase individually; each title is edited to be contextually exclusive from any other title within the series. For pricing and ordering information please visit http://www.igi-global.com/book-series/advances-human-social-aspects-technology/37145. Postmaster: Send all address changes to above address.

Titles in this Series

For a list of additional titles in this series, please visit: www.igi-global.com/book-series

Technological Influences on Creativity and User Experience
Joshua Fairchild (Creighton University, USA)
Information Science Reference • © 2022 • 305pp • H/C (ISBN: 9781799843542) • US $195.00

Human Factors Issues and the Impact of Technology on Society
Heather Lum (Embry-Riddle Aeronautical University, USA)
Information Science Reference • © 2021 • 333pp • H/C (ISBN: 9781799864530) • US $195.00

Ubiquitous Technologies for Human Development and Knowledge Management
Hakikur Rahman (Institute of Computer Management and Science, Bangladesh)
Information Science Reference • © 2021 • 396pp • H/C (ISBN: 9781799878445) • US $195.00

Technological Breakthroughs and Future Business Opportunities in Education, Health, and Outer Space
Angus Hooke (Australian Institute of Higher Education, Australia)
Business Science Reference • © 2021 • 353pp • H/C (ISBN: 9781799867722) • US $225.00

Machine Law, Ethics, and Morality in the Age of Artificial Intelligence
Steven John Thompson (University of California, Davis, USA & University of Maryland Global Campus, USA)
Engineering Science Reference • © 2021 • 266pp • H/C (ISBN: 9781799848943) • US $295.00

Examining the Socio-Technical Impact of Smart Cities
Fenio Annansingh (York College, City University of New York, USA)
Information Science Reference • © 2021 • 231pp • H/C (ISBN: 9781799853268) • US $190.00

Human-Computer Interaction and Technology Integration in Modern Society
Hakikur Rahman (Ansted University Sustainability Research Institute, Malaysia)
Engineering Science Reference • © 2021 • 347pp • H/C (ISBN: 9781799858492) • US $195.00

Information Technology Applications for Crisis Response and Management
Jon W. Beard (Iowa State University, USA)
Information Science Reference • © 2021 • 294pp • H/C (ISBN: 9781799872108) • US $195.00

ICT Solutions for Improving Smart Communities in Asia
Noor Zaman (Taylor's University, Malaysia) Khalid Rafique (AJK Information Technology Board, Pakistan) and Vasaki Ponnusamy (Universiti Tunku Abdul Rahman, Malaysia)
Engineering Science Reference • © 2021 • 377pp • H/C (ISBN: 9781799871149) • US $235.00

IGI Global
PUBLISHER of TIMELY KNOWLEDGE

701 East Chocolate Avenue, Hershey, PA 17033, USA
Tel: 717-533-8845 x100 • Fax: 717-533-8661
E-Mail: cust@igi-global.com • www.igi-global.com

Table of Contents

Detailed Table of Contents

Detailed Table of Contents

The chapter focuses on the effects of using digital technology amongst Zimbabwean adolescents. The research was necessitated by the fact that adolescents are a vulnerable group still in the self-discovery age. COVID-19 pandemic resulted in the widespread use of digital learning to facilitate learning, communication, and social interaction among adolescents. Focus groups and interviews were used to gather data in primary schools. There is a disparity in technological device use and access to the internet between urban and rural areas dwellers. Ninety percent of students in rural areas did not have access to technological devices, and during the pandemic, they hardly used them for learning purposes but mostly used them for communicating with relatives. In Zimbabwe, digital technology devices have not yet replaced social relationships as most adolescents are restricted to 2-3 hours on their devices. A model for effective use of digital technologies in adolescents can be developed and implemented.

In this pandemic circumstance, the Bangladesh government started distance learning using broadcasting and online classes. This is a comparative study exploring the impact of online classes on adolescent learning during COVID-19 lockdown in Bangladesh. The study conducts a telephonic interview with 10 headteachers of urban and rural government schools through a semi-structured questionnaire. The study uses descriptive and documentary methods for analysis. The study finds the limitations and challenges facing by the teachers and the students to attend online classes. However, online examinations and assignments are more impactful than online classes to engage students and practice lessons but not enough to compensate for the learning gaps. Even parents are not satisfied with these. Many parents admitted their children to madrasah, and many parents want to continue their children's education after reopening the schools.

Due to the COVID-19 pandemic, governments around the world closed all the educational institutions to control the spread of disease, which is creating a direct impact on students, educators, and institutions. The purpose of this study was to analyze the perception of academic stress experienced by students during current online education and coping strategies using emotional intelligence adopted by them. The study aims to conduct a timely assessment of the effects of stress due to COVID-19 pandemic on the mental health of college students. The authors conducted interview surveys with 227 students at a private university in India to understand the effects of online education during pandemic on their mental

its peak, easy access to e-platform and exponential growth of the knowledge available on social media networks has made it intricate to differentiate between right and wrong information, which has caused large effects on the offline society already. A crucial goal in improving the trustworthiness of data in online social networks is to spot fake news so the detection of spam news becomes important. For sentiment mining, the authors specialise in leveraging Facebook, Twitter, and Whatsapp, the most prominent microblogging platforms. They illustrate how to assemble a corpus automatically for sentiment analysis and opinion mining. They create a sentiment classifier using the corpus that can classify between fake, real, and neutral opinions in a document.

Preface

Today digital technology covers the digital information in every form. We live in an information age in which massive amounts of data is being produced to improve our daily lives. The intelligent digital work will incorporate interconnected people, robots, gadgets, content, and services all determined by digital transformation. The role of digital technologies in adolescent lives is significantly increasing across the world. Checking homework assignments online, playing interactive games, chatting with parents and friends on messaging apps, watching dance videos, and asking for information via smart home devices are just a few examples of the daily ways in which digital technologies have become embedded in the lives of adolescents across the world. New and emerging devices and services promise to make their lives easier, as they create new ways of connecting, creating and relaxing. They also promise to support learning at home and school by enabling ready access to information, and new and exciting pathways for adolescents to follow their interests. Yet alongside these conveniences come trade-offs, with implications for privacy, safety, health and wellbeing.

This book aims at providing deeper understanding on the impact and role of digital technologies on adolescent lives and how they navigate developmental tasks and issues utilizing these technologies. We hope that this book will be helpful to global research community to find out research gaps for their research work.

1. Effect of Digital Technology on Adolescent

Digital technology has become more advanced and it transforming human life so rapidly. Not long ago, digital technology was only meant for adults in the working environment. Nowadays, the assimilation of technology has transformed our life into a new digital culture across all ages. The adolescent has to keep up with the current trend in digital technology as they use it as a platform for education, entertainment, social as well as sport. Technology use among adolescents has so many positive sides such as virtual classrooms, widening social circles, and increasing visual reasoning. Apart from that, it also increases creativity, improves technical skills and self-esteem.

2. Effects of Digital Technology on Adolescent: Pros and Cons

Nowadays there is a widespread use of digital technologies such as social media like Facebook, twitter, WhatsApp and mobile phones by adolescents. They are spending an average of 2 to 3 hours a day by texting, gaming and watching irrelevant videos. Due to this excessive use, it may lead to some health issues, it also helps to encourage the adolescents to improve their communication and knowledge. It thus follows that Digital technology has both positive as well as negative effect on the young minds., however there is no denying the fact that negative impact is much more vigorous than its positive repercussions. As far as the harmful effects of, digital technology is concerned, Social media and mobile devices may lead to a Psychological and physical issues. On the positive side, it empowers learning and communication. The need of the hour is to help and nurture our young minds in a way that they learn to focus on the positivity of the digital technology so that they can use it to the optimum and get benefitted in the best possible manner.

3. An Adverse Effect of Social, Gaming, and Entertainment Media on Overall Development of Adolescents

Digital technology in general and social media in particular entails both positive as well as negative impact on the psyche of the adolescents. The current generation of gen- z has grown up in a media-saturated world. However, it is pretty difficult to judge how media exactly influences their lives. It's commonly observed that social media platforms help saving time establishing contact with our acquaintances. The excessive use of it however leads to various negative repercussions. This may include cyber bullying, low self-esteem, gaming addiction, and less time devoted for doing physical activities. Playing indoor and outdoor games makes adolescents physically fit and mentally healthy. There are however various video games on social media gaming platforms that affect their creativity and logical thinking in an adverse manner. Negative Consequences of video game addiction can lead to many problems such as severe body-ache, skin blisters, weakened eye sight and insomnia. Long-term addiction could lead to obesity, weakness or numbness in the hands and even blood clots.

4. An Intersectionality Perspective on Rural Adolescent Access to Digital Technology in South Africa

The recent Covid-19 Pandemic has brought to the fore compounding disparities in the South African education system. The world at large has suddenly been faced with the challenge of blended learning given that Covid-19 has reconstituted the traditional form of education delivery. Much emphasis has been placed on global education, yet due to compounding inequalities stemming from the effects of apartheid, the South African education system has been left further entrenched in the digital divide. It is for this reason that the authors of this chapter illuminate the lack of digital technology, its subsequent effect on adolescents in rural areas and how it impacts on their ability to learn and compete in the global education sphere. The perspective of Intersectionality theory highlighted in this chapter is used to address the inequalities perpetuated by digital technology on adolescents in rural areas.

5. Harnessing Social Media to Improve Educational Performance of Adolescent Freshmen in Universities

Social Media use in different industries and commerce is high. Little exists on harnessing Facebook and YouTube to improve the educational performance of adolescent students. Researchers support that the use of social media has a positive correlation with educational performance. However, 'little' or 'no' research has been done on how social media platforms are customized to improve the academic performance of adolescents. Data was collected using interviews in the qualitative case study. The book

chapter proffers solutions like coming up with a private resource center hosted by university departments. The resource center is solely accessible for accessing research content and making contributions by academics, students, and industry. Further research might include a framework for the coalition of universities in sharing educational content on a global scale and designing local social media platforms.

6. Media and Parental Communication: Effects on Millennial Value Formation

Both popular media and scholarship have attributed adolescents' antisocial values and behaviors in part to media usage. Nevertheless, many scholars posit that media usage can be positive in certain contexts, and that parent-child communication may abate negative media effects. An analysis of the Educational Longitudinal Study data from 2002 to 2006 examines the effects both parent-child communication and various forms of media had on Millennial teenagers' values development during a time of tremendous digital evolution. Results support that not all media was negative in effect on Millennials' values development and that parent-child communication may counteract some negative effects of media.

7. Obesity Levels of Individuals With Intellectual Disabilities: Prediction for Intervention

Individuals with intellectual disabilities (ID) have considerable health inequalities including higher levels of unmet health needs and a shorter life expectancy compared to the general population. The prevalence of obesity, a commonly accepted measure of health inequalities, is higher in people with ID than in the general population and the factors leading to the increased prevalence among people with ID have not been well understood yet. This has become worse during the COVID-19 pandemic due to nationwide full and partial curfews. In this study, based on a dataset that comprises a set of parameters related to eating habits and physical conditions of a number individuals, the use of classification algorithms for predicting obesity levels of individuals with ID is proposed, and a performance analysis is made using well-known performance metrics. The results could be used by researchers and practitioners in this field to choose the best classifier for their mobile application solutions. Opportunities, research challenges and future research directions in this topic are also presented.

8. Complexity of Digital Media Crowning the Mental Health of Adolescents: Impact of Digital Media on Adolescent Mental Health.

Digital media is working as a different planet showing the disparities between the fantasies of what everyone thought about their lives, and the reality of how they are actually living. It is important to have hedonic and eudemonic happiness in the life of an adolescent which contributes to overall well-being and to flourish with achievements, but 75 percent of 12-22 years are on digital media and spend on average two hours a day there, and this issue needs to be addressed. The first section of the chapter deals with the disruptions created with the digital media in order the way adolescents compare their lives with everyone's highlighted on media. The other section targets the direct impact of the same on an adolescent's life and analyses the various recovery measures and stages required and various techniques the parents and peers can use to deal with such situations. The basic purpose of this is to add value in the world of 'Economy of Attention and how to outgrow it without hurting oneself and not turning out micro-moments into macro moments of Digital Media.

9. Stress and Anxiety in the Digital Age: The Dark Side of Digital Technology

While digitalization, undoubtedly has improved the access to information and communication with the world, increased cases of cyberbullying, harassment, cyber abuse and suicide numbers has also surfaced especially among teens. The hype of the online world, the celebrity culture and their high status seldom generate the fear of inadequacy, dissatisfaction and isolation which ultimately aggravate the feeling of anxiety and depression. This chapter will underline the causes of the digital stress and technology related anxiety among adolescents, their deteriorating psychophysical behavior, factors disintegrating their mental fitness and physical well-being and most importantly the answer to the problem. Various small steps can be taken to avoid serious problems such as talking with your family and friends. Some other approaches include cutting off from the negative digital surroundings, limiting the watch time etc. With the growing online world, there is an urgent need to control the exposure one gives to the social world.

10. The Impact of Mindfulness Practices and the Implementation of Technology in Higher Education

Maintaining efficient time management and learning how to balance different facets of life can be difficult for undergraduate students. Students can have difficulty with maintaining their schedules and academic responsibilities, which can disrupt daily living functionality. Mindfulness is an evidence-based practice that can help students establish a self-care routine. Additionally, establishing a mindfulness regimen can help to improve students' academic performance. With the inclusion of technology, mindfulness mobile applications have gained popularity over the past decade and have provided a convenient method for students to engage in guided meditations.

11. Empowering Adolescents Through Instilling Information Security Culture in Zimbabwean Schools

There is an excellent opportunity to ensure that information security culture (ISU) is embedded in in schools children before they are employed in the industry. However, for the majority of time spent in primary and secondary school education, pupils are alienated from supervised use of technology, making it difficult to teach pupils proper use of technology. Thus, there is no deliberate effort to empower and impart ISC to school pupils in Zimbabwe. The purpose of this study is to develop a framework for instilling Information security culture in secondary school pupils. Based on the literature, the first version of the framework was developed and subjected to a focus group for review. Data from this focus group was analysed, resulting in a second improved version of the framework. Consequently, it was shown that the framework was relevant, useful, and applicable within Zimbabwean settings.

12. Perceptual Difference Between Teens and Their Parents About Internet Usage

This chapter is designed to interpret and understand the difference between parents and their children with reference to their usage of internet and its influence in their daily lives. knowledge in the use of interactive technology. Teenagers between the age of 13 and 16 are the most vulnerable and most prone to cyber bullying, and other crimes. Children's effective use of Internet relies heavily on the factors surrounding them that not only includes their parents but also their peer and school and neighbourhood. Parents believe that depriving their children of internet access or tools can stop them from accessing the web or restrict their skills and learning about social media. They do not believe that Internet use effects family relations negatively. The chapter presents study undertaken of parents and their children . This present differences in perception and understanding of internet usage.

13. Reform of Education Services for Adolescents Through Digital Technology: HIV/AIDS Prevention

The problem of HIV/AIDS cases affects the number of human resources in education such as teachers, administrators, principals who are reduced due to illness or death. Increasing cases of HIV/AIDS that spread among adolescents makes the government should pay more attention to the increase in these cases. Empirical facts found during the study that one of the stakeholders in the world of education stated that learning about HIV/AIDS prevention has not been considered an urgent matter. The research is qualitative research using a descriptive approach based on the perspective of public policy. In this case, researchers have developed a basic health promotion system about HIV / AIDS, and have been introduced to sufferers newly infected with HIV / AIDS. The results of studies in the city of Banjarmasin explored policy issues in education for the prevention of HIV/AIDS in the city of Banjarmasin. Study findings to assist policymakers in making decisions regarding HIV/AIDS prevention policies.

14. Effects of Digital Technologies on Zimbabwean Adolescent Lives During the COVID-19 Era: Pros and Cons

The book chapter focuses on the effects of using digital technology amongst Zimbabwean adolescents. The research was necessitated by the fact that adolescents are a vulnerable group still in the self-discovery age. COVID-19 pandemic resulted in the widespread use of digital learning to facilitate learning, communication, and social interaction among adolescents. Focus groups and interviews were used to gather data in primary schools. There is a disparity in technological device use and access to the Internet between urban and rural areas dwellers. Ninety percent of students in rural areas did not have access to technological devices and during the pandemic, they hardly used them for learning purposes but mostly used them for communicating with relatives. In Zimbabwe, digital technology devices have not yet replaced social relationships as most adolescents are restricted to 2-3 hours on their devices A model for effective use of digital technologies in adolescents can be developed and implemented.

15. Impact of COVID-19 on Adolescent Online Learning in Bangladesh: Insights From Government School Teachers

In this pandemic circumstance, the Bangladesh government started distance learning using broadcasting and online classes. This is a comparative study exploring the impact of online classes on adolescent learning during COVID-19 lockdown in Bangladesh. The study conducts a telephonic interview with ten headteachers of urban and rural government schools through a semi-structured questionnaire. The study uses descriptive and documentary methods for analysis. The study finds the limitations and challenges facing by the teachers and the students to attend online classes. However, online examinations and assignments are more impactful than online classes to engage students and practice lessons. But not enough to compensate for the learning gaps. Even parents are not satisfied with these. Many parents admitted their children to madrasah, and many parents want to continue their children's education after reopening the schools.

16. COVID-19: The Effects and Implications on Students Mental Health and Wellbeing

Due to COVID-19 pandemic, the government around the world has closed all the educational institutions to control the spread of disease, which is creating a direct impact on students, educators and institutions. The purpose of this study was to analyze the perception of academic stress experienced by students during current online education and coping strategies using emotional intelligence adopted by them. Our study aims to conduct a timely assessment of the effects of stress due to COVID-19 pandemic on the mental health of college students. We conducted interview surveys with 227 students at a private university in the India to understand the effects of online education during pandemic on their mental health and well-being. The data were analyzed through quantitative and qualitative methods. Of the 227 students, 71% indicated anxiety and stress due to ongoing pandemic. Further, the stressors facing behavioral stress of losing efficiency and effectiveness to perceive their academic studying during this period of COVID-19 pandemic.

17. Web-Based Learning for Adolescents: Innovation and Challenges

This chapter assesses the innovations and challenges of web-based learning for adolescents. The Technology Acceptance Model (TAM) was used to establish the determinants for the challenges and innovations of web-based learning for adolescents. Challenges coined include virtual engagement, adaptability, motivation, technical issues, time management, learning styles, destructions, communication skills, and the digital divide. Web-based learning is an emerging and innovative approach to teaching and learning that meets the evolving needs of adolescent learning. Adolescents should embrace the transformation from traditional brick-and-mortar classroom practices to a web-based environment. Stakeholders must continuously overcome the challenges of web-based learning. Future research can concentrate on developing well-defined web-based course content that supports the holistic learning needs of adolescents and the development of a web-based learning model for adolescents.

18. Artificial Intelligence and Blockchain as Disruptive Technologies in Adolescent Lives

COVID-19 and future pandemics drastically change the way of life globally. Research has predominantly focused on the use and integration of disruptive technologies in industry and commerce. Little of the recent studies focused on the implementation of artificial intelligence and blockchain technologies in educational applications. The book chapter focuses on how these can be implemented, from development, deployment, use, and maintenance of applications. A computer program's lifespan is usually spent during its use. The qualitative case study was administered using a digital learning platform that provides interactive learning for primary and secondary learners. The disruptive technologies inform new teaching methodologies and the development of student-centered algorithms for learning. Further research includes privacy issues in the implementation of disruptive technologies and data sharing governance issues and evaluating the effectiveness of artificial intelligence and blockchain-based learning platforms.

19. Fake News Polarization for Sentimental Analysis

The popularity of the internet has increased the use of e-commerce websites and news channels. Currently, fake News has been around for many years, and with the arrival of social media and modern-day news at its peak. easy access to e-platform and exponential growth of the knowledge available on social media networks has made it intricate to differentiate between right and wrong information. which has caused large effects on the offline society already. a crucial goal in improving the trustworthiness of data in online social networks is to spot fake news timely So the detection of spam news becomes important nowadays. For sentiment mining, we specialise in leveraging Facebook, Twitter, and Whatsapp, the most prominent microblogging platforms. We illustrate how to assemble a corpus automatically for sentiment analysis and opinion mining. We create a sentiment classifier using the corpus that can classify between fake, real, and neutral opinions in a document.

Shaveta Malik
Terna Engineering College, India

Rohit Bansal
Department of Management Studies, Vaish College of Engineering, Rohtak, India

Amit Kumar Tyagi
Vellore Institute of Technology, Chennai, India

Acknowledgment

First of all, we would to extend our gratitude to our Family Members, Friends, and Supervisors, which stood with us as an advisor in completing this book. Also, we would like to thanks our almighty “God” who makes us to write this book. We also thank IGI Global (who has provided their continuous support during this COVID 19 Pandemic) and our colleagues with whom we have work together inside the college/ university and others outside of the college/ university who have provided their support.

Also, we would like to thank our Respected Madam, Prof. G Aghila, and our Respected Sir Prof. N Sreenath for giving their valuable inputs and helping us in completing this book.

Shaveta Malik

Rohit Bansal

Amit Kumar Tyagi

Chapter 1
Effect of Digital Technology on Adolescents

Nurul Naimah Rose
Universiti Malaysia Perlis, Malaysia

Nor Hafizan Habib Sultan
Universiti Malaysia Perlis, Malaysia

Aida Shakila Ishak
Universiti Malaysia Perlis, Malaysia

Fauziah Ismail
Universiti Malaysia Perlis, Malaysia

Adi Fahrudin
https://orcid.org/0000-0001-7140-6006
University of Muhammadiyah, Jakarta, Indonesia

ABSTRACT

Digital technology has become more advanced, and it is transforming human life rapidly. Not long ago, digital technology was only meant for adults in the working environment. Nowadays, the assimilation of technology has transformed our lives into a new digital culture across all ages. The adolescent has to keep up with the current trend in digital technology as they use it as a platform for education, entertainment, social, as well as sport. Technology use among adolescents has so many positive sides such as virtual classrooms, widening social circles, and increasing visual reasoning. Apart from that, it also increases creativity and improves technical skills and self-esteem.

INTRODUCTION

Recent development in digital technology has heightened the need for our society to carefully examine the effects among adolescents. Over a year, the world experiences a great change in the way of living due to the outbreak of covid-19 pandemic. At the early outbreak, people around the world were forced to undergone a movement control order and practise social distancing to minimize the spread of the virus infection. Thus, everyone needs to change their normal way of doing things by utilizing the digital technology in almost all aspects of lives. Digital technology help them to go through their lives such as

DOI: 10.4018/978-1-7998-8318-0.ch001

working and schooling remotely from home, buying food, household necessities, getting medications supply and also as a medium of communication. Digital media technology has been a useful medium as it provides many advantages to users especially adolescents.

The process of growth and development among adolescent may be influences by internal and external factors and both are interrelated. As child growing up and enter the adolescence, they will quickly absorb everything around them and later on will shape them as an adult individual. Affective Neuroscience Theory (ANT) suggested that social emotions and learning strongly influenced by social and cultural experience (Immordino Yang, M. H., & Singh, V., 2011). Digital technology affected adolescence growth and development process in the sense that it regulates the emotion control. By right, digital technology such as computers and smart phone does not have any emotions but it contains data and algorithms provided to them. The users somehow put the emotion and later establishing close engagement to the technology (Liberati, 2019). Otherwise, if users fail to establish connections, technology becomes meaningless to them. In the case of adolescent, letting the digital technology as a sources of social and emotional learning may resulted to adolescent involves in prosocial or anti-social behaviour depending on how they process the digital media content from any websites, social media sites or gaming site (Erreygers. et al, 2017).

This chapter will discuss about the effect of digital technology on adolescent, including the current trend and types of digital technology mostly used among adolescent. The chapter also touches on the issue of overusing the internet during pandemic covid-19 among the adolescent. It can be seen that even though the digital technology has helped adolescent in many ways to get through the pandemic time but it also carries other risks such as behavioural and psychological problems. As the pandemic is still on going, the current situation is at alarming and in need of certain control mechanism especially from parents. The chapter has been organised in the following way, the first part deals with the current trend of digital technology and the types of social media and digital use among adolescent. The second part deals with the adolescent development in the digital age and the effects of digital technology among adolescent.

CURRENT TREND IN DIGITAL TECHNOLOGY

Going into the 2020 century, the increasingly popular digital technologies consist of the internet of things (IoT), artificial intelligence (AI), big data, block chain technology and telecommunication networks 5G. These technologies are interconnected with each other and benefit human life in general. The following is a description of some examples of digital technology applied according to the current developments. Internet of things (IoT) defined as dynamic global network infrastructure with self-configuring capabilities based on standards and interoperable communication protocols, physical and virtual things. While IoT also have identities and attributes and using intelligent interfaces and being integrated as information network (Shancang Li & Li Da Xu & Shanshan Zhao, 2015). IoT also known as an inter-connected worldwide network based on sensory, communication, networking, and information processing technologies. Moreover, IoT is able to interact without human intervention. This is a new version of information and communication technology (ICT) (Kranenburg 2013; Marry 2013). Several technologies are applied in IoT include wireless sensor networks (WSNs), barcodes, intelligent sensing, RFID, NFC, low energy wireless communications, cloud computing, and so on (Shancang Li & Li Da Xu & Shanshan Zhao, 2015). The sectors involved in the adoption of IoT technology consist of the healthcare, transportation, banking, automotive industries, education, and others (Ri et al., 2014)

Most of the equipment used in daily life can be operated and monitored using IoT. Many processes are carried out with the help of sensors on IoT. The sensor is transmitted to a specific part and this sensor converts the raw physical data into a digital signal. In this way, one can monitor environmental changes remotely for each region of the world via the internet. The architecture of the system will be based on the context of operations and processes in the scenario occurring at the time. In home automation, each electrical switch box will be connected to a smartphone or via a remote control so that it can be operated remotely (Apri Junaidi, 2015).

In health sector, the IoT provides a platform that allows public-health agencies access to data for monitoring the COVID-19 pandemic. For example, the 'Worldometer' provides a real-time update on the actual number of people known to have COVID-19 worldwide, including daily new cases of the disease, disease distribution by countries and severity of disease (recovered, critical condition or death) (Rita Komalasari, 2020). In addition, the use of IoT technology in the health sector can also help in monitoring the health of patients through wireless sensors attached to the patient's body. Among the things that can be monitored using IoT technology such as patient psychology, blood pressure and heart rate. All these activities can be performed remotely through internet-connected equipment and still be able to maintain the confidentiality of patient data (Apri Junaidi, 2015). Besides that, the use of IoT in the health sector also allows long-distance consultation between doctors and patients to be done through WLAN networks and the internet (Y. Wang, 2011).

Artificial intelligence (AI) is a program used to see the ability of machines to learn human behaviour and provide responses or reactions like humans (Shabbir & Anwer, 2015). The program is incorporated into software, machines, or robots so that it can act like what humans do and think in accordance with what humans want (Aliff nawi, n.d). AI technology is a part of computer science that makes it easier for humans to work. In general, AI is a science used to mimic the intelligence possessed by living beings and applied through machines with the aim of solving problems. Teahan (2010) defined that AI is the science of how to build a computer system that shows intelligence in various ways. Several applications used through this AI technology is the natural language processing. Through this application, Artificial intelligence can learn human-spoken language all day long so that humans can communicate with computers. Next is an expert system application, this application refers to a computer that can store all information and can imitate the expertise possessed by humans. Another application is speech recognition which is a technology that can read the human voice into a command and can be understood by computers (Dardya Putra Hastungkara, Endah Triastuti & Anglo-Saxon, 2019).

Most AI today is designed not only to prove the intelligence of AI, but also to be able to match humans and be able to function much better. The google company has produced many AI programs. One example of an AI program run by Google is called AlphaGo. AlphaGo has managed to beat the world champions in the Go game. The silicon fibres used in AI as compared to the carbon fibres found in the human brain have given AI the advantage to act fast, not forgetful and have unlimited energy. Google translate is an example of AI technology that helps people convert to the right language. In addition, AI technology has successfully prompted Google to produce a variety of publicly usable programs such as Waymo (driverless vehicle) and RankBrain (AI logarithms that help generate the most accurate information for use by users searching through Google).

Big data is defined as a collection of data that is very large (volume), very fast changing (velocity), comes in various forms / formats (variety), has a certain value (value) and comes from the right source (veracity) (Hapnes Toba, 2015). Big data is a term for a set of data that is so large or complex that it can no longer be handled by conventional computer technology systems (Hurwitz, et al., 2013). In fact, big data is the idea of a system that can gather information about different things and look for useful features in those edicts (Irhamni Ali. 2015). In general, big data is the result of the process of moving an event which is then used as information as needed, so that it can be used as material for making decisions (policies). Big data obtained may come from various forms such as transactions, devices, and web pages that produce data such as data on weather, finance, political movements, ATMs, food, number of visitors, gender, preferences, hobbies, etc. Big data is different from ordinary data because big data has a factor of 3V (volume, velocity, variety). The size of big data itself cannot be determined, but processing big data requires a lot of effort in terms of analyser (data analyst), software, and hardware. 5G is the fifth-generation wireless connection technology. This technology is a continuation of the previous generation from 1G to 4G. However, 5G technology has different advantages compared to previous technologies. In the age of 1G and 2G technology, connections were made for the purpose of delivering voice services, followed by 3G for mobile data connections when 4G technologies was developed for mobile internet. 5G technology also has an effort to change the function of transferable technology to society.

The concept of digital technology refers to a wide range of devices, services, and applications, whereas among adolescent digital technology refers to the heavy use of social media application (Dienlin, T., & Johannes, N., 2020). Smart phones devices are currently used by the majority of adolescent and it plays a significant part in their lives, since they provide the features and functionalities of a number of different media. Nowadays, social media has changed the way adolescent learn to socialize and shaping their identity (Khusnnul and Helmi, 2017). Social media from any website or software allows them to chat, share knowledge, videos, photos, socialize, receive information and disseminate information interactively (Boone, 2011; Wright and Ross, 1997; Jain, Gupta and Anand, 2012; Suriati et al. 2012). The next part will elaborate on the various types of media and digital platform commonly used by adolescent.

TYPES OF SOCIAL MEDIA AND DIGITAL PLATFORM USED AMONG ADOLESCENT

The use of digital technology in our society has become handy and helpful in managing our daily live more efficiently. People nowadays not only using digital technology for the purpose of working but they also applied it at home and in handling personal business. So as among the adolescents, they had benefited from the digital technology in various aspects of such as in education, social relationship, character building and entertainment. Having one smart phone device is seen as a major requirement for adolescents. Apart from that, they need to master the skill of using the information technology especially social media applications in teaching and learning (Noorhadi and Zurinah, 2017). The following discussion is about the types of digital technology and social media applications most commonly used among the adolescent around Asian countries.

A study saw Malaysia ranked third after the Philippines and Hong Kong in the case of social media addiction (Mohd Azul et al, 2019). This is because the average Z generation adolescent suffers from lack of socialization because they tend to play virtually with their friends. Adolescents no longer tend to play physically with their peers. In fact, they are more interested in surfing the internet and playing smartphones on their own. This attitude can generate a rather aggressive and dominant introvert personality. According to Shela and Shafie (2020), the social media network site Facebook is the application that occupies the highest chart and is frequently visited by the Malaysian community, followed by twitter which gathered a relatively large number of followers on social media sites. In addition, in the context of daily interaction, the majority of people in Malaysia use WhatsApp, Facebook, TikTok, Instagram, Zoom, and Hangout Meet. Facebook and WhatsApp applications have become the choice that dominates 80% to 90% of social media usage among adolescents as these applications are able to speed up communication relationships and be able to reach other users at a time (Subramanian, 2017). The results of Tariq and Mattoo (2013) study have classified social media into four categories, namely online networks and ecosystems such as Facebook, LinkedIn, Instagram and Twitter. The second category of online publications such as YouTube, flickers, RSS and slide share is a medium to share information and reference materials more easily. The third category is social media which is an online collaboration platform such as mediawiki and blogs. While the fourth category is the online-based feedback system aims to interact, submit reviews, related to surveys and others.

Facebook is a virtual social networking site created by Mark Zuckerberg and friends in 2004 has allowed one to stay in touch with each other, share information more easily and increase business marketing opportunities. In the year 2019, 79% of Malaysians regularly use Facebook as the main medium of social media (Mohd Azul. et al, 2019). In addition, in 2019 also showed the tendency of Malaysians in other applications, especially Twitter, YouTube and Instagram. In fact, Malaysians are seen to be able to spend time every 10 minutes just to check on their news feed on social media. Facebook is the most popular social media among adolescents and society (Boyd and Ellison, 2008). A study has proven that adolescents nowadays are more comfortable to interact through the Facebook than face to face (Robin, B., 2008). This is because the use of Facebook allows a person to interact with anyone around the world without being limited by time and geographical distance. Through the Facebook social site, all information related to current issues, education and technology can be accessed quickly and easily. Facebook has become a well-known communication media as a medium to communicate and get information quickly. Therefore, it is not surprising that the number of Facebook users has increased, especially among teenagers. In addition to the old social Facebook used to find contacts, interact with friends, Facebook is also used by adolescents to share information and academic materials especially adolescents who are a student. Through the Facebook page, they can share articles and reference materials related to learning.

In line with the outbreak of the pandemic Covid-19 that hit the whole country in March 2020, the teaching and learning process at the school and university level was conducted using Facebook page as a platform of interaction. More about the use of digital technology among adolescent during pandemic will be discussed in the next part of this chapter. Facebook page makes it easier for students to connect with teachers and lecturers. The Facebook social site becomes a medium for students to access teaching and learning information and references. Reference materials and learning information can also be referenced continuously by using the Facebook page without any time restrictions. According to Rahimi and Zawawi (2005), indirectly it can improve the teaching and learning system with better quality. Adolescents often use the social media application Facebook to carry out various activities. Among them are aimed at learning, entertainment information and social activities such as traveling, interesting places to eat, finding a job and connecting with friends.

WhatsApp is the most popular social media among adolescent nowadays. WhatsApp is a messaging application that assists in the communication process through the highest used mobile communication technology platform in Malaysia (Mohd Azul et al, 2019). According to Akhmarisha and Rusniyati (2017) WhatsApp is derived from the English word "what's up" has become one of the mediums for sending and receiving messages via smartphones by allowing users to exchange pictures, videos, voices or messages in writing. In fact, these social media features provide space for users to channel information more clearly about current issues and become the focus of society. Ali and Siti Minanda (2017) stated that social media applications play a role as a medium to establish relationships between individuals. WhatsApp has features like convenience and privacy compared to other social media to connect with family and friends (Jamiah et al, 2016). Communication that takes place through this application allows a user to chat or tell a story as if there is a real presence among the users who interact.

According to Azul et al (2019), WhatsApp application allows users to create profiles to interact with others, build and maintain relationships, invite users to join a virtual community or group to share information content more easily and quickly. Cetinkaya (2017) stated that WhatsApp has a diverse platform feature and multimedia support that helps interaction among individuals and groups of mobile social networks. The WhatsApp application is also function as a communication platform for any group of people. The formation of groups through this application benefits group members as a medium of discussion related to various topics such as family issues, education, information, safety, community, business, teaching learning and so on. Thus, this application provides a space for continuous social interaction to support the benefits to users.

In October 2010, Instagram was launched as a social networking application. In Malaysia, Instagram is popular among adolescents as they love to upload and share theory own personal photos and videos (Rohana and Faisal, 2018). Instagram works as an alternative that allows its users to store their photos and videos digitally as well as share with other users (Norazlah and Normaliza, 2017). According to Dyah Ayu (2016), Instagram is an application that allows users to share photos, edit using the 20 filters provided and then upload to the Instagram site, to be shared with other users. According to Dellavale (2014), Instagram is a social media channel where each of its users consists of individuals, organizations, sellers and merchants who will upload their photos for the views of followers on the social media. Since its launch in October 2010, Instagram ranks fifth in the top ten list of active social media networks. In the year of 2017, Instagram has about 400 million users per month, 75 million users per day, and 40 billion who share photos and 3.5 billion who sign "like" per day (Norazlah and Normaliza, 2017).

Instagram prioritizes pictures over text. This situation creates a visually oriented state in providing new experiences to store living memory (Lee et al, 2015). A study showed that the use of Instagram can be seen as one of the successful social media in disseminating story-shaped material (Normaliza et al., 2017). This is because the information in Instagram can be read and referenced by users at all times and is easy to understand as the written captions are not long. Photo sharing on Instagram has become a daily routine that need to be done in completing the daily lives of adolescents (Lee et al, 2015).

The findings of the study by Hafiza Abas et al (2017) that 48.8% of generations Y and Z have used Instagram as a social medium to upload their daily lives. According to a study, there are photo categories with various types of photos shared in Instagram (Hu et al, 2014). While the study conducted by Dorsch et al (2017) divided the pictures on Instagram into 10 categories namely (i) friends, (ii) food, (iii) gadgets, (iv) the category of pictures with captions, (v) the category of pets, (vi) activity category, (vii) selfie category, (viii) fashion category, (ix) architecture category and (x) landscape category. According to Rohana and Faisal (2018), the use of photos uploaded in Instagram directly and indirectly can have an impact on users.

Next social media application that also among the most popular used by the adolescent is YouTube. YouTube is a video sharing site that allows users to upload, download, view, comment on video recordings and share video clips including video blogs. According to Norhayati and Suriati Ghazali (2015), YouTube is one of the social media that disseminates information, entertainment and news in the form of video. The history of YouTube was first introduced in Malaysia on March 22, 2012 and the average user watching YouTube recorded a total of 1 million a day (Shela and Shafie, 2020). Although YouTube is not seen as a new phenomenon, but lately the social media is gaining attention from the younger generation. According to Meyerson (2010), a total of 72 hours of video duration were uploaded per minute via YouTube.

According to Norizah et al (2016) apart from getting information, YouTube also serves as a platform that encourages the younger generation to contribute and share their ideas. Therefore, the younger generation uses YouTube as one of the references or tutorials to get information. According to Tan and Carol (2013) YouTube application is also used as a medium of teaching and learning. Furthermore, YouTube application is also a medium that can be linked to website network applications such as blogs, wikis and so on in providing fast information (Mohamed, 2011). Through the study of Norhayati and Suriati Ghazali (2015) the community uses the YouTube medium to change around the world. YouTube has over 78 million users and a total of 150,000 videos are uploaded per day (Sawyer, 2011). Furthermore, YouTube subscribers also do not take into account geographic distance (Thelwall, Sud and Vis, 2012). People across the world are free to join the conversation and give comment in the display section.

Social media platform offers a various advantages for adolescent users and it should be used properly and with caution. Adolescents are also responsible for using social media carefully and not abusing the technological facilities that available for them. Social media should not be a major source for adolescent to socialize. Now, adolescent may rely solely on the social media as a main platform to make friends and establish networking. Furthermore, they also rely on social media to help them solve problem in real life. The following part will continue with the discussion about adolescent development in the digital age. It can be seen that, the digital technology has it impacts on the growth and development among adolescent. Thus, it cannot be taken lightly especially by parents in making sure that their youngsters able to be well developed well despite the challenges of living in the digital technology world.

ADOLESCENT DEVELOPMENT IN THE DIGITAL AGE

Digital technology has become more advanced and it transforming human life so rapidly. Not long ago, digital technology was only meant for adult in working environment. Nowadays, the assimilation of technology has transformed our life to a new digital culture across all ages. A study found that there is a relationship between adolescent well-being and digital technology use (Orben, A., & Przybylski, A. K., 2019). Even the relationship found was relatively small but adolescent still has to keep up with the current trend in digital technology as they use it as a platform for education, entertainment, social relationship as well as sport. Technology use among adolescent has so many positive side such as virtual classroom, widening social circle and increasing visual reasoning. Apart from that, it also increases creativity, improve technical skills and self-esteem. But, the over use of the digital technology may bring more harmful effects to adolescent well-being such as privacy loss, health related issues and changing social norm (Odgers, C. L., & Jensen, M. R., 2020). However, a study examined the technology readiness among adolescent is depends on the acceptance of their parents (Mishra, A., Maheswarappa, S. S., & Colby, C. L., 2018). The acceptance of technology among adolescent is not on the concept-oriented but merely the socio-oriented. It can be suggested that, adolescents generally ready to adapt the digital technology but at the same time they are not really understand the effects of technology to their life. Thus, guidance and restriction from parents is essential to put limit and protect the well-being of their young children.

There are also other potential risks such as cyberbullying, addiction and other psychological effects (Odgers, C. L., & Jensen, M. R., 2020). There is a need to balance the technology use so as to reduce the harmful effects. There is a need to understand that technology is an excellent tool that helps adolescent to learn, grow and develop. Adolescents need to know how to control the technology before technology takes control over their life. Adolescence is a stage where human beings experience changes and development in terms of physical, cognitive, emotional and psychological (Gutgesell & Payne, 2004). These experienced changes pose a great challenge to individuals as very significant changes occur to children who are beginning to enter adolescence. During this period of development, adolescents need to learn and adapt by making choices in terms of self-image and lifestyle that are biologically as well as sociocultural. However, digital technology has making an entrance into the process of learning about the world, to establish their independence and identities, and to socialize with their peers (Giedd, J. N., 2012).The pressure while making these choices creates pressure on teens in determining their lifestyle that will become a habit and routine until adulthood. Each option needs to be scrutinized and observed as it will have a lifelong impact. The period of adolescence is likened to a period of transition where they are expected to behave and think like adults while their behaviour is still like that of children (Giedd, J. N., 2012).

Adolescence age is the most critical time in addressing issues such as self-image, mental health, interest tendencies and talents. The transition process that takes place requires proper adaptation to help teens form a healthy and well-functioning lifestyle. Traditionally, some agents of socialization such as parents, school, family and friends have a very important role in influencing adolescents to form their personalities (Erikson, 1994). However, the socialization process that occurs now has a new influencing factor that is the digital media content such as YouTube (Balleys, C. et al, 2020). The new world of digital creates other factors that have an equally strong influence on the formation of adolescent self -image. Being a content creator and receiver, adolescent establish strong bond with digital community such YouTube (Balleys, C. et al, 2020). In fact, digital media content such as social media site is able to be a major factor in the process of change and self-development of adolescents. In managing the risk of harmful effects of digital technology particularly the internet of thing, adolescent need to know how to manage the risk rather than to avoid its functionalities (Smith, J., Hewitt, B., & Skrbiš, Z., 2015).

Digital media includes video games, streaming, social media, augmented reality and virtual reality used by teens for communication, education, and entertainment. Other than that, some adolescent also involves in online gambling via website, video games, smart phone application and social media site (King, D. L., et al, 2014). These applications and platforms provide advantages and satisfaction through the optimal use of media. Excessive media use has a variety of health risks to adolescent including developmental problems, musculoskeletal problems, physical inactivity, obesity, and inadequate sleep quality (Mustafaoğlu, R., Zirek, E., Yasacı, Z., & Özdinçler, A. R., 2018). Nevertheless, digital media offers adolescent benefits as well as risks at the same time. The risk of excessive media use also involves cybersecurity due to the collection of data and personal information that is difficult to control (Guan, J., & Huck, J., 2012). The information collected can then be analysed and used for targeted and profitable advertising and marketing for some parties.

Excessive use of digital media can have an impact on the well-being and mental health of adolescents (Meates, J., 2020). The use of digital media technology is inseparable in the lives of mostly Z generation. It has become a part of the lives of adolescent and most of them have smartphones and spend a lot of time on the internet. Adolescent uses a wide range of social media platforms such as Instagram and Snapchat, games, and various applications to communicate, do schoolwork, and engage in activities of daily living. Today, digital media has become part of the process of development and growth of adolescents. Healthy activities such as exercise, adequate sleep, and face to face social interactions are no longer daily activities and as a result they are capable of damaging the mental health and well -being of adolescents. Therefore, better and healthier use of media needs to be promoted among adolescents so that digital media can function better and help the development of healthier and healthier adolescents mentally and physically.

THE POSITIVE EFFECTS OF DIGITAL TECHNOLOGY ON ADOLESCENT

The development of technology has changed the conventional society way of life to rely on the devices that have been created. The birth of gadgets such as computers and smartphones has a huge impact on each individual, whether for personal needs or organizational needs, especially for youngsters who are still in school, (Muhammad Sarwar and Tariq Rahim Soomro, 2013). There is no doubt that the advancement in digital technologies has had some positive consequences (Meenakshi Sigh, 2020). A virtual classroom is an internet-based tool for conducting live classroom-style sessions where students and tutors may interact at the same time (Hana, 2020). Students can access coursework and keep updated with the virtual classroom from anywhere at any time. They have the ability to research and complete coursework at any time and from any location that fits their busy schedule. They can also do assignments, send homework, view presentations, participate in student discussions, perform study group, communicate with teachers and peers, seek help from student support services, receive input, and view test results. There are also features such as application sharing, voting, breakout sessions, and quizzes available. It enables both the teacher and the students to take part in live lessons and discussions. Students will engage in breakout sessions, ask questions, and draw on the whiteboard. Digital classrooms can do almost everything that can be done in a physical classroom. Furthermore, the entire classroom session can be recorded (Kim, 2020)

In addition, the student also can do asynchronous discussions with classmates. Asynchronous discussion allows students to read and respond after that. This form of online discussion, as experienced in a college literacy course, creates a text of talk that has the potential to be reflective given the freedom participants have in their response time (Sarah, 2019). Besides that, adolescent can manage their time effectively. An online education provides a welcome environment for students who need to manage time and be more organized (Tracey & Jeanne, 2004). They will be able to save hours every week not having to commute back and forth to classes on a campus (Alexander, M. W., Truell, A. D., & Zhao, J. J., 2012). Practising online distance learning also benefits in sharpens your time-management skills because you have to be disciplined and find the time to study. Since the study mostly takes place at home, students may need to balance the demand from the school work and daily routine. Adolescent may learn to be more responsible as they are studying alone and be more independent and not relying so much on teachers, lecturers and fellow friends (Dumford, A. D., & Miller, A. L., 2018).

The others positive side of digital technology for adolescent is they can get immediate feedback on tests. By switching to the online distance learning, the waiting period for test results and feedback may be shortened as they can get immediate results for the test (Alruwais, N., Wills, G., & Wald, M., 2018). Taking the tests online, and they are usually scored when the test finish. Student can quickly see if they are doing well or they need improvement on certain subject matter. While increasing knowledge and skills in study, student also is improving their visual reasoning in digital skills on the most sophisticated online learning technology. Visual reasoning is the ability to analyse, process and mentally manipulate visual information to solve problems, (Cowan, 2014). Visual reasoning is the key in school settings and is also important in many real-life situations. Graphs, charts, and maps involve a combination of verbal, statistical, and visual interpretation to interpret the data. The impact of digital technology can improve visual reasoning and assist students in problem solving by analysing, processing, and manipulating visual data. Students can solve problems by recalling the relative position of objects and patterns in their heads. Providing children with the ability to improve their spatial skills is important in school and helps them manage day-to-day functions effectively.

Technology is often blamed for causing us to withdraw into our own virtual worlds, but the internet is actually assisting people in forming new friendships (Amichai Hamburger, Y., Kingsbury, M., & Schneider, B. H., 2013). Continue reading to learn how to build a social circle in just a few clicks. Nobody could have expected the internet's rapid development, and the advent of smartphones has ensured that almost everyone now has access to online knowledge at their fingertips, no matter where they are. People should use technology to jump-start a social life when move to a new city where or simply want to expand the existing social circle. Adolescents should know how to use the internet to help meet new people and widen social circle for educational purpose.

THE NEGATIVE EFFECTS OF DIGITAL TECHNOLOGY ON ADOLESCENT

The rapid development of digital media technology has changed the human lifestyle so significantly. Adolescents, in particular, are experiencing significant lifestyle changes as a result of the influence of digital media technology in their lives. The influence of digital media technology on the lives of teenagers begins as early as childhood. A child had been exposed to the use of gadgets which contains digital media and its technology at their early age. Digital media technology then greatly influences the development of children to adolescence especially in the aspects of education, behaviour and personality. On average, children in the US spend eight hours a day on digital media technology, with 65% of them having a television in the bedroom, and 50% of them having a television in the living room (Morowatisharifabad, M. A., Karimi, M., & Ghorbanzadeh, F., 2015). The use of smartphones for the purpose of surfing the internet and social media applications is the beginning of significant changes in the life and social environment of a family. In Malaysia, children spend 19 hours a week surfing the internet and 49.2% of students were addicted to the internet (WS, W. I., Bahar, N., & Mahadevan, R., 2018). Meanwhile, parents in Malaysia think that their children surfing the internet around 11 hours a day. This showed that parents are not aware that their children actually spend more time on the internet. They were doing activities on the internet that do not receive monitoring and control from parents. This situation is very dangerous and is based on the Norton Online Family Report 2010 dated 7 July 2010, which found that nine out of ten children in Malaysia had negative experiences while surfing the internet alone.

The Z generation around the age between 18 and 25 year olds are more at risk for internet addiction especially university and college students. When wireless internet networks are proliferated, the situation worsens. Given that the internet is today's most dominant source of digital technology in the form of the world's largest computer network, harmful topic such as of pornographic websites has become easy access to children and adolescents (D'Alberton, F., & Scardovi, A., 2021). Digital media content on pornographic websites can be viewed not only by adults but also by adolescents and worst by children. The problem of obesity or 'epidemic obesity' among children and adolescents are also discussed by health experts around the world and it is very difficult for them to separate the factors of excessive gadgets and media use especially TV watching as one of the contributors to the problem (Bener, A., et al, 2011). The influx of gadgets has eliminated children's physical activities outside the home, such as sports, in addition to limiting communication between them.

The obsession with surfing the internet in Malaysia that leads to addiction syndrome should be investigated further in the area of pathological psychology, which is a branch of psychology that deals with mental health issues. There has been a rise in heavy internet usage over time, which causes depressive symptoms and social relationship issues (Twenge, J. M., 2020). The effect of increased intense internet browsing on adolescents subjectively impacted the well-being since many adolescent seem to be engaged in the virtual world but are quiet in the real world. This research on the issue of internet addiction syndrome is backed up by a report by Johari and Raja Shahrina (2012) which discovered that adolescents use internet highly at all-time. Their results revealed that in 2010, only two students out of 149 samples were found to be addicted to the internet, but by 2012, that number had risen to 20. Adolescents who spend so much time with technology have less real engagement and their communication skills suffer as a result. This is due to the fact that children and adolescents spend the majority of their time using new technical devices solely on their own initiative. As a result, their communication and social interaction activities with others around them are severely limited, despite the fact that communication skills are critical for acquiring social skills such as watching, listening, and conversing with others at a young age.

Adolescent growth is influenced by media such as the internet, television, magazines, and newspapers. Violent scenes and unethical actions shown on television and in the media make children and teenagers' minds aggressive and hard. They then incorporate the violent actions to everyday interactions, causing problems in their social relationships with friends and family. Azizi, Goh, and Halimah et al. (2014) acknowledged the problem of aggressive media influence, suggesting that mass media has become one of the causes of illegal behaviour among children and adolescents. According to their conclusions, the average mean for the item 'Negative media effect' was 2.80, with a 1.01 standard deviation. This suggests that viewing negative media has negative consequences. Cyberbullying attitudes are also common among adolescent nowadays. A study by Balakrishnan (2015) explored the relationship between internet frequency and cyberbullying among adolescents in Malaysia. Based on the study finding, the frequency of internet engagement significantly predicts cyberbullying activities among adolescents. The study also concluded that those spending 2 to 5 hours on the internet daily reported engage more in cyberbullying as a victim compared to those spending less than an hour on the internet in a day.

Today, adolescents generally spend more hours in front of a computer simply to satisfy their need for self-satisfaction by surfing the internet. It is unfortunate when students who are supposed to use the digital technology for the purpose of learning and education but they engage in activities such as video games, cyberbullying and other unnecessary leisure activities. They will drop out of school once they do not participate, and their academic performance will decrease as a result. Excessive internet activity was found to increase absenteeism symptoms for high school students in a study by Austin and Totaro (2011) on 10,170 students aged 13-17 years in the United States. This means that the unhealthy use of digital media, such as smartphones, is a threat. The excessive use of digital media technology may have a negative social and academic effect on users, especially children and adolescents in school.

Since the beginning of pandemic covid-19 around December 2019, people all around the world shifted all works to online platform. The potential impact of the Covid-19 pandemic had changed the perception of work in terms of occupational status, work from home effectiveness, and occupational mobility (Kramer & Kramer, 2020). Digital technology suddenly became essential to everyone ranging from young children to older people. There is an increase digital technology overuse and excessive screen time among children and adolescents since the school and college had to be closed during pandemic

(Iyer & Sharma, 2020). The problem of overusing the digital technology mostly point at the excessive screen time resulted to psychological problem such as stress, anxiety and depression (Jones, Mitra, & Bhuiyan, 2021). Apart from that, some behaviour problems may occur such as substance abuse, suicide, relationship problems, poor academic performance and absenteeism from work (Meherali, et al, 2021). Some of the factors to excessive screen time including the shift of face to face class to virtual classroom and parental neglects.

According to Montag and Elhai (2020), the outbreak of covid-19 pandemic has forced the authority to implement compulsory lockdown and social distancing which then lead to the closure of day care and school. The situation has led parents to pay less attention to their children as they need to work from home. Parents are preoccupied with work because of fear of losing the job as it will affect the financial securities. Children who are left unguarded will spend more time on screen and using digital technology out of boredom, to play and seeking pleasure. Without parental guardian, children especially adolescent takes control of their own screen time. Adolescent will become preoccupied with digital technology more than before the time of pandemic (Paschke, 2021). Adolescent who experience gaming addiction, compulsive use of internet and social media will resulted having high level of depression, loneliness, escapism, lack of sleep and pandemic related anxiety (Fernandes, 2020). Anywhere around the world, the pandemic outbreak has had significant effect on adolescent overuse of digital technology and psychological well-being.

CONCLUSION

Digital technology has become an important part of everyday life for many adolescents between the ages of 13 and 17 all over the world. This technology is not an alien to them as they are born with it. In fact, they are more comfortable with the use of technology compared to older people. Compared to adult, adolescence use it for multipurpose such as education, entertainment, social interaction and it is a place where they turn to for almost about anything. So it is no surprise that majority of adolescents are heavy user of the digital technology especially social media sites. Despite the negative perceptions on the digital world that had been created on social media, still it offers some benefit such as a place of collaborative learning, meaningful connections, and a support system for those who needed. Digital technology have had impacted the live of adolescent in so many ways, either good or bad depending on how they use the technology. The adaptation of digital technology into our world has begun long time ago and it continues blooming ever since the time of pandemic. Most of the digital technologies are easy to use and adolescent can use it on their own. So, parents need to take the efforts to guard, control and limit the screen time to avoid the risk of behaviour and psychological problems. Therefore, adolescents need more parental supported and direction as they gradually develop their own digital literacy skills. To become a responsible digital citizen, adolescent need to learn how to navigate and control the effects that could either benefits or damages their life in future.

REFERENCES

Abas, H., Yusoff, R. L., & Tukiman, Z. (2017). Apps: Platform Dakwah untuk Generasi Y dan Z. Dalam Hasanah Abd Khafiz, Zainab Ismail (penyunting). Prosiding Seminar Antarabangsa Dakwah ke 3, Hlm. 58-67. Penerbit Jabatan Pengajian Dakwah dan Kepimpinan Universiti Kebangsaan Malaysia: Bangi Selangor

Abd Rahim. (2017). Tahap Penggunaan Instagram dalam kalangan pelajar Universiti Putra Malaysia. *Journal of Business and Social Development, 5*(2).

Al Nuaim. (2020). The Use of Virtual Classrooms in E-learning: a case study in King Abdulaziz University. Academic Press.

Alexander, M. W., Truell, A. D., & Zhao, J. J. (2012). Expected advantages and disadvantages of online learning: Perceptions from college students who have not taken online courses. *Issues in Information Systems, 13*(2), 193–200.

Ali, I. (2015). Big Data: Apa dan pengaruhnya pada perpustakaan? (What is Big Data and its Influence to Library). Perpustakaan Nasional, 22(4).

Alruwais, N., Wills, G., & Wald, M. (2018). Advantages and challenges of using e-assessment. *International Journal of Information and Education Technology (IJIET), 8*(1), 34–37. doi:10.18178/ijiet.2018.8.1.1008

Amichai-Hamburger, Y., Kingsbury, M., & Schneider, B. H. (2013). Friendship: An old concept with a new meaning? *Computers in Human Behavior, 29*(1), 33–39. doi:10.1016/j.chb.2012.05.025

Andonia, M., Robua, V., Flynna, D., Abramb, S., Geachc, D., Jenkinsd, D., McCallumd, P., & Peacock, A. (2019). Blockchain technology in the energy sector: A systematic review of challenges and opportunities. *Renewable & Sustainable Energy Reviews, 100*, 143–174.

Aripin, N. (2016). "YouTube" dan Gneerasi Muda Islam: Satu Pendekatan Kelompok Fokus dalam Kalangan Pelajar Universiti. *Jurnal Komunikasi Malaysian Journal of Communication Jilid, 32*(1), 371–400.

Asha'ari. (2018). Pola Agama dalam penggunaan Instagram di kalangan mahasiswa. *Jurnal Pengajian Islam.*

Austin, W. A., & Totaro, M. W. (2011). Gender differences in the effects of internet usage on high school absenteeism. *Journal of Socio-Economics, 40*(2), 192–198. doi:10.1016/j.socec.2010.12.017

Ayu, D. (2016). Motif dan kepuasan penggunaan Instagram. In *Kesenjangan Antara Motif Dan Kepuasan Penggunaan Media Sosial Instagram pada Mahasiswa Universitas Muhammadiyah Surakarta Angkatan 2013*. Universitas Muhamaddiyah Sukarta.

Balakrishnan, V. (2015). Cyberbullying among young adults in Malaysia: The roles of gender, age and Internet frequency. *Computers in Human Behavior, 46*, 149–157. doi:10.1016/j.chb.2015.01.021

Balleys, C., Millerand, F., Thoër, C., & Duque, N. (2020). Searching for oneself on YouTube: Teenage peer socialization and social recognition processes. *Social Media + Society, 6*(2), 2056305120909474. doi:10.1177/2056305120909474

Bener, A., Al-Mahdi, H. S., Ali, A. I., Al-Nufal, M., Vachhani, P. J., & Tewfik, I. (2011). Obesity and low vision as a result of excessive Internet use and television viewing. *International Journal of Food Sciences and Nutrition, 62*(1), 60–62. doi:10.3109/09637486.2010.495711 PMID:20645888

Boyd, D. M., & Ellison, N. B. (2008). Social network sites: Definition, history, and scholarship. *Journal of Computer-Mediated Communication, 13*(1), 210–230. doi:10.1111/j.1083-6101.2007.00393.x

Cetinkaya, L. (2017). An educational technology tool that developed in the natural flow of life among students: WhatsApp. *International Journal of Progressive Education, 13*(2), 29–47.

Cholissodin & Riyandani. (2016). *Analisis Big Data (Teori & Aplikasi) "Big Data vs Big Information vs Big Knowledge"*. Versi 1.01

Cowan, N. (2014). Working Memory Underpins Cognitive Development, Learning, and Education. *Educational Psychology Review, 26*(2), 197–223. doi:10.100710648-013-9246-y PMID:25346585

D'Alberton, F., & Scardovi, A. (2021). Children Exposed to Pornographic Images on the Internet: General and Specific Aspects in a Psychoanalytic Perspective. *The Psychoanalytic Study of the Child, 74*(1), 131–144. doi:10.1080/00797308.2020.1859277

Daniel, S. W. T., Carin, L., Dzau, V., & Wong, T. Y. (2020). Digital technology and COVID-19. *Nature Medicine, VOL, 26*, 458–464.

Dellavalle, R. R. (2014). Dermatology on Instagram. *Dermatologi Online Journal, 20*(7).

Dienlin, T., & Johannes, N. (2020). The impact of digital technology use on adolescent well-being. *Dialogues in Clinical Neuroscience, 22*(2), 135–142. doi:10.31887/DCNS.2020.22.2/tdienlin PMID:32699513

Dumford, A. D., & Miller, A. L. (2018). Online learning in higher education: Exploring advantages and disadvantages for engagement. *Journal of Computing in Higher Education, 30*(3), 452–465. doi:10.100712528-018-9179-z

Embi. (2011). *Aplikasi Web 2.0 Dalam Pengajaran dan Pembelajaran*. Pusat Pembangunan Akademik Universiti Kebangsaan Malaysia.

Erikson, E. H. (1994). *Identity: Youth and Crisis* (7th ed.). WW Norton & Company.

Erreygers, S., Vandebosch, H., Vranjes, I., Baillien, E., & De Witte, H. (2017). Nice or naughty? The role of emotions and digital media use in explaining adolescents' online prosocial and antisocial behavior. *Media Psychology, 20*(3), 374–400. doi:10.1080/15213269.2016.1200990

Fernandes, B., Biswas, U. N., Mansukhani, R. T., Casarín, A. V., & Essau, C. A. (2020). The impact of COVID-19 lockdown on internet use and escapism in adolescents. *Revista de psicología clínica con niños y adolescentes, 7*(3), 59-65.

Ghazali, S., Mapjabil, J., Nor, A. M., Samat, N., & Jas, L. S. J. (2012). Dufusi Ruangan Budaya Transeksualisme dan Imaginasi Geografi Pelajar Lelaki Berpenampilan Silang di Universiti Tempatan Malaysia. *Journal of Social Sciences and Humanities, 7*(1), 252–266.

Ghazali. (2015). Tindak Balas Pengguna YouTube Terhadap Kes Buli dalam Kalangan Remaja di Malaysia. *Sains Humanika, 6*(1), 9-17.

Giedd, J. N. (2012). The digital revolution and adolescent brain evolution. *The Journal of Adolescent Health, 51*(2), 101–105. doi:10.1016/j.jadohealth.2012.06.002 PMID:22824439

Guan, J., & Huck, J. (2012). Children in the digital age: exploring issues of cybersecurity. In *Proceedings of the 2012 iConference* (pp. 506-507). 10.1145/2132176.2132266

Gutgesell, M. E., & Payne, N. (2004). Issues of adolescent psychological development in the 21st century. *Pediatrics, 25*(3), 79–85. PMID:14993515

Hanafi & Rahim. (2017). Penggunaan media sosial dan faktor-faktor yang mempengaruhi terbentuknya gerakan sosial pemimpin pelajar universiti di Bandar Pekan Baru, Riau, Indonesia. *Journal of Social Sciences and Humanities.*

Hastungkara, D. P., & Triastuti, E. (2019). Application Of E-Learning And Artificial Intelligence In Education Systems In Indonesia. *Anglo-Saxon, 10*(2), 117–133.

Hu, Y., Mankikonda, L., & Kambhamati, S. (2014). What We Instagram: A First Analysis of Instagram Photo Content User Types. *Proceedings of the Eight International.*

Immordino-Yang, M. H., & Singh, V. (2011). Perspectives from social and affective neuroscience on the design of digital learning technologies. In *New perspectives on affect and learning technologies* (pp. 233–241). Springer. doi:10.1007/978-1-4419-9625-1_17

Iyer, M., & Sharma, R. (2020). Impact of internet overuse on psychological well-being of adolescents during covid 19 shutdown. *J Crit Rev, 7*(9), 2832–2839.

Jain, M. R., Gupta, P., & Anand, N. (2012). Impact of Social Networking Sites in the Changing Mindset of Youth on Social Issues – A Study of Delhi-NCR Youth. *Journal of Arts, Science & Commerce, 2*(2), 36–43.

Johari & Shahrina. (2012). Ketagihan Penggunaan Internet Di Kalangan Remaja Sekolah Tingkatan 4 Di Bandaraya Johor Bahru. *Journal of Technical, Vocational & Engineering Education*, 23-43.

Jones, E. A., Mitra, A. K., & Bhuiyan, A. R. (2021). Impact of COVID-19 on mental health in adolescents: A systematic review. *International Journal of Environmental Research and Public Health, 18*(5), 2470.

Junaidi. (2015). Internet Of Things, Sejarah, Teknologi Dan Penerapannya: Review. *Jurnal Ilmiah Teknologi Informasi Terapan, 1*(3).

Keong, T. C., & Abu, C. (2013). Pengaplikasian Video YouTube: Bahan Bantu Mengajar (BBM) Dalam Proses Pengajaran dan Pembelajaran Mata Pelajaran Sains Sosial. Seminar Pendidikan Sejarah dan. Universiti Malaysia Sabah: Sabah.

Kim, J. (2020). Learning and Teaching Online During Covid-19: Experiences of Student Teachers in an Early Childhood Education Practicum. *International Journal of e-Collaboration, 52*, 145–158. https://doi.org/10.1007/s13158-020-00272-6

King, D. L., Delfabbro, P. H., Kaptsis, D., & Zwaans, T. (2014). Adolescent simulated gambling via digital and social media: An emerging problem. *Computers in Human Behavior, 31*, 305–313.

Komalasari, R. (2020). Manfaat Teknologi Informasi Dan Komunikasi Di Masa Pandemi Covid 19. Jurnal Teknologi Informasi Dan Komunikasi, 7(1).

Kramer, A., & Kramer, K. Z. (2020). *The potential impact of the Covid-19 pandemic on occupational status, work from home, and occupational mobility*. Academic Press.

Lee, H. E., Mustaffa, N., & Salman, A. (2015). Faktor-Faktor yang Mempengaruhi Remaja Lembah Klang untuk Terus Membaca dan Membeli Melalui Pengiklanan Facebook. *Jurnal Komunikasi*, *30*(1), 229–250.

Li, S., Da Xu, L., & Zhao, S. (2015). The internet of things: A survey. *Information Systems Frontiers*, *17*(2), 243–259. doi:10.100710796-014-9492-7

Liberati, N. (2019). Emotions and Digital Technologies. *HUMANA MENTE Journal of Philosophical Studies*, *12*(36), 292–309.

Manap (2020). Ancaman Tort Kecerdasan Buatan Terhadap Manusia: Retorik Atau Realiti? (The Threats Of Artificial Intelligence In Tort: Rhetoric Or Reality?). *Dewan Bahasa dan Pustaka, 32*(1).

Meates, J. (2020). Problematic Digital Technology Use of Children and Adolescents: Psychological Impact. *Teachers and Curriculum*, *20*(1), 51–62.

Meherali, S., Punjani, N., Louie-Poon, S., Abdul Rahim, K., Das, J. K., Salam, R. A., & Lassi, Z. S. (2021). Mental health of children and adolescents amidst CoViD-19 and past pandemics: A rapid systematic review. *International Journal of Environmental Research and Public Health*, *18*(7), 3432.

Meyerson, M. (2010). *Success Secrets Of The Social Media Marketing Superstars*. Entrepreneur Media Inc.

Michael, J. W. (2018). Blockchain Technology And Regulatory Investigations. Thomson Reuter. *The Journal Litigation*.

Mishra, A., Maheswarappa, S. S., & Colby, C. L. (2018). Technology readiness of teenagers: A consumer socialization perspective. *Journal of Services Marketing*.

Montag, C., & Elhai, J. D. (2020). Discussing digital technology overuse in children and adolescents during the COVID-19 pandemic and beyond: On the importance of considering Affective Neuroscience Theory. *Addictive Behaviors Reports*, *12*, 100313.

Morowatisharifabad, M. A., Karimi, M., & Ghorbanzadeh, F. (2015). Watching television by kids: How much and why? *Journal of Education and Health Promotion*, 4.

Mustafaoğlu, R., Zirek, E., Yasacı, Z., & Özdinçler, A. R. (2018). The negative effects of digital technology usage on children's development and health. *Addicta: The Turkish Journal on Addictions*, *5*(2), 13-21.

Nawi, A. (2019). Penerokaan Awal Terhadap Isu Dan Impak Penggunaan Teknologi Kecerdasan Buatan Terhadap Kehidupan Manusia. *Asian Journal of Civilizational Studies.*, *1*(4), 2682–8022.

Odgers, C. L., & Jensen, M. R. (2020). Adolescent development and growing divides in the digital age. *Dialogues in Clinical Neuroscience*, *22*(2), 143.

Orben, A., & Przybylski, A. K. (2019). The association between adolescent well-being and digital technology use. *Nature Human Behaviour*, *3*(2), 173–182.

Paschke, K., Austermann, M. I., Simon-Kutscher, K., & Thomasius, R. (2021). Adolescent gaming and social media usage before and during the COVID-19 pandemic. *Sucht: Zeitschrift für Wissenschaft und Praxis*.

Patil, & Patil, & Bhat. (2012). A Review on 5G Technology. *International Journal of Engineering and Innovative Technology*, *1*(1), 2277–3754.

Praveen, Chamola, Hassija, & Kumar. (2020). *Blockchain for 5G: A Prelude to Future Telecommunication*. IEEE Network.

Putrawangsa & Hasanah. (2018). Integrasi Teknologi Digital Dalam Pembelajaran Di Era Industri 4.0. Kajian dari Perspektif Pembelajaran Matematika. *Jurnal Pemikiran dan Penelitian Pendidikan, 16*(1).

Rahardja, U., Harahap, E. P., & Christianto, D. D. (2020). Pengaruh Teknologi Blockchain Terhadap Tingkat Keaslian Ijazah. Technomedia Journal, 4(2).

Rahim, N. A., Affendi, N. R. N. M., & Awang, A. A. P. (2017). Dissemination of Values and Culture through the E-Folklore. *The Turkish Online Journal of Educational Technology*, *16*(1), 32–36.

Robin, B. (2008). The Effective Uses Of Digital Storytelling As A Teaching And Learning Tool. Handbook of Research on Teaching Literacy through the Communicative and Visual Arts, 2.

Saad & Ismail. (2005). Pengajaran dan Pembelajaran Bahasa Arab Berasaskan Internet. In Z. Ismail & M. S. Othman (Eds.), *Pendidikan Bahasa Arab di Malaysia: Cabaran dan Inovasi*. Minda Imtiyaz Publications.

Saari & Mahiyaddin. (2017). Analisis linguistik terhadap penggunaan media sosial dalam kalangan pelajar di IPTS. *Advances in Language and Literary Studies*, *1*(1), 151–165.

Saleha, N. S., & Rosli, M. S. (2020). *Tinjauan Literatur Sistematik: Interaksi Media Sosial Atas Talian dalam Kluster Pendidikan dan Pekerjaan. In Sains Humanika*. Universiti Teknologi Mara.

Salleh. (2019). Mudah Guna dan Manfaat Aplikasi WhatsApp terhadap hubungan antara Peribadi masyarakat dan Polis. *Malaysian Journal of Communication Jilid*, *35*(4), 135–154.

Salman, A., & Pulungan, S. M. (2017). Pendedahan diri, motivasi dan kepuasan penggunaan Facebook dalam menjalinkan persahabatan. *Jurnal Komunikasi: Malaysian Journal of Communication*, *33*(1), 438–459.

Sarah, F. N. (2019). *Understanding Students' Self-Regulation in Asynchronous Online Learning*. University of Minnesota.

Sarwar & Soomro. (2013). Impact of Smartphone's on Society. *European Journal of Scientific Research, 98*(2), 216-226. http://www.europeanjournalofscientificresearch.com

Sawyer, R. (2011). *The Impact of New Social Media on Intercultural Adaptation*. Senior Honors Projects: Honors Program at the University of Rhode Island. http://digitalcommons. uri.edu/cgi/viewcontent.cgi ?article=1230&context=srhonorsprog

Singh. (2020). Impact of Technology in Indian Education. *International Journal of Advance and Innovative Research, 7*(1).

Smith, J., Hewitt, B., & Skrbiš, Z. (2015). Digital socialization: Young people's changing value orientations towards internet use between adolescence and early adulthood. *Information Communication and Society*, *18*(9), 1022–1038. doi:10.1080/1369118X.2015.1007074

Subramanian, K. R. (2017). Influence of social media in interpersonal communication. International. *Journal of Scientific Progress and Research*, *38*(2), 70–75.

Tahir. (2017). Kepentingan Penggunaan Media Sosial Teknologi Maklumat dalam pendidikan IPTA. *Journal of Social Sciences and Humanities*, *12*(3).

Tariq Banday, M., & Mattoo, M. M. (2013). Social Media in e-Governance: A Study with Special Reference to India. *Social Networking*, 47–56.

Toba, H. (2015). *Big Data: Menuju Evolusi Era Informasi Selanjutnya*. https://www.researchgate.net/publication/274712686

Tracey Crosbie & Jeanne Moore. (2004). *Work-Life Balance and Working from Home*. Cambridge University Press.

Twenge, J. M. (2020). Why increases in adolescent depression may be linked to the technological environment. *Current Opinion in Psychology*, *32*, 89–94.

Wright, J. W., & Ross, S. D. (1997). Trial by media? Media reliance, knowledge of crime and perception of criminal defendants. *Communication Law and Policy*, *2*(4), 397–416.

WS, W. I., Bahar, N., & Mahadevan, R. (2018). Internet Addiction among Secondary School Students in Klang Valley, Malaysia: What is the Association with Depressive Symptoms, Anxiety Symptoms, and SelfEsteem? *IIUM Medical Journal Malaysia, 17*(2).

Yaga, Mell, Roby, & Scarfone. (2018). *Blockchain Technology Overview*. National Institute of Standards and Technology Internal Report 8202.

Yahaya, A., Gooh, M.L., & Ma'alip, H. (2014). Pengaruh media berunsur agresif dan kesannya terhadap tingkah laku pelajar di sekolah menengah. *National Action Research Conference 2014, Official Conference Proceeding*.

Yudhanto. (2007). *Apa itu IOT (Internet of Things)?* Komunitas eLearning IlmuKomputer.Com.

Chapter 2
Effects of Digital Technology on Adolescents:
Pros and Cons

Kavita Ajit Saptasagar
SSR College of Arts, Commerce, and Science, Silvassa, India

ABSTRACT

There is a widespread use of digital technologies such as social media like Facebook, Twitter, WhatsApp, and mobile phones by adolescents. They are spending an average of 2-3 hours a day texting, gaming, and watching irrelevant videos. Due to this excessive use, it may lead to some health issues; it also helps to encourage the adolescents to improve their communication and knowledge. It thus follows that digital technology has both positive as well as negative effects on the young minds. However, there is no denying the fact that negative impact is much more vigorous than its positive repercussions. As far as the harmful effects of digital technology is concerned, social media and mobile devices may lead to psychological and physical issues. On the positive side, it empowers learning and communication. The need of the hour is to help and nurture our young minds in a way that they learn to focus on the positivity of the digital technology so that they can use it to the optimum and get benefitted in the best possible manner.

INTRODUCTION

In present day life, digital technologies such as social media platforms viz. Facebook, twitter, WhatsApp and mobile phones are widely used by adolescents. These digital technologies exercise negative impact on them as they spend an average of 2 to 3 hours a day texting, gaming, posting and sharing messages, photos and watching irrelevant videos. These activities may affect their physical and mental health. However, every cloud has a silver lining. some forms of these digital technologies undoubtedly have made positive influence on the youngsters by improving their communication skills, helping them gain knowledge of the recent trends and using E-Learning Digital platforms in their education.

DOI: 10.4018/978-1-7998-8318-0.ch002

NEGATIVE IMPACT OF DIGITAL TECHNOLOGIES ON ADOLESCENTS

The following are the negative effects of using digital technologies

- Psychological Disorders.
- Physical Exertion.
- Violence and Aggression.
- Cyberbullying and sexting.
- Risky sexual behavior

Psychological Disorders

- Isolation
- Depression and anxiety

Isolation

Digital technologies, such as social media platforms like Facebook, Twitter, Snapchat, and Instagram are designed to bring people together, yet they may have the opposite effect in some cases because most of the time they are active on social media platforms for posting and sharing messages, photos, videos and gaming. This behavior may isolate them from the social life. It also does not allow them to spend enough time with their family members.

Depression and Anxiety

Social Media, with the aid of using its nature, actively encourages social comparisons, as its miles plagued by records which could effortlessly be used as metrics of obvious social success (e.g., friends, likes, shares, fans and so forth). These metrics are problematic in themselves, because if adolescents don't get enough likes to a comment or picture they have posted, or if someone has more likes or friends than them, it can make them feel inferior. Furthermore, the disparity between real life and what people actually post on social media means that they tend to only see an extremely edited 'highlight reel' of other people's lives. This effectively gives the misunderstanding that others lead a more interesting / perfect /exciting life than our own, which, actually has its justifiable share of ups, middles and downs for everybody. So, this kind of negative social comparisons may experience higher levels of depression and anxiety in adolescents.

Physical Exertion

- Eyestrain
- Sleep problems
- Poor posture
- Reduced physical activity

Eyestrain

The adolescents can face the eyestrain problems like blurred vision, dry eyes and it may also lead to pains in other areas of the body, such as the head, neck, or shoulders, due to the over use of digital technologies, such as computers, smartphones and handheld tablets.

The following factors may lead to eyestrain, such as:

- screen brightness
- viewing too close or too far away
- poor sitting posture
- underlying vision issues
- screen time
- screen glare

Sleep Problems

Keeping a television, computer, or mobile phone in the bedroom during early childhood is associated with less sleep. Adolescents who make excessive use of social media or who sleep with mobile devices in their bedrooms are at increased risk of experiencing sleep disturbances because of the blue lights from cell phones, computers and it also stimulates their brain. The use of electronic devices during the daytime also can affect sleep quality.

Poor Posture

Most of the time adolescents are using the computers and mobile devices for playing games, and texting that to sitting in one place may contribute to incorrect posture. Over time, this might cause musculoskeletal issues.

Many technologies promote a "down and forward" user position, meaning the person is hunched forward and looking down at the screen. This will put an unnecessary amount of pressure on the neck and spine.

Reduced Physical Activity

Most of the adolescents are playing games in mobiles and computers rather than physical games like cricket, volly ball, kabaddi etc. and most of the time they are online with their devices in social media like Facebook, twitter, Instagram etc. which allegedly discourages them from participating in more traditional physical activities, so, this may lead to a childhood obesity and also obstruct their physical development.

Violence and Aggression

Violence shown on social media platforms does not always depict the truth and successively glorifies and downplays the negative aspects and to some extent may even depict it as humorous, which the adolescents will find difficult to untie from reality. Repeated exposure to media violence can cause the acceptance of violence as an appropriate means of solving conflict and desensitization, which can cause a rise in aggression and reduce in altruism. Most of the web interactive games may contain violence and playing these sorts of games can increase the aggressiveness in adolescents. "A recent analysis of video games by Anderson *et al.* revealed that more than half of all games contain violence, including >90% rated as appropriate for children aged 10 years and above. According to Bushman *et al.*, the relationship between exposure to media violence and real-life aggression is nearly as strong as the relationship between cigarette, smoking and lung cancer" (Pasam, 2019). Also, if the adolescents are not getting enough resources to using digital technologies then they become more aggressive due to the addiction.

Cyberbullying and Sexting

"**Cyberbullying** has been defined as 'deliberately using digital media for false communication, embarrassing or hostile information about another person'" (O'Keefe and Clarke-Pearson, 2011).

Cyberbullying is seen in adolescents while using social networking websites by sending harassing messages, posting humiliating pictures, or threatening/intimidating someone electronically due to this someone can experience negative feelings such as embarrassment, self-harm/suicide and fear and also adolescents can experience these kinds of negative feelings due to cyberbullying. There are so many accounts of suicide in youngsters related to cyberbullying.

Sexting means sending sexual videos, photos or messages through mobile phones or through social media. A major portion of adolescent have admitted to receiving and sending text messages of a sexual nature, such as nude or seminude photos, and provocative videos. It is also one kind of cyberbullying. If adolescents are engaged in these kinds of activities then they are charged under criminal offences.

Suppose you may be dating the person now, but after four months you may not be. Pictures and messages, videos which are shared with them they will still exist and can be forwarded to others.

Sexting and having these images an on your cellphone is dangerous. If your phone or computer gets into the wrong hands, then these photos, videos can be exposed on social media and can damage your reputation.

Keep in mind it may be fun to flirt, but these messages carry with them an expectation and affect the impression others have of you. The best way to keep yourself safe from hackers and wrongdoers sharing your personal and private photographs and conversations is simply not to have them on your phone or in your email correspondence.

Risky Sexual Behavior

Digital technology comes first for providing sexual information to adolescents. Older technology had restricted access to pornography, but with newer technologies it is very easy to access sexual related information like sexual videos, images, stories and messages etc. "A study by Brown *et al.* found that adolescents' exposure to sexually explicit content predicted the perpetration of sexual harassment, more permissive sexual norms, having oral sex, and engaging in sexual intercourse while in high school. Social networking websites enable adolescents to present themselves publicly, sometimes in very sexually aggressive ways" (Pasam, 2019).

POSITIVE IMPACT OF DIGITAL TECHNOLOGIES ON ADOLESCENTS

If the adolescents are using the digital technologies in the right way then it, exercises' a positive impact on them such as.

- Improves Communication and Building the Relationship:
- Improve the Problem-Solving Skills:
- Enhanced Education and Digital Learning Tools:
- Increases Interactive Creativity and Digital Expression:
- e-health related benefits

Improves Communication and Building the Relationship

Digital technologies such as social media platforms like Facebook, Twitter, Instagram etc. and Popular messaging apps like WhatsApp, Telegram etc. are connecting thousands of adolescents with each other. These kinds of digital technologies allow adolescents to keep in touch with their family members, relatives, friends, through group messages or through online video calls. These kinds of activities will build the relationship between them even if they are staying thousands of miles in distance. It also improves the communication between them. Now a days, it is very easy to communicate with each other because of these technologies.

Improve the Problem-Solving Skills

There are so many digital technology apps like YouTube, internet, Braingle, Eidetic and also some gaming apps improve the problem solving skills in the adolescents. If the adolescents are facing any issues, be technical or otherwise, they can easily get solutions using various search engines and applications such as YouTube etc. The social media platforms help them come in contact with thousands of their acquaintances. They can also use Google Maps to navigate a destination if they are stuck in some place. Also, there are so many gaming apps like Lumosity, CogniFit Brain Fitness etc. available online that help them enhance their memories and improve their problem-solving skills.

Enhanced Education and Digital Learning Tools

The digital technologies also enhance the performance of the education sector. Varieties of digital learning tools are available through which adolescents can engage in self-directed learning, collaborate on group projects or exchange ideas between them. Moreover, many e-Books are available for free that help them add to their knowledge. Even the google or social networking sites are helpful for them to avail of plenty of study materials online. These platforms also offer them access to different discussion forums through which they can clear their doubts and also contact their teachers any time. YouTube too, offers numerous educational videos that help them strengthen their grip over the concepts.

Adolescents are using various digital learning tools. Applications like Duolingo helps them to learn languages. Audible app facilitates them to listen to the eBooks. SoundNote app helps them significantly in recording lectures, whereas, remind app enables them to share photos and handouts, submit assignments, Google classroom apps, MOOCS concepts help them continue to work productively even while at home and completing virtual assignments. These kinds of learning apps help adolescents avail of better education prospects.

Increases Interactive Creativity and Digital Expression

There are so many interactive and creative apps, smartphone applications and interactive websites that help adolescents to share their artistic passions to the world. For example, adolescents can record their talents like musical, dancing and any other creative activities and upload the same on YouTube, Photo editing apps like Sumopaint, Photo Editor by Aviary etc. to show their creativity, or apps like MXTakaTak, Instagram Reel to record their talents and post it on their account. So, there are so many different types of digital tools which are helpful the adolescents to show their creativity and they can easily share their creations. It also offers the chance for encouragement and artistic feedback.

e-Health Related Benefits

There are many digital technologies like Social media, mobile health apps, Mental Health apps like (Moodfit, Talksapce, headspace) will provide information about psychological and physical health problems to adolescents, which may cause a positive benefit in terms of promoting engagement with health-care services. There are wide variety of apps designed for adolescents to encourage their behaviors such as healthy activity, enhanced medication adherence and smoking cessation, as well as specific mental health support. Also, adolescents can use social media to express their views and problems with their friends to reduce their depression.

CONCLUSION

There is no denying the fact that the usage of Digital technology entails both good as well as evil effects.

If adolescents use the digital technology appropriately, they can use it to the best of their advantage. If they learn to use these platforms rationally and self- restrict its excessive use, they may easily overcome the anxiety and isolation associated with it.

If adolescents had more positive interactions and social support on the social platforms then it may be helpful to reduce the spread of depression amongst them. Besides, they can use mental health apps to overcome their day-to-day life stress.

If adolescents are taking regular break away from the screen, it may reduce the eyestrain problem and also, they can use 20-20-20 rule for digital viewing. While using digital technology, taking a small break for a walk may help them overcome the problem linked to poor posture. The young people should also avoid using digital devices in the hour or two before bedtime to overcome the sleep problems. while surfing on internet they need to be extra cautious to avoid cyberbullying and sexting. If the adolescents are taking these kind of safety measures then the digital technology will help them in a positive way.

REFERENCES

Hughes, N., & Burke, J. (2018). Sleeping with the frenemy: How restricting 'bedroom use'of smartphones impacts happiness and wellbeing. *Computers in Human Behavior, 85*, 236–244. doi:10.1016/j.chb.2018.03.047

Kopecký, K. (2014). Cyberbullying and Sexting between Children and Adolescents - Comparative Study. *Procedia: Social and Behavioral Sciences, 149*, 467–471. doi:10.1016/j.sbspro.2014.08.292

Mustafaoğlu, R., Zirek, E., Yasacı, Z., & Razak Özdinçler, A. (2018). The negative effects of digital technology usage on children's development and health. *Addicta: The Turkish Journal on Addictions, 5*(2), 13-21.

O'Keefe, G. S., & Clarke-Pearson, K.Council on Communications and Media. (2011). The impact of social media on children, adolescents and families. *Paediatrics, 127*(4), 800–804. doi:10.1542/peds.2011-0054 PMID:21444588

Pasam, R. S. (2019). Advances in digital technology: Effect on child and adolescent behavior. *Archives of Mental Health, 20*(2), 37. doi:10.4103/AMH.AMH_25_19

Richards, D., Caldwell, P. H. Y., & Go, H. (2015). Impact of social media on the health of children and young people. *Journal of Paediatrics and Child Health, 51*(12), 1152–1157. doi:10.1111/jpc.13023 PMID:26607861

Small, G. W., Lee, J., Kaufman, A., Jalil, J., Siddarth, P., Gaddipati, H., Moody, T. D., & Bookheimer, S. Y. (2020). Brain health consequences of digital technology use . *Dialogues in Clinical Neuroscience, 22*(2), 179–187. doi:10.31887/DCNS.2020.22.2/gsmall PMID:32699518

Chapter 3
An Adverse Effect of Social, Gaming, and Entertainment Media on Overall Development of Adolescents

Sonali Sagar Kharade
Department of Computer Science, SSR College of Arts, Commerce, and Science, Savitribai Phule Pune University, Silvassa, India

ABSTRACT

Digital technology in general and social media in particular entail both positive as well as negative impacts on the psyche of the adolescents. The current generation of gen-z has grown up in a media-saturated world. However, it is pretty difficult to judge how media exactly influences their lives. It's commonly observed that social media platforms help saving time establishing contact with our acquaintances. The excessive use of it however leads to various negative repercussions. This may include cyber bullying, low self-esteem, gaming addiction, and less time devoted for doing physical activities. Playing indoor and outdoor games makes adolescents physically fit and mentally healthy. There are however various video games on social media gaming platforms that affect their creativity and logical thinking in an adverse manner. Negative consequences of video game addiction can lead to many problems such as severe body-ache, skin blisters, weakened eyesight, and insomnia. Long-term addiction could lead to obesity, weakness or numbness in the hands, and even blood clots.

INTRODUCTION

Many parents wonder how digital technology would affect adolescent's development. Various researches have, however, proved that majority of people reported social media playing a positive role in the lives of their wards. Many of them uses social media to connect with their family, friends and also for the sake of entertainment. The use social media has both negative and positive impacts on our youths and teens today. Young people spend a lot of time on social media. They're also more susceptible to peer pressure, low self-esteem and mental ill-health. A number of studies have found associations between increased

DOI: 10.4018/978-1-7998-8318-0.ch003

social media use and depression, anxiety, sleep problems, eating concerns, and suicide risk. In this paper, I tried to bring your attention towards the impacts of social media specifically to the adolescence of this generation. These impacts are both positive and negative.

Positive Impact of Social Media

- The social media keeps youngsters connected to their friends
- Social media keeps you abreast of important global events. Getting worldwide information, political, economic or otherwise is now just a click away.
- The social media also provides youngsters with platform to express themselves more conveniently and thereby helps increasing their self-esteem.
- These platforms also help people connect with others better.one can establish contact with like-minded strangers and thus befriend them.
- One more interesting impact of social media is that its far easier to interact with peers than talking to them in person

Negative Impact of Social Media

- Social media tops the list of priority of millions of adolescents across the world. Young People love being on the social media sites and become forgetful of their responsibilities
- The social media sites at times may prove to be deceptive as they do not reveal the true side of an individual. many people portray themselves differently on theses media sites. They never reveal their true colors
- Social media sites can also make young people fall prey to cyberbullying. It may lower their confidence and affect their emotional and psychological health adversely. At times, such issues may lead to depression and trigger suicidal thoughts in the minds of the adolescents.
- The social media platforms can also make the youngsters very sensitive to their looks and appearance. It may lead them to anxiety disorders.
- The youngsters are found to be getting addicted to using the social media sites to the extent that their lifestyle largely becomes sedentary. It may cause obesity and hypertension among them and thus impair both their physical as well as mental health. There is a very strong temptation in social media. It can become an addiction for the youth and teens.

Like any form of social engagement, social media too involves many risks and problems. Some of the most common risks are:

- spending too much time online and being disconnected from the real world
- being the victim of online bullying
- damaging your online reputation
- having your personal information shared online
- being harassed or annoyed by someone you do not want attention from
- being the victim of an online scam
- Having reduced self-esteem

VARIOUS SOCIAL MEDIA PLATFORMS

Social media refers to any digital platform, system, website or app that enables people to create and share content, and connect with each other. Following are some popular social media platforms among adolescents.

Facebook

A free site where registered users (aged 13 and up) can share pictures, links, videos, and other content with their 'friends'. Most of the time, anything you share will only be seen by the people you have 'friended'. Facebook changes its privacy settings regularly so it's important to stay up-to-date with their settings and policies.

Instagram

A free image sharing service that is used mainly on mobile devices. On Instagram, it's common to connect with people you don't know but share common interests. You can have a public account or a private account where only friends can see your posts. Nowadays adolescences are very much attracted to the new service provided by the Instagram that is "Instagram Reels". Adolescence is attracted more towards to make reels due to ban on TikTok app. Instagram also has their own privacy settings to stay up-to-date with.

Snapchat

A mobile phone messaging application that allows you to send video or picture messages to one or more people at a time. The messages are only viewable for a few seconds at a time, as determined by the sender. However, there are ways of saving the images using other phone functions.

Tumblr

A blogging website where people can publish their own articles and share it with people who follow their blog or profile. They also view the content of people they follow. On Tumblr, it is more common to connect with people you don't know in real life than people you do.

WhatsApp

WhatsApp and other messaging applications let you contact friends using internet data (or Wi-Fi) instead of your mobile carrier, which reduces the cost of messaging. It can be used without a sim card if connected to a Wi-Fi network. It is popular and easy to use because it allows you to send text-messages, audio-messages and video-messages to a group of friends at one time.

Other Games and Sites

YouTube, Club Penguin, or online games like World of Warcraft, Pub G, League of Legends, Counter Strike nay many more are also considered social media because of the way they enable social interaction. Many adolescents create their e-content and upload on the YouTube to get the fame.

Among many social media platforms YouTube, Instagram and Snapchat are the most popular online platforms among adolescences. (Kaur & Bhatt, 2016) According to "Pew Research Centre" approximate 95% of adolescence has access to a smartphone, and 45% say they are online almost all the time.

(Akram & Kumar, 2017) New data from "YouGov" finds that almost 17% of teenagers, defined as 13-17 years old, report spending 4 hours or more online every day. YouGov conducted the study among 8-to-17-year-olds and gained the children's parental permission before asking the questions on a range of topics, including online habits. Overall, teens report spending more time online than do children: 45% of teens say they spend 2 hours or more browsing the internet, watching videos, playing games, chatting with friends, and other online activities. Fewer children approximate 39% report spending the same amount of time online.

ADDICTIVE EFFECTS OF SOCIAL MEDIA IN VARIOUS WAYS

Using social media platforms become a risk to adolescents more than most adults realize.

The Effect of Social Media on Education: Work/School

- Social media helps the students to cheat on school assignments and for those who work can get some idea about their work
- If social media is used on a light scale, it will help the students to improve their grades. However, for those students who are heavy users, their grades tend to fall incredibly. A study shows that students who are using Facebook have an average GPA of 3.06 while those who don't use Facebook have an average GPA of 3.82.
- On average of 106 minutes in every 90 minutes that is spent on Facebook college students drop their grades.
- The social media platform affects the productivity of the employees. These workers are generally between the age of 25-34 and check their social media during work, especially teachers use these social media while teaching which degrades their teaching methods and ultimately it effects on students learning. This is the reason why many schools had ban to bring mobile phones in the classroom or school during working hours.

Lacking Privacy

- One thing that is seen common among the young people is they often get too open in the public and share their personal information. Most of the people don't read the privacy policies and are unaware that their information is being read by the insurance companies, advertisers and fraudster's.

- The main risk to preadolescents and adolescents online today is risks from each other, risks of improper use of technology, lack of privacy, sharing too much information, or posting false information about themselves or others.
- When Internet users visit various Web sites, they can leave behind evidence of which sites they have visited. This collective, ongoing record of one's Web activity is called the "digital footprint". One of the biggest threats to young people on social media sites is to their digital footprint and future reputations.
- Preadolescents and adolescents who lack an awareness of privacy issues often post inappropriate messages, pictures, and videos without understanding that "what goes online stays online". Indiscriminate Internet activity also can make children and teenagers easier for marketers and fraudsters to target.

Cyber Bullying

- One of the very prime negative impact of social media on youth as some of the negative minded individuals use it for bullying someone
- People use it for sending intimidating messages to others that sometimes create trauma for the others as well
- Misinformation or false propaganda is another negative impact of social media on adolescences which lead to cyberbullying.
- Cyberbullying is deliberately using digital media to communicate false, embarrassing, or hostile information about another person. It is the most common online risk for all teens and is a peer-to-peer risk.
- Although "online harassment" is often used interchangeably with the term "cyberbullying," it is actually a different entity. Current data suggest that online harassment is not as common as offline harassment and participation in social networking sites does not put most children at risk of online harassment. On the other hand, cyberbullying is quite common, can occur to any young person online, and can cause profound psychosocial outcomes including depression, anxiety, severe isolation, and suicide.

The Effect of Social Media on Health Issues

- Social media also effects the health of adolescences in one or the other way like sleeping disorders, bad eating habits, rude behaviours, change in attitudes and hiding unwanted materials etc. which will bring very dangerous effect on adolescents' overall development.
- Social media negatively effects sleep because the artificial light given off by smartphones activates arousing neuron which is present in the brain. Hence, these chemicals disrupt the body's ability to produce melatonin which is a sleep-inducing hormone.
- Another 2019 study of more than 12,000 13- to 16-year-olds in England found that using social media more than three times a day predicted poor mental health and well-being in teens. Other studies also have observed links between high levels of social media use and depression or anxiety symptoms (Bhardwaj et al., 2017).

The Effect of Social Media on Identity Formation

- Nowadays adolescents are very much aware of their identity or can say formation of unique identity and the effect of social media on them extend to an important part of their development.
- Many social media platforms provide various ways to improve their skills and attitude but much use of social media can adversely effect on their psychological health and distract them and that make them worse mentally. Because the more effective way to build good identity of yourself is to explore real world challenges and not to explore virtual world on the social media platforms.

The Effect of Social Media on Friendships

- Friends play a vital role in overall development of adolescences. Nowadays adolescents make many known and unknown friends on social media platforms. According to "Pew Research Centre" around 81% of teen's ages 13 to 17 years says social media makes them feel more comfortable and connected to what's going on in their friends and family lives (**Kaur & Bhatt, 2016**).
- Aside from cyberbullying, oversharing and sexting issues, social media can also put negative effect and pressure on friendships, especially when one friend is very active about posting pictures, status updates and opinions that hurt others. Here I tried to describe how social media effects on adolescences friendships:
    - Your child feels left out when one of his/her friend posted some images or videos of any party or event on social media platforms in which he/she was not invited.
    - Your teen's disagreements are publicly posted. If your teen friend regularly posts about the disagreements they have, this is called fake friendship. Those friends who posts bad comments, engages in rumours, gossips and reveal your child's secret or any disability he/she have, publicly that make your teens moral down and then they feel depressed and anxiety comes into their mind.
    - Communicating too much time online doesn't make a healthy and life long friendship. But adolescents nowadays prefer to communicate online rather than face-to-face, because they believe communicating online make them feel comfortable and can express their feeling in better way by texting or sending images and emojis.

The Effect of Entertainment Media on Adolescence

Under the name of entertainment, the social media is challenging the minds of adolescence. Television exposure and total media exposure in adolescence are associated with increased mental illness in them. Both TV viewing and mobile phone use contribute to the development of depressive symptoms in adolescence. Moreover, sexually explicit materials, pornography and semi-nudes' images or videos is now indiscriminately available freely on internet and social media which have significant negative effects on mental and physical health of adolescence.

The Effect of Gaming Media on Adolescence

- Nowadays many adolescences or teens believe that interactive screen such as gaming or using social media is better than watching TV or playing outside with some friends. But in reality, it is not true. Playing online games or using social media is often worse for them, as it has many adverse effects on adolescence brain which can lead to emotional outbursts, meltdowns and distractibility. Furthermore, it can cause in insomnia, problem at work, academic failure, lack of concentration, low confidence, aggression, poor wellbeing, feeling loneliness and it can also lead to memory loss.

- Many researchers have found that playing excessive online or video games can lead to high feeling of social isolation that makes adolescence depressed, anxious and alone. Online video game addiction leads to a loss of interest in after school activities, including social clubs and indoor-outdoor sports. Addiction of online game can distract the adolescence towards mobile gambling in very teen age which can affect them in their entire life.

TIPS TO USE SOCIAL MEDIA IN A RIGHT WAY

Social media has given the power to the adolescents in one or the other way like:

1. Educate yourself by using social media correctly because there are lot of study material or e-Contents available through which you can obtain information and gain knowledge.
2. Use the social media for spreading awareness in the society and provide a global exposure to express their thoughts.
3. Social media act as a platform where people can able to showcase their talents. Adolescences can post their creativity on the social media which will be acknowledged by the people.
4. A social media platform boosts your skills. It offers many opportunities to the teens like they can hone their multiple and unique skills through posting blogs, photos, articles, videos, audios and many more.
5. Blogging is the way to express the feelings or emotions about any topic which you like. Adolescence can write any type of blogs like they can write about life experiences, poetry, quotes, recipes, travelling blogs and can express their thoughts in any way. It is a good idea to become a blogger.

CONCLUSION

Just like in the offline world, there are things you can do to protect your child from all these online risks. It's important to prepare adolescences for how to respond if these things do occur with them. Parents can play an important role in this. They need to be act like a friend of their teen so their child can share their feelings what happen with them.

Parents can help their teens to develop positive habits like taking regular breaks from checking social media, and making sure notifications are turned off so that their mobile device isn't distracting can help reduce the compulsion to check in with social media so often. This can lead your adolescence to live in a better way with the limited access of social media and they can use the social media platforms in a positive way.

REFERENCES

Agarwal, V. (2012). Harmful effects of media on children and adolescents Burden of care of principal caregivers of autistic children and adolescents visiting health facilities in Lucknow city.

Best, P., Manktelow, R., & Taylor, B. (2014). Online communication, social media and adolescent well-being: A systematic narrative review. *Children and Youth Services Review*, *41*, 27–36.

Hellström, C., Nilsson, K. W., Leppert, J., & Åslund, C. (2015). Effects of adolescent online gaming time and motives on depressive, musculoskeletal, and psychosomatic symptoms. *Upsala Journal of Medical Sciences*, *120*(4), 263.

Kaur, R., & Bashir, L. (2015). Impact of social media on mental health of adolescents. *International Journal of Education*, *5*, 22–29.

Kaur, R., & Bhatt, L. (2016). Impact of stress on mental health of students: Reason and intervention. *International Journal of Education*, *5*, 30–35.

Eijnden, R., Lemmens, J., & Valkenburg, P. (2016). The social media disorder scale: Validity and psychometric properties. *Computers in Human Behavior*, *61*, 478–487. doi:10.1016/j.chb.2016.03.038

Bhardwaj, A., Avasthi, V., & Goundar, S. (2017). Impact of social networking on Indian youth: A survey. *I.J. of Electronics and Information Engineering*, *7*(1), 41–51.

Akram, W., & Kumar, R. (2017). *A Study on positive and negative effects of social media on society*.

McCrae, N., Getting, S., & Purssell, E. (2017). Social media and depressive symptoms in childhood and adolescence. *Systematic Reviews*, *2*(4), 315–330.

Anderson, M., Jiang, J., & Smith, A. (2018). *For media or other inquiries* (Vol. 28).

Anderson, M., & Jiang, J. (2018). *Teens' Social Media Habits and Experiences _ Pew Research Center*.

Anderson, M., & Jiang, J. (2018). *Teens*. Social Media & Technology.

Bekalu, M. A., McCloud, R. F., & Viswanath, K. (2019). Association of Social Media Use With Social Well-Being, Positive Mental Health, and Self-Rated Health: Disentangling Routine Use From Emotional Connection to Use. *Health Education and Behavior, 46*(2_suppl), 69–80.

Ganguly, N., & Kumaraguru, P. (2019). The positive and negative effects of social media in India. *Communications of the ACM*, *62*(11), 96–99.

Beyens, I., Pouwels, J. L., van Driel, I. I., Keijsers, L., & Valkenburg, P. M. (2020). The effect of social media on well-being differs from adolescent to adolescent. *Scientific Reports*, *10*(1).

Singh, A., Kumar, H., Kumari, S., Scholar, R., & Professor, A. (2020). Impact of Social Media on Adolescent's Mental Health. *International Journal of Creative Research Thoughts*, *8*(7).

Cipolletta, S., Malighetti, C., Cenedese, C., & Spoto, A. (2020). How can adolescents benefit from the use of social networks? The igeneration on instagram. *International Journal of Environmental Research and Public Health*, *17*(19), 1–15.

Boer, M., van den Eijnden, R. J. J. M., Boniel-Nissim, M., Wong, S. L., Inchley, J. C., Badura, P., Craig, W. M., Gobina, I., Kleszczewska, D., Klanšček, H. J., & Stevens, G. W. J. M. (2020). Adolescents' Intense and Problematic Social Media Use and Their Well-Being in 29 Countries. *The Journal of Adolescent Health*, *66*(6), S89–S99.

Sandhya Keelery (2020). Social media usage in India- statistics & facts.

Boer, M., Stevens, G. W. J. M., Finkenauer, C., de Looze, M. E., & van den Eijnden, R. J. J. M. (2021). Social media use intensity, social media use problems, and mental health among adolescents: Investigating directionality and mediating processes. *Computers in Human Behavior*, *116*, 106645.

Pelphrey, K. A., Foell, J., Zerwas, S., Esposito, G., Cataldo, I., Lepri, B., Jin, M., & Neoh, Y. (2021). Social Media Usage and Development of Psychiatric Disorders in Childhood and Adolescence. *RE:view*.

Austermann, M. I., Thomasius, R., & Paschke, K. (2021). Assessing Problematic Social Media Use in Adolescents by Parental Ratings: Development and Validation of the Social Media Disorder Scale for Parents (SMDS-P). *Journal of Clinical Medicine*, *10*(4), 617.

Chapter 4
An Intersectionality Perspective on Rural Adolescent Access to Digital Technology in South Africa

Amy Sarah Padayachee
https://orcid.org/0000-0002-6021-6642
University of the Free State, South Africa

Fumane Portia Khanare
https://orcid.org/0000-0002-1707-9179
University of the Free State, South Africa

Delin C. Louw
University of the Free State, South Africa

Ntombizandile Gcelu
University of the Free State, South Africa

ABSTRACT

The recent COVID-19 pandemic has brought to the fore compounding disparities in the South African education system. The world at large has suddenly been faced with the challenge of blended learning given that COVID-19 has reconstituted the traditional form of education delivery. Much emphasis has been placed on global education, yet due to compounding inequalities stemming from the effects of apartheid, the South African education system has been left further entrenched in the digital divide. It is for this reason that the authors of this chapter illuminate the lack of digital technology, its subsequent effect on adolescents in rural areas, and how it impacts on their ability to learn and compete in the global education sphere. The perspective of Intersectionality theory highlighted in this chapter is used to address the inequalities perpetuated by digital technology on adolescents in rural areas.

DOI: 10.4018/978-1-7998-8318-0.ch004

INTRODUCTION

The demand for basic and higher education on a global level has been elevated since the recent pandemic which took the world by storm. According to Bonilla-Molina (2020: 2), "in just few months, educational neoliberalism has mainstreamed tensions between face-to-face education in schools and virtual education at home in global public discourse." The glaring disparities became more evident when a significant number of schools fell behind the wheel in delivering the curriculum due to lack of technological infrastructure.

The effects of globalisation on education brings rapid developments in technology and communications (Mitchell & Nielsen, 2012) placing emphasis on the shift in education meeting global standards. The outbreak of pandemics, such as the recent Covid-19 pandemic has shown that teaching and learning are not confined to the traditional four walls of a classroom but have shown how education and technology are becoming increasingly intertwined (Wargo, Chellman, Budge, & Davis, 2020). Yet despite the shift to blended learning, the digital divide in the South African education system remains apparent. According to Steele (2019):

[d]igital inequality is evident between communities living in urban areas and those living in rural settlements; between socioeconomic groups; between less economically developed countries and more economically developed countries; between the educated and uneducated population. Individuals with access to a broadband connection can be digitally split. How? Low-performance computers, limited broadband speeds and limited access to subscription-based content widen the gap.

A quarter century later, yet the digital inequalities still continue. Several of the initial schisms that were identified in the year 1995 are still relevant today (Robinson et al., 2020). This is especially true in numerous schools across South Africa. Access to digital communication technologies and the ability to use those technologies remain a challenge for many schools. The lack of resources, specifically in township schools, has failed to redress the historical marginalisation of learners who attend these schools and Ndimande (2016) argues that the lack of resources in these schools directly contribute towards poverty in the lives of the learners.

The effect of apartheid on the South African education system can further be seen in the inequities in accessing online learning platforms and software which are compounded by differences in cultural capital (Bourdieu & Passeron 1977). Despite numerous attempts by the Departments of Education and Higher Education to narrow the digital gap, South African rural adolescents are still susceptible to falling behind the academic wheel. The infrastructure of a school also affect the sustainable use of mobile technology, as some institutions, especially those in rural areas in South Africa, lack basic equipment like desks and many computer facilities at these schools are under-serviced and dysfunctional (Terhoven & Fataar, 2018:24). Whilst tertiary institutions attract students from various demographic areas, including rural and township areas, where students might lack digital fluency as they might not have prior exposure to information and communication technologies in education (Mbodila 2020).

However, lack of access to digital technology and it subsequent requirements to be an effective tool for teaching and learning are not the only problem faced by rural students. Petrie et al. (2020) maintain that it is also essential for teachers to have the necessary technical and pedagogical skills to integrate digital devices into instruction and also develop creative initiatives like play-based collaborative learning. Notwithstanding, it is imperative that teachers be trained in the skilful use of fourth industrial revolution tools in order to make learning meaningful.

In this chapter, we aim to present our findings from a scoping review to identify the intersectional experiences of adolescents living in the rural setting in the age of digital technology.

BACKGROUND AND CONTEXT OF THE PROBLEM

Lack of Access to Fourth Industrial Revolution (4IR) Tools

As the world moves from knowledge economy to digital economy, there is need for all sectors including education to embrace information and communication technology (Akintolu and Uleanya, 2021: 713). The current COVID-19 crisis has revealed egregious disparities of compounding inequalities faced by the Global South. In his response during a *Virtual Imbizo*, President Ramaphosa stated that the South African government aims towards "building a more just and equal society." Regrettably, the response to the pandemic by the Department of Basic Education (DBE) has further contributed to South Africa being touted as a country with ostensibly high levels of inequality. The DBE's Neoliberal approach to "save the academic year" has propelled the Global South to live up to the expectations of the Global North, particularly in terms of expediting the use of Fourth Industrial Revolution tools, despite an evident digital divide, poor infrastructure and limited resources.

Much emphasis has been placed on the benefits of blended learning and how in times of crises, such an approach could be seen as a paradigm shift from traditional pedagogical method to technology based teaching and learning (Mahaye, 2020). However, for rural adolescents, the lack of resources leads is congruent with an accumulation of limitations. These limitations subsequently impact on their ability to keep pace with the rest of the world who has shifted to blended learning. As the Global North progresses to digitized education platforms, the Global South, inter alia South Africa remains tangled in a web of limitations. Maurya (2020:177) illuminates the reality of digitization by stating that; it is a very paradoxical situation that while on one hand we are moving towards digitization of education on the other hand the digital divide in rural areas is also increasing.

South Africa's Digital Divide Factors

With underdeveloped infrastructure and education budget cuts, South African educational institutions struggle to reconceptualise a digital space for its youth. As much as South African Departments of Education have established plans to assist adolescents in vulnerable communities, it is reported that the economically disadvantaged South African adolescents cannot access these provisions adequately (Mahaye, 2020). Furthermore, in a country wrought with disparities such as South Africa, there is unequal access to technologies used in online learning as well as unequal access to data and connectivity (De Preez & le Grange, 2020:92). As a result of these disparities, more vulnerable adolescents, such as for instance those from rural backgrounds, are especially likely to fall behind during this emergency period (Di Pietro et al., 2020:4). Walters (2020:8) states that in many ways, this is not new information; we have been talking about educational inequality and the achievement gap for a generation now.

The digital divide can be characterised by two fundamental problems; firstly a limited and costly infrastructure and secondly a limited digital literacy (Chetty et al., 2018). Schools in rural areas often have limited access to digital technologies due to high costs as well as a lack of infrastructure and security, at times they might not even have access to electricity. This is linked to the limited digital literacy, because if people do not have access to digital technologies they cannot educate themselves on its usage.

In order to bridge the digital divide and encourage remote learning, investing in infrastructure should be a priority. This would lower the cost of consumers with regards to gaining access to online material. Montoya (2020) argues that closing the digital divide would require government to invest in supporting learners early on in the foundation phase. However, this would require better measures of access to as well as the use of digital technologies.

Gómez-García et al. (2020) state there are a lot of benefits of using ICT for educational purposes and this has been supported by various studies. For instance, it allows educators to move beyond the classroom size restrictions set out in the Covid-19 protocols. However, this can only be achieved if all learners have access to computing devices. Sadly, this is not always the case when it comes to learners from rural areas. According to Power, Musgrove and Nichols (2020) schools face challenges with implementing a learner-computing ratio of 1:1, however these challenges are greater in rural school districts with scarcer resources. Power et al. (2020) further highlights that rural districts often face challenges regarding replacing aging technology. You might find that schools are equipped with ICT devices, but they are not able to support online learning.

Hargittai and Dobransky (2017) suggests that digital inequality research has established that individuals from underprivileged backgrounds are less likely to be internet users. This could be as a results of the costs associated with internet enabled devices as well as the training on using the devices that go a long with it. In places where learners do not even have access to their own textbooks, ICT devices often don't even make it on the school's budget. Most researchers on public education are noting the pressing need for a focus on access to the internet and investing in the technology needed to close the digital divide in online learning for all students (Walters, 2020:8). The year 2020 has further illuminated the inequities in the struggles faced by parents to assist their children with virtual learning. For single-parent families, large families living in small and crowded spaces, families where one or both parents are poorly educated, and the large percentage of child-headed households, this task is Herculean (Schneider, 2020).

Exercises from the past recommend that as subsidizing becomes limited, disadvantaged schools endure the most difficulties, with the overall impact on such schools remaining long-term. It is therefore important to note that these schools are already in a position of disadvantage. Be that as it may, there is an occasion to re-examine and redesign teaching and learning. Significantly, it's an opportunity to address the gross disparities and deficiencies in South African education and training.

LITERATURE REVIEW

Placing Rural Adolescents on the Digital Map

Rural areas are defined as "farms and traditional areas characterized by low population density, low level of economic activity and low level of infrastructure" (Department of Basic Education, 2017: 20). To put the definition of adolescents into perspective, the Cambridge dictionary describes adolescents in the following way: "We know and accept that adolescents are neither children nor adults and that they have specific developmental needs." In the context of the study at hand, embracing the Fourth Industrial Revolution tools is an equally imperative developmental need that deserves much attention. Furthermore, social inclusion is regarded as a national desirable goal, which means that individuals can fully participate in political, social and economic life (Ye and Yang, 2020:2), however, rural adolescents are subject to a myriad of social exclusionary conditions.

The Department of Basic Education (DBE) Rural Education Draft Policy of 2017 states that "it is difficult to recruit, retain and develop qualified teachers in a rural setting due to the distances of schools from towns, poor infrastructure and limited service delivery" (Department of Basic Education, 2017: 18). Poor infrastructure and poor limited service delivery have therefore been identified as contextual factors negatively affecting effective teaching and learning in rural areas.

According to UNESCO (2020), "it is envisaged that the education of an African child would be the most affected post COVID-19 era due to economic and technological backwardness of most African countries." In South Africa, a vast majority of schools are situated in rural areas and subsequently, learners in rural schools are have less opportunity to access quality education (Du Plessis, 2014). "With only 37% of South African households having consistent access to the internet through cell phones or computers, according to Statistics South Africa, the departments of basic education and higher education and training were faced with the near impossible task of continuing the academic year" (Hanekom, 2020).

The availability of, as well as access to digital technology is often perceived as a social equalizer and a means of sustainable development in education (Wang, Tigelaar and Admiraal, 2019). However, rural areas comprising students from low-income families, who have little or no motivation and generally lower academic performance, remain digitally disadvantaged due to many constraints such as the availability of financial and social capital and the limits posed by geographical isolation (Salas-Pilco, 2018; Warschauer, Knobel, and Stone, 2004). The need to place rural adoloescents at the forefront of the educational endeavour is now more important than it has ever been. Krönke (2020) suggests that government's efforts to redress socio-economic inequalities need to be improved radically to avoid the persistent widening of an education gap among their citizens, especially between the rural and urban areas.

Parental Involvement in Rural Learning Settings

In developed countries, learners have access to access to a variety of digital tools at home where they learn to create content, solve problems, engage, collaborate, and network which helps them develop 21st-century skills (Papadakis & Kalogiannakis, 2017). Despite the limitations caused by economic recession and lack of infrastructure in a developing country such as South Africa, a further contribution to the digital divide can be dated back to the era of apartheid. The lack of cultural capital stemming from apartheid is seen today where educated parents who are wealthy generally have the knowledge, competence, resources and skills to support online and home learning (Sayed & Singh, 2020). Moreover, blended learning also requires active mediation of learning, which is disproportionately unequal between wealthy and poor rural households (Sayed & Singh, 2020). The digital divide exacerbates already existing learning gaps in South Africa where blended learning can only be enjoyed by the privileged few (Bashman 2020:10). Not only is the shift to blended learning and teaching expected to cause greater inequality in cognitive abilities, it may exert a similar effect as regard students' emotional well-being and motivation (Di Pietro et al., 2020:4).

There are also many other challenges that have negatively impacted on parental involvement in rural learning settings. These include economic resources (Hohlfeld et al., 2010); lack of internet access (Hollingworth et al., 2011); lack of interest of in using technology (Beckman et al., 2019); and having low digital self-efficacy (Povey et al., 2016).

Compounding Inequalities That Further the Digital Divide

The COVID-19 crisis has deepened the disparities of the compounding inequalities of the educational system across basic and tertiary education level. Most schools in South Africa in underdeveloped rural areas lack basic necessities such as running water and sanitation. In addition, this pandemic has highlighted the enormous digital divide amongst South African educational institutions. According to Jantjies (2020) there are many schools that have little to no access to technology facilities. In addition, not all educators are computer literate or even have the basic digital skills to make use of e-learning. To make matters worse, South Africa has some of the highest data cost on the continent which makes access to the internet a costly endeavour and unattainable method of teaching.

O'Connell and Clarke (2020, p. 9) postulate that "there may now be scope for refocusing schooling away from its traditional academic orientation, towards creating more learning opportunities engendering such skills and attributes as critical thinking, communication, self-management, research and social analysis, which may contribute to forging a successful and cohesive future society." The repositioning of teaching approached have now become more apparent than ever before as traditional teaching methods have now become outdated. Despite the myriad of negative challenges that Covid-19 brought on, it has forced us to introspect and adopt future research-validated management practices.

Research shows that many governments in developing countries have initiated the inclusion of digital technology in schools by having crafted policies that are aimed at supporting mobile learning implementation in an effort to support 21st-century teaching and learning (Lam & Duan, 2012). However, Mangwaya et al. (2013) posit that many countries in the Global South still heavily rely on traditional face-to-face teaching. This may be due to the poor infrastructure which makes the shift to digital teaching and learning difficult, if at all possible.

Educational institutions have further contributed to the digital inequality given that students from more advantaged backgrounds may be more likely to attend primary, secondary and tertiary institutions with better Information and Communication Technology (ICT)-based infrastructure and where educational practitioners have higher levels of digital skills (Di Pietro et al., 2020:4). Immediate short-term plans that cater for students with access to online resources and those without access had to be developed (Mbodila 2020). To address the issue of making resources available to students, educational institutions have negotiated with several cellular networks to make data available to students whilst also providing devices to disadvantaged students (Hedding et al., 2020:1). However, despite such attempts to narrow the digital divide, online learning is particularly challenging in Africa, 'where less than a third of the population has access to broadband connectivity' (Ngalomba 2020). Furthermore, rural schools are continually exposed to practical limitations. One such limitation is that educators are responsible for teaching multiple subjects and different grades in a single classroom while also expected to take on responsibilities such as coordinating extracurricular and co-curricular activities whilst also engaging in support programs and infrastructure for student achievement (Willis and Templeton, 2018). The taxing workload of teachers in rural schools leaves little or no time to capacitate themselves with the required technological skills needed.

Students in urban areas have access digital technology in nearby public spaces such as schools, libraries, internet cafes etc., where wireless internet access is available to children for schooling purposes. However, rural residents face further challenges due to distance required to travel to public access points that are often situated in nearby urban areas (Lai and Widmar, 2020). The lack of access to digital technology has compounding effects on students living in these areas. Rural areas without sufficient internet connectivity and services face challenges including a brain drain of skills and limitations to distance learning (Hennessy, Läpple, and Moran 2016; Thompson 2020). It is for these reasons that students in rural areas need to be afforded the opportunity to acquire 21st century skills required to be active citizens of a global village.

THEORETICAL FRAMEWORK

According to Rutherford and Davidson (2019:1), Intersectionality provides a framework for examining how multiple social categories combine in systems characterized by both oppression and privilege to affect the experiences of those occupying the intersections of these social categories.

In the context of crises, the intersectional theory is used to address how multiple social differences are (re)produced in responses to crises (Kuran et al., 2020). Rural adolescents are subject to a myriad of barriers that stem from inadequate access to digital technologies, thereby negatively impacting on their ability to learn. The intersectional theory thereby deepens the understanding of how different social groups are both affected by, and also have their own effect on, the impacts of various crises (Kuran et al., 2020). An additional perspective of intersectionality theory is its address of social inequalities that are perpetuated by institutional oppression (Shields, 2008; Walby, Armstrong, & Strid, 2012) which remains evident in the South African education system. Intersectionality perspective has become an invaluable lens used by scholars in the field, to explore the various experiences of marginalized groups who over time have become almost invisible to the 'social scientific gaze,' (Rutherford and Davidson, 2019). To this extent, Purdie-Vaughns and Eibach (2008) have referred to this unique phenomenon as "intersectional invisibility." In the context of this study, rural adolescents have over time developed intersectional invisibility to the education sector.

Warner (2008) opines that individuals' identities are informed by the social context and structures influencing them. In the context of this chapter, rural adolescents and their ability to participate in the global education context are greatly influenced by the limited access to a global village which has been most evident during the COVID-19 crises. Subsequently, individuals' identity development is largely influenced by their context, which can also influence their understanding of privilege and oppression (Chan, Cor and Band, 2018: 61). The intersectionality framework therefore addresses the oppression of rural adolescents in relation to their social contexts and structures. Therefore the Intersectionality perspective can be used to guide multicultural frameworks within educational redress. Using this theoretical lens challenges educational institutions to search beyond interpersonal interactions to form knowledge of social inequalities connected to their institutions (Chan et al., 2018). Furthermore, this framework offers a perspective from which we can begin to co-construct the narratives of digital technology. Building participatory, democratic interpretive communities across differences of experience, expertise, and resources has been the hallmark of intersectional projects (Collins, 2019: 692).

The Intersectionality Wheel (Simpson, 2009) is helpful in the adoption of Intersectionality as an approach. According to Simpson (2009), "Societal forces are expressed through types of discrimination that affect individuals according to their personal social dimensions in ways that intersect for each individual in their unique circumstances and specific social location." In this regard, social forces adversely affecting rural adolescents, which are further underpinned by discriminations must be approached head-on. Moreover, the government will need to go beyond service provision, to address social determinants of accessing digital technologies and to empower them to overcome the unique barriers they experience in this regard (McCollum et al., 2019).

Figure 1. Intersectionality Wheel as described by Simpson (2009)

MATERIALS AND METHODS

This chapter employed the use of literature review approach. The authors further integrated document analysis and conceptual analysis of a variety of secondary sources of data. Secondary sources of data included: online newspaper articles, peer-reviewed journals articles, and research reports. The authors were fortunate that since the Covid-19 pandemic, a host of recent published journals and reports were found related to the study at hand. Since last year, much research has been centred on accessing digital technologies, more especially for students in rural areas as a means to bridge the digital divide gap. Hence, secondary sources of data were readily available encompassing all aspects that this study addresses.

Listed below are approximate number of online newspaper articles, peer-reviewed journal articles and research reports that were used by the authors to contribute to this study:

- Online newspaper articles 7
- Peer-reviewed journal articles: 26
- Research reports: 19

RESULTS AND DISCUSSION

The common thread that runs through the number of articles as aforementioned is how rural adolescents lack access to digital technologies. For reasons as discussed in the literature of this chapter, rural students remain marginalized due to geographical isolation. It is also undeniable that in order to eliminate poverty, reduce inequality and spur economic development, South Africa's National Development Plan (NDP) identifies and illuminates education, training, life-long learning and innovation as key priorities (National Planning Commission, 2012). However, in rural areas, such priorities have been a bone of contention due to the challenges that face those who live there. At grass roots, rural schools in South Africa are still struggling to implement the White Paper on e-education (Department of Basic Education, 2004). This is largely due to contextual issues, lack of basic resources and infrastructure as well as the ability to translate the country's national policies to provincial policies (Herselman et al., 2020:24).

For decades, rural education advocates and scholars alike have argued that rural students constitute a forgotten minority (Azano & Stewart, 2015). Hence, the Intersectionality perspective provides a framework for examining how multiple social categories combine in systems characterized by both oppression and privilege to affect the experiences of those occupying the intersections of these social categories (Rutherford and Davidson, 2019:1).

Whilst much literature is presented on the compounding inequalities that these students face, more literature is needed regarding how to change the course of these students' lives. Authors such as Maurya (2020: 177) state that the digital divide can be eradicated by providing various technical facilities in this direction such as technical skills education and awareness for rural people. Taking all this into consideration, "it is worth exploring the extent to which the pandemic has forced us to examine discourses of social justice afresh, as pertinent to the possibilities that online learning can proffer to achieve the goals of democratic teaching/learning contexts in future research endeavours" (Du Preez and le Grange, 2020: 101).

ADDRESSING THE COMPOUNDING INEQUALITIES FACED BY RURAL ADOLESCENTS

Partnerships That Matter

As discussed in "South Africa: Broken and unequal education" (2020) the government enacted the Minimum Norms and Standards for educational facilities in 2013, which required them to ensure that by November 2016 all schools ought to have access to sanitation and electricity. However the National Education Infrastructure Management System (NEIMS) standard report (2019) indicated that there were still schools without electricity (DBE, 2019). This is a testament to the state in which a number of schools in the rural areas are. In order for schools in South Africa to compete with schools across the globe it is time that our government addresses these issues and come up with feasible solutions to these problems. One such way is through partnerships and collaboration between government and private companies. The partnership between the Vodacom Foundation and the Department of Basic Education (DBE) is one such example. The partnership established the Mobile Education Program which provided digital literacy to learners and teachers (Advertising Supplement, 2019). It is through partnerships like these that schools in the rural area can be digitally empowered, because it would be pointless to equip schools with ICT devices and not train teachers on how to use those devices.

Additionally, the South African government needs to team up with projects such as "Computer for Every Child" which main goal, according to Zilka (2016) is to reduce the digital divide between the wealthy and underprivileged populations. It is time that the government reach out and seek assistance with regards to narrowing the digital divide. Without immediate intervention, the digital gap would surely increase and worsen the compounding inequalities in the education system.

Digitalization

Educational researchers in countries like Sweden and certain Nordic countries maintain that schools in rural contexts (Pettersson, 2017) are promising in terms of remote and distance teaching. This can be particularly beneficial to overcoming challenges including, but not limited to providing equal teaching and learning opportunities and inclusion for children facing challenges, to learn in schools (Stenman and Petterson, 2020; Hilli, 2020). To put the context of digitization into perspective, Siljebo (2020: 33) postulates that: digitalization in rural schools corresponds with the synchronous use of digital technologies within rural school organizations for human learning and development, where teachers and pupils are separated in space. As with other countries, South Africa can also adopt digitization for learning achievements and pedagogical experimentation in rural adolescents.

Collaboration of Government Sectors

Despite the limited funding resources allocated to the South African Department of Education, there still remains opportunity for improving access to digital technologies. The establishment and implementation of collaborative efforts across the various government sectors including that such as as the Departments of Communications, Science and Technology, Rural Development and Landform, can foster highly productive relations that can assist the Education sector.

Developing Platforms for the Use of Technology

There remains a great need for researchers in the field to continuously and extensively respond to the growing body of research regarding the need for equal access to digital technologies, more especially by developing viable platforms for the use of technology in the improvement of skills and knowledge delivery to rural adolescents.

Policy as a Major Tool to Overcome the Digital Divide

Many scholars have highlighted the benefits of using policy as a tool to narrow the digital divide. Shenglin, Simonelli, Zhang and Li (2017) assert that policy can be used as a major tool which is instrumental in narrowing the digital divide. This is most beneficial for contributing towards the reduction or overall possible eradication of the existing digital divide and the lack of infrastructure contributing to the problem. This can be in the area of inclusion of ICT content in the curriculum of students in various school levels, policies promoting the provision and availability of computers for students, availability of internet facilities in public domains, amongst others (Akintolu and Uleanya, 2021: 717).

CONCLUSION

In this chapter, the authors illuminated the challenges faced by rural adolescents in accessing digital technologies. The extent to which rural students as marginalized groups remain deprived in the midst of crises are is also widely discussed. The authors recognize the disparities of the education system that continue to further marginalize our rural youth, yet also maintain that it should be used for the primary purpose of improving the educational endeavour rather than to recover an equilibrium of social justice in South Africa. There still however, remains a gap that further needs to be researched. One in which critical emancipatory research can create a disquisition around, provoke dialogue, inform policy, reconstruct and empower the principles of democracy in South Africa.

REFERENCES

Akintolu, M., & Uleanya, C. (2021). Ensuring Sustainable Development Goal in Rural Africa through Adult Literacy Programme: A Case Study of Technology Usage in Developing Nations. *Universal Journal of Educational Research, 8.*

Azano, A. P., & Stewart, T. T. (2015). Exploring place and practicing justice: Preparing preservice teachers for success in rural schools. *Journal of Research in Rural Education, 30*(9), 1–12. https://jrre.psu.edu/

Bashman, L. (2020). *Online learning punishes the poor*. Independent Online. Available athttps://www.iol.co.za/news/opinion/online-learning-punishes-the-poor-46963674

Beckman, K., Bennett, S., & Lockyer, L. (2019). Reproduction and transformation of students' technology practice: The tale of two distinctive secondary student cases. *British Journal of Educational Technology, 39*(6), 346. doi:10.1111/bjet.12736

Bonilla-Molina, L. (2020). Covid-19 on Route of the Fourth Industrial Revolution. *Postdigital Science and Education*, 1-7.

Bourdieu, P., & Passeron, J. C. (1977). Reproduction in Education, Society and Culture. *Sage (Atlanta, Ga.).*

Chan, C. D., Cor, D., & Band, M. P. (2018). Privilege and Oppression in Counselor Education: An Intersectionality Framework. *Counselor Education Faculty Publications and Presentations.*, 50. https://archives.pdx.edu/ds/psu/24227

Chetty, Qigui, Gcora, Josie, Wenwei, & Fang. (2018). Bridging the digital divide: measuring digital literacy. *Economics: The Open-Access, Open-Assessment E-Journal, 12*(23), 1-20.

Collins, P. H. (2019). *Intersectionality as critical social theory*. Duke University Press.

Department of Basic Education. (2004). *White paper on E-Education: Transforming learning and teaching through information and communication technologies*. Available: http://www.education.gpg.gov.za/Legislation/White%20paper%207%20on%20e-Education.pdf

Department of Basic Education. (2017). *Rural Education Draft Policy*. Retrieved from http:// pmg-assets.s3-website-eu-west-1.amazonaws.com/180126draftruraledupolicy.pdf

Department of Education South Africa. (2019, August 30). *National Education infrastructure Management System Standard Report August 2019*. Author.

Di Pietro, G., Biagi, F., Costa, P., Karpiński, Z., & Mazza, J. (2020). *The likely impact of COVID-19 on education: Reflections based on the existing literature and recent international datasets* (Vol. 30275). Publications Office of the European Union.

Du Plessis, P. (2014). Problems and complexities in rural schools: Challenges of education and social development. *Mediterranean Journal of Social Sciences*, *5*(20), 1109–1117. doi:10.5901/mjss.2014.v5n20p1109

Du Preez, P., & le Grange, L. (2020). *The COVID-19 pandemic, online teaching/learning, the digital divide and epistemological access*. Unpublished paper.

Gómez-García, M., Hossein-Mohand, H., Trujillo-Torres, J. M., & Hossein-Mohand, H. (2020). The Training and Use of ICT in Teaching Perceptions of Melilla's (Spain) Mathematics Teachers. *Mathematics*, *8*(10), 1641. doi:10.3390/math8101641

Hanekom, P. (2020). Covid-19 exposes South Africa's digital literacy divide. *Mail & Guardian*. Available at: https://mg.co.za/opinion/2020-09-08-covid-19-exposes-south-africas-digital-literacy-divide/

Hargittai, E., & Dobransky, K. (2017). Old dogs, new clicks: Digital inequality in skills and uses among older adults. *Canadian Journal of Communication*, *42*(2). Advance online publication. doi:10.22230/cjc.2017v42n2a3176

Hedding, D. W., Greve, M., Breetzke, G. D., Nel, W., & Van Vuuren, B. J. (2020). COVID-19 and the academe in South Africa: Not business as usual. *South African Journal of Science*, *116*(7-8), 1–3. doi:10.17159ajs.2020/8298

Hennessy, T., Doris, L., & Brian, M. (2016). The Digital Divide in Farming: A Problem of Access or Engagement? *Applied Economic Perspectives and Policy*, *38*(3), 474–491. doi:10.1093/aepp/ppw015

Herselman, M. E., Botha, A., Maremi, K. J., Dlamini, S. B., & Marais, M. A. (2020). *Mobile technology affecting teaching and learning in rural schools*. Academic Press.

Hilli, C. (2020). Distance teaching in small rural primary schools: A participatory action research project. *Educational Action Research*, *28*(1), 38–52. doi:10.1080/09650792.2018.1526695

Hohlfeld, T. N., Ritzhaupt, A. D., & Barron, A. E. (2010). Connecting schools, community, and family with ICT: Four-year trends related to school level and SES of public schools in Florida. *Computers & Education*, *55*, 391–405. doi:10.1016/j.compedu.2010.02.004

Hollingworth, S., Mansaray, A., Allen, K., & Rose, A. (2011). Parents' perspectives on technology and children's learning in the home: Social class and the role of the habitus. *Journal of Computer Assisted Learning*, *27*, 347–360. doi:10.1111/j.1365- 2729.2011.00431.x

Jantjies, M. (2020), April 29. How South Africa can address digital inequalities in e-learning. *Conversation*. Available at: https://theconversation.com/how-south-africa-can-address-digital-inequalities-in-e-learning-137086

Krönke, M. (2020). Africa's digital divide and the promise of e-learning. *Afrobarometer Policy Paper*, (66).

Kuran, C. H. A., Morsut, C., Kruke, B. I., Krüger, M., Segnestam, L., Orru, K., Nævestad, T. O., Airola, M., Keränen, J., Gabel, F., Hansson, S., & Torpan, S. (2020). Vulnerability and vulnerable groups from an intersectionality perspective. *International Journal of Disaster Risk Reduction, 50*, 101826. doi:10.1016/j.ijdrr.2020.101826

Lai, J., & Widmar, N. O. (2020, October 12). (2020). Revisiting the Digital Divide in the COVID-19 Era. *Applied Economic Perspectives and Policy*. Advance online publication. doi:10.1002/aepp.13104 PMID:33230409

Lam, J., & Duan, G. (2012). A review of the mobile learning environment in the higher education sector of Hong Kong: Technological and social perspectives. In S. K. Cheung, J. Fong, L. Kwok, K. Li, & R. Kwan (Ed.), Hong Kong (pp. 165-173). Springer. doi:10.1007/978-3-642-32018-7_16

Mahaye, N. E. (2020). The impact of COVID-19 pandemic on education: navigating forward the pedagogy of blended learning. *Research Online*.

Mangwaya, E., & Ndlovu, E. (2013). Rural Secondary School Teachers' Capacity to Respond to Hiv and Aids: The Case of Shurugwi District in Zimbabwe. *Asian Social Science*, *9*(2), 129. doi:10.5539/ass.v9n2p129

Maurya, P. (2020). Need for Digital Learning in Rural Areas: Special reference to Higher Education. *Research Review Journals.*, *5*(9), 175–177. Advance online publication. doi:10.31305/rrijm.2020.v05.i09.044

Mbodila, M. (2020). *Online learning – The Pandemic cannot Change Reality*. University World News: Africa Edition. Available at: https://www.universityworldnews.com/post.php?story=20200420130222745

McCollum, R., Taegtmeyer, M., Otiso, L., Tolhurst, R., Mireku, M., Martineau, T., & Theobald, S. (2019). Applying an intersectionality lens to examine health for vulnerable individuals following devolution in Kenya. *International Journal for Equity in Health, 18*(1), 1–15. doi:10.118612939-019-0917-2 PMID:30700299

Mitchell, D. E., & Nielsen, Y. S. (2012). *Internationalization and Globalization in Higher Education*. Academic Press.

National Planning Commission. (2012). National Development Plan 2030: Our future–make it work. Presidency of South Africa.

Ndimande, B. S. (2016). Pedagogy of Poverty: School Choice and Inequalities in Post-Apartheid South Africa. *Global Education Review*, *3*(2).

Ngalomba, S. (2020). *Using WhatsApp to Enhance Online Learning*. University World News: Africa Edition. Available at: https://www.universityworldnews.com/post.php?story=20200421102812987

O'Connell, A., & Clarke, S. (2020). A School in the Grip of COVID-19: Musings from the Principal's Office. *International Studies in Educational Administration*, *48*(2), 4–11.

Papadakis, S., & Kalogiannakis, M. (2017). Mobile educational applications for children. What educators and parents need to know. *International Journal of Mobile Learning and Organisation*, *11*(3), 256–277. doi:10.1504/IJMLO.2017.085338

Petrie, Aladin, Ranjan, Javangwe, Gilliland, Tuominen, & Lasse. (2020). *Spotlight: Quality education for all during Covid 19 crisis*. Retrieved from https://hundred-cdn.s3. amazonaws.com/uploads/report/file/15/hundred_spotlight_covid-19_digital.pdf)

Pettersson, G. (2017). *Inre kraft och yttre tryck: perspektiv på specialpedagogisk verksamhet i glesbygdsskolor* [Inner power and outer pressure: perspectives on special education in rural schools] (Doctoral dissertation). Umeå Universitet.

Povey, J., Campbell, A. K., Willis, L.-D., Haynes, M., Western, M., Bennett, S., Antrobus, E., & Pedde, C. (2016). Engaging parents in schools and building parent-school partnerships: The role of school and parent organization leadership. *International Journal of Educational Research*, *79*, 128–141. .Ijer.2016.07.005 doi:10.1016/j

Power, J. R. A. T. M., & Nichols, B. H. (2020). Teachers Bridging the Digital Divide in Rural Schools with 1: 1 Computing. *Rural Educator*, *41*(1), 61–76. doi:10.35608/ruraled.v41i1.576

Purdie-Vaughns, V., & Eibach, R. P. (2008). Intersectional invisibility: The distinctive advantages and disadvantages of multiple subordinate-group identities. *Sex Roles*, *59*(5-6), 377–391. doi:10.100711199-008-9424-4

Robinson, L. J., Schulz, J., Blank, G., Ragnedda, M., Ono, H., Hogan, B., Mesch, G. S., Cotten, S. R., Kretchmer, S. B., Hale, T. M., Drabowicz, T., Yan, P., Wellman, B., Harper, M.-G., Quan-Haase, A., Dunn, H. S., Casilli, A. A., Tubaro, P., Carvath, R., ... Khilnani, A. (2020). Digital inequalities 2.0: Legacy inequalities in the information age. *First Monday*, *25*(7). Advance online publication. doi:10.5210/fm.v25i7.10842

Rutherford, A., & Davidson, T. (2019). Intersectionality and the history of psychology. In Oxford Research Encyclopedia of Psychology. doi:10.1093/acrefore/9780190236557.013.468

Salas-Pilco, S. Z., & Law, N. W. Y. (2018). ICT curriculum planning and development: Policy and implementation lessons from small developing states. In *ICT-Supported Innovations in Small Countries and Developing Regions* (pp. 77–98). Springer. doi:10.1007/978-3-319-67657-9_4

Sayed, Y., & Singh, M. (2020). Evidence and education policy making in South Africa during Covid-19: Promises, researchers and policymakers in an age of unpredictability. *SARE*, *26*(1), 20–39.

Schneider, J. (2020). Perspective: How COVID-19 laid bare the vast inequities in U.S. public education. *Washington Post*. Retrieved from https://www.washingtonpost.com/education/2020/04/14/how-covid-19-has-laid-bare-vast-inequities-us-public-education/

Shenglin, B., Simonelli, F., Zhang, R., & Li, W. (2017). *Digital infrastructure: Overcoming Digital Divide in Emerging Economies.* Available from https://economic-policy-forum.org/wp-content/uploads/2017/03/Policy-Brief_ZJUCEPS_Digital-infrastructure_Over coming-Digital-Divide-in-Emerging-Economies_2017022 8.pdf

Siljebo, J. (2020). Digitalization and digital transformation in schools: A challenge to educational theory? *Education in the North, 27*(2), 24–37.

Simpson, J. (2009). *Everyone belongs: a toolkit for applying intersectionality. Ottawa. South Africa: Broken and unequal education perpetuating poverty and inequality. 2020, February 11.* Amnesty International. Available at: https://www.amnesty.org/en/latest/news/2020/02/south-africa-broken-and-unequal-education-perpetuating-poverty-and-inequality/

Steele, C. (2019). *What is the Digital Divide?* Digital Divide Council. Available at: http://www.digitaldividecouncil.com/what-is-the-digital-divide/

Stenman, S., & And Pettersson, F. (2020). Remote teaching for equal and inclusive education in rural areas? An analysis of teachers' perspectives on remote teaching. *The International Journal of Information and Learning Technology, 37*(3), 87–89. doi:10.1108/IJILT-10-2019-0096

Supplement, A. (2019). Vodacom has partnered with government for teacher development. *Mail & Guardian.* Available at: https://mg.co.za/article/2019-10-11-00-vodacom-has-partnered-with-government-for-teacher-development/

Terhoven, R., & Fataar, A. (2018). The role of leadership practices in establishing a curriculum policy platform at working-class schools. *South African Journal of Education, 38*(2), 1–10.

UNESCO. (2020). *COVID-19 Educational Disruption and Response.* https://en.unesco. org/news/covid-19-educational-disruption-and-response

Walters, A. (2020). Inequities in access to education: Lessons from the COVID-19 pandemic. *The Brown University Child and Adolescent Behavior Letter, 36*(8), 8–8. doi:10.1002/cbl.30483

Walters, A. (2020). Supporting youth and families during COVID-19. *The Brown University Child and Adolescent Behavior Letter, 36*(6), 8–8. doi:10.1002/cbl.30472

Wang, J., Tigelaar, D. E. H., & Admiraal, W. (2019). Connecting rural schools to quality education: Rural teachers' use of digital educational resources. *Computers in Human Behavior, 101*, 68–76. doi:10.1016/j.chb.2019.07.009

Wargo, E., Chellman, C., Budge, K., & Davis, C. (2020). On the digital frontier: Stakeholders in rural areas take on educational technology and schooling. *Journal of Research on Technology in Education*, 1–19. doi:10.1080/15391523.2020.1760753

Warschauer, M., Knobel, M., & Stone, L. A. (2004). Technology and equity in schooling: Deconstructing the digital divide. *Educational Policy, 18*(4), 562–588. doi:10.1177/0895904804266469

Willis, J. C., & Templeton, N. R. (2018). Investigating the establishment and sustainability of professional learning communities in rural east Texas: The principals' perspectives. *Rural Educator*, *38*(1), 30–37. doi:10.35608/ruraled.v38i1.233

Ye, L., & Yang, H. (2020). From digital divide to social inclusion: A tale of mobile platform empowerment in rural areas. *Sustainability*, *12*(6), 2424. doi:10.3390u12062424

Zilka, G. C. (2016). Reducing the digital divide among children who received desktop or hybrid computers for the home. *Journal of Information Technology Education*, *15*, 233–251. doi:10.28945/3519

Chapter 5
Harnessing Social Media to Improve Educational Performance of Adolescent Freshmen in Universities

Munyaradzi Zhou
https://orcid.org/0000-0002-2184-0290
Midlands State University, Zimbabwe

Cyncia Matsika
Midlands State University, Zimbabwe

Tinashe Gwendolyn Zhou
Midlands State University, Zimbabwe

Wilfreda I. Chawarura
Midlands State University, Zimbabwe

ABSTRACT

Social media use in different industries and commerce is high. Little exists on harnessing Facebook and YouTube to improve the educational performance of adolescent students. Researchers support that the use of social media has a positive correlation with educational performance. However, 'little' or 'no' research has been done on how social media platforms are customized to improve the academic performance of adolescents. Data was collected using interviews in the qualitative case study. The chapter proffers solutions like coming up with a private resource center hosted by university departments. The resource center is solely accessible for accessing research content and making contributions by academics, students, and industry. Further research might include a framework for the coalition of universities in sharing educational content on a global scale and designing local social media platforms.

DOI: 10.4018/978-1-7998-8318-0.ch005

INTRODUCTION

Social media platforms play a crucial role in education through seamless access of information in learning groups and other educational systems (Greenhow & Robelia, 2009). The use of social media platforms like Facebook and YouTube for learning has received attention even though customization of these services received 'little' or 'no' research to endow soft and practical skills in freshmen adolescents. Several pieces of research support that the use of social media has a positive correlation with academic performance (Kolhar et al., 2021; Lambić, 2016; Lavuri et al., 2019). This research solely focuses on students who are aged 19 or below according to UNICEF's definition of adolescents (UNICEF, 2019). A methodological gap is addressed in this research since others have focused on university students in general and including adults (Kolhar et al., 2021). Most researchers have focused on the effects (positive and negative) of social media networks on academic performance (Kolhar et al., 2021; Osharive, 2015). The impact of electronic media use has received much attention but less is known about techniques to harness social media to attain academic performance are (Kolhar et al., 2021). This study prospectively informs how social media can be customized to improve academic performance in late childhood and tries to apply semantic analysis of data in coming up with academic resource centers.

The book chapter on Harnessing Social Media to improve Educational Performance of Adolescent Freshman in Universities will fulfil the following objectives.

1. To identify general strategies to improve the educational performance of adolescent freshmen
2. To enquire how to use Facebook and YouTube to boost students' academic performance.
3. To endow Industry technical skills and 4IR skills industry technical skills using Social Media
4. To come up with strategies to develop department centered educational resource centers in universities

The upcoming sections focus on the background, the main focus of the book chapter, solutions and recommendations, future research directions, and the conclusion.

BACKGROUND

First-year students are essential in the university system since they're new in the academic journey and they are the lifeblood for the continuity. The new generation of students are techno-savvy, and they are accustomed to entertainment hence engaging them on Facebook and YouTube is essential (Duverger & Steffes, 2012; Moghavvemi et al., 2018). Most researches mainly focus on the bearing, and use of social media on academic performance. 'Little or 'no' research was done on the processes to follow to endow or aimed at enhancing academic performance (Abbas et al., 2019; Altawalbeh, 2021; Ansari & Khan, 2020; Faraj & Al-Abdallat, 2018; González et al., 2016; Kulidtod & Pasagui, 2017; Moghavvemi et al., 2018; Rouis et al., 2011; Talaue et al., 2018). Faraj & Al-Abdallat (2018) researched Jordan University first-year students studying the English Language. They focused on the impact of YouTube and Facebook on their performance. The research did not explicitly focus on the process of customizing these technologies to improve academic performance. Leal-Rodrígueza & Albort-Morant (2017) focused on experiential learning and did not focus on social media customization to attain this. The book chapter dovetails the concepts of expanding the social media platforms like Facebook's new pages for academic purposes. The pages contain material like specific modules, student-centered materials, learning or discussion groups linked within university subject matter domains (Bich Diep et al., 2021). Ultimately, this results in enhanced academic performance or activities rather than students being addicted to social activities (Mim et al., 2018).

The customization of Facebook pages linked with Learning Management Systems (LMS) like Google Classroom proffers a high level of interaction among lecturers, students, and other stakeholders. The interaction results in bonding among the stakeholders concerned. A programme and or single module-specific Facebook accounts are used to connect lecturers and students so that they can seamlessly interact in the academic journey (Bich Diep et al., 2021). The andragogic learning model which is mainly used in universities supports learning through facilitation. Facilitators can assign students to create videos to complement existing materials, thus supporting collaborative learning (Olasina, 2017). Social media is a networking digital technologies tool to connect people in different sectors for communication, sharing of content (videos, music, pictures, documents, and so on), interactive engagement (Krajcovic & Cabyova, 2020). This book chapter refers to Facebook and YouTube as Social Media. Academic performance is the attainment of the ultimate goal of learning outcomes during study and post-graduation (employment in the industry) (Lamas, 2015).

MAIN FOCUS OF THE CHAPTER

The book chapter mainly focuses on techniques to harness social media to improve the educational performance of first-year students in universities. The United Nations International Children's Emergency (UNICEF) focuses on supporting education. Adolescents are children between the ages of 10 and 19 years of age (UNICEF, 2019). The research primarily focuses on freshmen students aged between 17 and 19 years inclusive. Little research on the customization of Facebook to enhance academic performance exists. Customization can include the launching of free resource centers with course or module-specific content. Interactive content can be sorted and searched according to preferred parameters for a specific module and or cohorts of students (Aydin, 2012). The dovetailing of elements includes improving the educational performance of students; the use of Facebook and YouTube to improve the educational performance of first-year students; endowing Industry technical skills and 4IR skills industry technical skills and 4IR skills using Social Media; and the process of developing department-centered educational resource centers in universities.

Improving Educational Performance of Students

Educational performance is directly affected by the framework that defines the existing education systems. The framework includes elements, namely, the availability of learning materials and academic staff, the motivation and involvement of the academic staff, students, and other stakeholders such as guardians and industry captains. Student motivation results in the full utilisation of resources and skills available for learning. The culture of owning the education system by all universities stakeholders is critical. Industry captains participate in the transforming teaching methodologies and content used in universities (Masino & Niño-Zarazúa, 2016). The book chapter proposes the use of Facebook and YouTube to bridge the gap of inadequate educational content and support the seamless transfer of knowledge among community members. The learner is motivated to learn when full support is received from the community.

All stakeholders' involvement synergically promotes innovation which can be spearheaded by industry experts. The process supports new knowledge generation and experiential learning practices (Leal-Rodrígueza & Albort-Morant, 2017). Facebook and YouTube improve information searching strategies and the betterment of the learning environment. The content accessed is personalised according to the user's frequency of access. The community complements highly engaging chatbots on Facebook to help the students to understand the searching strategies. The chatbots can deflect the conversation and link it to a specific human operator (community member) thus improving students' experiences (Tsai & Tsai, 2003). Facebook use greatly increases the interaction among students and ultimately academic performance (Al-Dheleai & Tasir, 2017). The use of Facebook mainly for socialization results in poor academic performance (Kirschner & Karpinski, 2010). Disincentivizing the use of Facebook mainly for socialization and monitoring its use to avoid addiction is key (Azizi et al., 2019; Dontre, 2020). The pros and cons of using Facebook depend on the individual's use, duration, frequency, purpose, and so on.

Students believe academic performance is attained through efforts by both the instructor (64%) and the learner (36%). The remaining 2% relate to environmental and other factors. Instructors' responsibilities are giving more examples or exercises; moderating the teaching speed; avoiding 8 am classes; giving students extra work; making presentations more interesting, and improving teaching style; providing important materials for exam preparation; exercising control to the class; and giving special attention to weak students. Students' responsibilities include doing more revisions or exercises; being hard workers; having a positive attitude towards learning; prior preparation for classes; studying smart, and focusing during lectures or tutorials. Digital technologies help in monitoring students' activities through plugins that monitor the students' behaviour during the use of resources on YouTube and Facebook. Instructors can conform to their work processes with the aid of the applications. In addition, the applications and agents help students to memorize formulae; manage time; initiate group discussions, and solve past exam papers and question banks (Sheng & Basaruddin, 2014).

Learning Management Systems (LMS) helps in improving learning experiences online. The integration of LMS with Facebook pages using artificial intelligence algorithms to generate exercises, tests, and more examples that are broadcasted to the students' profiles or Facebook pages improves the learners' experiences. Auto-generation of teaching materials help students to do more work like revising past exam papers, thus students remain classroom-bound virtually. The creation of time-lapse and slow-motion videos accommodates weak or slow learners. Time management during the use of the application must be student-centered and according to patterns of access times of the LMS and Facebook pages. Agents on the LMS and Facebook pages can act as reminders on the learner or lecturer's schedule. The academic community also plays a pivotal role in time management. However, time management must conform to the daily, weekly, monthly, and semester targets. Time management models must instill discipline in students. The process might include reminders on the Facebook pages or even integrated with the Short Message Service (SMS) or other real-time messaging technologies.

Videos and animations during presentations engage the students. Andragogic and pedagogical approaches to teaching and learning must match students' characteristics. The videos can include formulae stickers on each subject matter, and they can be flashcards-styled even when the student is not making use of them. Flashcards and stickers help students in memorizing the formulae. Hardworking students are motivated through a rewarding model with elements such as badges, monetary incentives, free mobile data, and so on. The model encourages students to do revisions or exercises regularly, aiming for recognition. Recognition might include internships and student-exchange programs with prestigious universities and other monetary or non-monetary incentives. Artificial intelligence might help to monitor students' behaviour on Facebook and YouTube during activities such as discussions, tutorials, lectures, and studies. Therefore, students remain focused and attentive. Pratama et al. (2018) concluded that YouTube provides motivation to students to learn more, and the application is very productive in practical modules (A. Buzzetto-More, 2014; Moghavvemi et al., 2018; Mustafa et al., 2020; Olasina, 2017).

Endowing Industry Technical Skills and 4IR Skills Industry Technical Skills and 4IR Skills Using Social Media

Facebook and YouTube are used for educational purposes without fully exploiting the fourth industrial revolution concepts. The fourth Industrial Revolution (4IR) is the integration of artificial intelligence-enabled automated educational processes and industrial processes through the acquisition and the use of various artificial intelligence technologies. The 4IR tenets include robotics, artificial intelligence, the internet of things, digitalization, and automation. The application of artificial intelligence is at an advanced stage, users and developers seek to come up with platforms that are partially human dependent (Deloitte, 2018; Doorsamy et al., 2020; Sima et al., 2020). The innovations in digital technologies have resulted in the adoption of the fourth industrial revolution (4IR). 4IR transforms the teaching models used globally and the way work processes are done in the industry. The teaching models might include the provision of on-job training, development of 4IR skills, and training of the curriculum through partnership with industry experts. The spread of COVID-19 like wildfire resulted in the full adoption of virtual learning by universities. In general 4IR results in innovations in the teaching methodologies, learners' experiences, and technical skills in the industry (Kamaruddin et al., 2016). 4IR is essential in this research since it informs the application of AI in the use of YouTube and Facebook for automation. Universities are preparing new graduates for the highly digitalized labour market and are enhancing the employability of the graduates (Ebehard et al., 2017).

The Process of Developing Department-Centered Educational Resource Centers in Universities

The synchronization of existing educational learning materials and knowledge bases with social media is essential for knowledge discovery and continuity in the learning landscape. The use of cloud computing services on the social media platform bridged the gap of inadequate technical infrastructure in universities. It further blends existing academic resources (Mishra et al., 2020). Education resource centers (ERC) are essential in knowledge retention. ERC can be developed by synergically merging different units such as the library resources, the lecturer and students resources, cloud computing-based resources, and relevant stakeholders (Ekanem, 2015).

Methodological Gap

Researchers mainly used quantitative data collection procedures using questionnaires to focus on the impact of social media on academic performance. This book chapter uses qualitative interviews to inform on how practically social media improves academic performance (Ainin et al., 2015; Al Zboon et al., 2018; Alkhawaja & Halim, 2019; Altawalbeh, 2021; Azizi et al., 2019; Bich Diep et al., 2021; Faraj & Al-Abdallat, 2018; Mustafa et al., 2020). A single case study in the Department of Information Systems was used as resulted in theoretical saturation in enquiring how social media can be used to improve the performance of adolescents (Boddy, 2016). The students will fit well in the industry if they become tech-savvy with 4th industrial revolution skills. Data was collected from 29 freshmen adolescents in the late ages of 17 to 19 years which satisfies the recommended range of adolescents (Dworkin, 2012).

SOLUTIONS AND RECOMMENDATIONS

The book chapter solutions and recommendations are presented in detail in this section. Incentives in the use of the social media platform on a quota-based system improve access by students. Using Facebook for socialization in a quota-based environment that is dependent upon the learning process is crucial. It supports high engagement and interaction among learners hence an improved learning experience through collaboration. The management of academic Facebook pages and YouTube channels derive value. For instance, indices are used to create patterns and match related learning material. Facebook custodians can be asked for personal statistics on the usage of the platforms by currently enrolled students. The social media platforms need to be monitored by lecturers, quality assurance members, and students. Evaluations twinned with benchmarking are critical in enhancing learning. An amalgamation of learning resources on social media platforms and other learning resources into Private Learning Resource Centers (PLRCs) hosted by University Departments is essential. PLRCs are solely accessible for accessing research content and collaboration by academics, students, and industry. Access to PLRCs is free and is defined to support authentic research re-use and analyses of data.

General Approach to Improve Educational Performance of University Freshmen

The involvement of all stakeholders is crucial since they feel in ownership. The provision of necessary resources is necessary for access to education, for instance, through providing data. Affordable mobile data acquisition is possible through the partnership of mobile network operators and universities. Virtual meetings support life-long learning as it allows students to discuss and run sessions as a class according to either proximity or defined parameters. Customized social applications networks students at all levels in the same department to interact for discussions. Cloud computing services integration results in the provision of necessary required resources for the students. The cloud supports the delivery of all services at one point on the cloud. Virtual libraries and LMS empowerment are critical for learning. However, cloud-based services need to be complemented by face-to-face lectures. The management of the transition from high school to university is essential during the first year at university. It will include training and carrying awareness campaigns on the use of virtual learning. Information and Communication Technologies tools provision to both the learners and lecturers enable utilization and use of the resources.

Improving Educational Performance of University Freshmen using Facebook and YouTube?

During the COVID-19 and other pandemics that deters gatherings, delivering lessons on social media promises to be fruitful as students connected on these sites. The platform supports the broadcasting of updates, and alerts, and live-streaming of lectures. Students also confirmed that creating Facebook pages and YouTube channels for the students is critical for learning. The use of YouTube enables students to access the content at any given time. Students use the videos on topics that they do not understand. Google Classroom supports live streaming and recording of the sessions. Lecturers and students uploads videos on YouTube channels or Facebook pages of the university according to the predefined parameters like student level, modules of study, and concepts. Pre-recorded videos on the cloud provide a favorable environment for revision. In addition, these videos generally are engaging in nature, more so if they're animated. Awareness programs are essential to encourage responsible use of cloud-based educational materials.

Developing Industry Technical Skills and the Fourth Industrial Revolution (4IR) Skills in Freshmen

Digitalization and COVID-19 have mainly resulted in employees working from home. Lockdowns call for online demonstrations on practical skills. Online networked hubs allow sharing of opportunities amongst individuals' networks, universities, and industry are a synergy to the development of the student. Students who have acquired the 4IR skills are prioritized during employment to encourage uptake of the skills. Employee development must continue to foster 4IR skills. YouTube supports visual learning. The application of YouTube in practical skills might tandem with the industry's evolving needs. The academic and industry experts collaborate continuously to keep the curriculum and skills of students in alignment with ever-changing industry needs. The demonstration of skills will include live demonstrations and a reintroduction to technology. Social media platforms support seamless submission of documentation or work among the industry, academic collaborators, and the student community. Social media platforms and plugins complement the process of reminding and notifying students and lecturers of pending work, assignments, lectures, and outstanding tasks.

Using Social Media Networking to Create Electronic Educational Resource Centers in Universities

Social media support the process of creating Electronic Educational Resource Centers (ERC). The library department is engaged in the process of integrating existing databases with the social media platform. Electronic learning systems resources are amalgamated by involving all the key stakeholders such as the academic staff and students. Repositories integration with eLearning platforms must support access through social media. Library links created on Facebook pages and YouTube channels embedded with filters like location identification help in the classification of students into groups according to predefined parameters. Students studying the same modules can easily discuss different concepts. Management of accounts that links the ERC and social media accounts must be semi-automated to improve the access and experience of users. Archived and real-time educational material on Facebook pages aggregated in educational resource centers and other formal learning management platforms like libraries and learning management systems must remain accessible. Online resource centers, hubs, and groups allow students to access learning materials from anywhere, and networking this enriching their learning experiences. Creating digital hubs and groups online will assist as well.

Machine learning agents crawls on social media to create customized educational resource centers. Institutions create pages on social media platforms and publish materials. Academics provide links to the resource centers. Forcing students to regularly switch to social media educational resources as if they are socializing is essential. Students provide research material and make new research materials based on their encounters and researches. Centers that offer and manage learning materials are must be provided with guidance, training to ensure proper usage of social media, and elaborating the communication and collaboration processes essential in sharing the bodies of knowledge.

Intelligent archiving in ERC defines the semantics of the resources to the user. Private Facebook pages created by lecturers for academic purposes like assessment and evaluation, communication (including automated reminders), supporting highly engaging interactions including hashtags, and stimulating discussions to improve the teaching and learning experiences. Auto-creation of flashcards by the system or students and lecturers improves teaching and learning experiences. The creation of flashcards algorithmically can be supported through the pinning of content. The pinning facility allows the appreciation of high-performing students using posters or tags on the Facebook pages. Flashcards created can be module-specific, topic-specific, program-specific, and so on, hence it can support micro-learning. A rewards and penalties model on the use of social media platforms can enhance student performance. Facebook pages integrated with institutional customized blogs which exploit cloud computing services and interacts with other universities' blogs. Data collected from social media platforms and the ERC can be cleaned and later shipped to customized Facebook pages. The bodies governing universities are fostering the implementation of a minimum board of knowledge (MBK). MBK allows seamless integration of knowledge by academics, students, and universities thus informing the building of social media platforms and ERC for educational purposes. The Facebook page content (student work) is profiled and aggregated after each academic semester for the specific modules, academic levels, degree programs, or a specific academic tenure and shipped to other institutions through the use of plugins that supporting collaboration.

An extension of the Facebook pages with Alumni Facebooks groups is critical to endow the industry linkages and new skills and technologies according to the curriculum on offer. The engagement of industry experts proffers extracurricular activities to bridge the skills gap. Furthermore, alumni contribute by indexing videos and flashcards in a bid to derive value from industry trends. Videos are crucial in demonstrating industry-practical skills. The interactive content on videos increases learners' engagement. Improving video engagement by gamification and animation enhances learning through the mastering of concepts and ideas. In addition, it boosts the cognitive reasoning of students and makes learning fun, and other students learn better using audio and visuals methods.

FUTURE RESEARCH DIRECTIONS

The researchers will seek ethical clearance from Facebook and stakeholders concerned to receive statistics on the use of Facebook. The statistics are used in comparing the academic performance of users and non-users of the platform or the frequency of use of Facebook groups and YouTube channels for educational purposes. After deployment of Facebook pages, study duration leading to boredom and addictions will be investigated. Stakeholders such as lecturers and industry experts audit trails and certify content so that information is contributed ethically by both educators and students (Nizami et al., 2017). Implementing pedagogical methodologies in institutional blogs, Facebook pages, and LMS improves academic performance. The partnership among parents and lecturers in monitoring social media use is essential so that academic performance is not compromised (Altawalbeh, 2021).

The effects of rewards and penalties to support social media use for academic purposes need to be identified. There is a need to develop an online learning management system that integrates all universities and conforms to university governing boards such as the Zimbabwe Council for Higher Education (ZIMCHE) and universities internationally on the board of knowledge regulations. The application of Natural Language Processing in chatbots needs to be studied further to improve students' experiences. It is critical to rollout YouTube channels and Facebook pages integrated with plugins to manage educational content. A model to address ethical issues and information security and privacy issues such as the use of plugins and chatbots is essential. In addition, a strategy to define the change management and collaboration metrics of industry and academia to capture the ever-evolving technologies during the training of graduates in alignment with social media use is important.

CONCLUSION

The customization of Facebook and YouTube undoubtedly results in improving the educational performance of students in universities. Digital technologies support knowledge retention via ERC, LMS, learning social media platforms, and endows soft and practical skills in freshmen adolescents. The use of digital technologies improves learning outcomes. The university-centered Facebook pages and YouTube channels are supported to remain highly engaging in an entertainment-like mode. System administrators must train students, lecturers, and other stakeholders to use customized virtual learning platforms. The management and monitoring of the social media platforms might include modeling a time management tool according to learners' characteristics. The modeling facility must switch users automatically from socializing to studying according to defined parameters. Monitoring and evaluation of system users can be both human or semi-machine-centered. Applications can switch engagement to humans to attain auditing and counseling on the learning platform. For example, agents automatically switch the user from personal accounts to learning groups or logout users to avoid addiction or ineffective use of the platform (Nizami et al., 2017; Umar & Idris, 2018). Providing students with promotional activities like free or discounted internet services for students solely using social media platforms for academic purposes is important. The algorithms applied to data allocation can segment more rewards to students who spend more time studying on social media platforms. The Facebook pages or YouTube channels need to be indexed and linked to existing educational resources. The virtual platform can be linked to lecturers and guardians to enhance performance. The lecturers, Internet service providers, and other stakeholders can award points to students based on an agreed model to encourage academic engagement. Parents and academia can approve or disapprove YouTube channels and Facebook groups in the process of knowledge management. The participation of community members in the educational system needs a rewarding policy to support the retention of knowledge in ERC and social media platforms. Students love social media thus it is prudent to blend it both as a tool for education and entertainment.

REFERENCES

Abbas, A., Gulzar, R., & Hussain, Z. (2019). Article. *Journal of Communication and Cultural Trends*, *1*(1), 26–44. doi:10.32350/jcct.11.02

Ainin, S., Parveen, F., Moghavvemi, S., Jaafar, N. I., & Shuib, N. L. M. (2015). Factors influencing the use of social media by SMEs and its performance outcomes. *Industrial Management & Data Systems, 115*(3), 570–588. doi:10.1108/IMDS-07-2014-0205

Al-Dheleai, Y. M., & Tasir, Z. (2017). Using facebook for the purpose of students' interaction and its correlation with students' academic performance. *The Turkish Online Journal of Educational Technology, 16*(4), 170–178.

Al Zboon, M. S., Al Ghammaz, S. A. D., & Al Zboon, M. S. (2018). The Impact of the Use of YouTube and Facebook on Students' Academic Achievement in Geography Course at the University of Jordan for the Bachelor's Degree. *Modern Applied Science, 12*(3), 164. doi:10.5539/mas.v12n3p164

Alkhawaja, M. I., & Halim, M. S. B. A. (2019). Challenges of E-Learning System Adoption in Jordan Higher Education. *International Journal of Academic Research in Business & Social Sciences, 9*(9), 487–494. doi:10.6007/IJARBSS/v9-i9/6317

Altawalbeh, M. A. (2021). The Impact Of Using Social Sites On Academic Performance For Students At Jordanian Universities. *Turkish Journal of Computer and Mathematics Education, 12*(7), 3264–3270.

Ansari, J. A. N., & Khan, N. A. (2020). Exploring the role of social media in collaborative learning the new domain of learning. *Smart Learning Environments, 7*(1), 9. Advance online publication. doi:10.118640561-020-00118-7

Aydin, S. (2012). A review of research on Facebook as an educational environment. *Educational Technology Research and Development, 60*(6), 1093–1106. doi:10.100711423-012-9260-7

Azizi, S. M., Soroush, A., & Khatony, A. (2019). The relationship between social networking addiction and academic performance in Iranian students of medical sciences: A cross-sectional study. *BMC Psychology, 7*(1), 1–8. doi:10.118640359-019-0305-0 PMID:31053171

Bich Diep, P., Minh Phuong, V., Dang Chinh, N., Thi Hong Diem, N., & Bao Giang, K. (2021). Health Science Students' Use of Social Media for Educational Purposes: A Sample from a Medical University in Hanoi, Vietnam. *Health Services Insights, 14*. Advance online publication. doi:10.1177/11786329211013549 PMID:34017181

Boddy, C. R. (2016). Sample size for qualitative research. *Qualitative Market Research, 19*(4), 426–432. doi:10.1108/QMR-06-2016-0053

Buzzetto-More, N. (2014). An Examination of Undergraduate Student's Perceptions and Predilections of the Use of YouTube in the Teaching and Learning Process. *Interdisciplinary Journal of E-Skills and Lifelong Learning, 10*, 17–32. doi:10.28945/1965

Deloitte. (2018). Preparing tomorrow's workforce for the Fourth Industrial Revolution. For business: A framework for action. *Global Business Coalition for Education*, 1–58. https://www2.deloitte.com/global/en/pages/about-deloitte/articles/gx-preparing-tomorrow-workforce-for-the-fourth-industrial-revolution.html?id=gx:2sm:3tw:4GBCed2018:6abt:20180919220600:SocialImpact-Report&utm_source=tw&utm_campaign=GBCed2018&utm_content=

Dontre, A. J. (2020). The influence of technology on academic distraction: A review. *Human Behavior and Emerging Technologies*, (August), 1–12. doi:10.1002/hbe2.229

Doorsamy, W., Paul, B. S., & Marwala, T. (2020). *The Disruptive Fourth Industrial Revolution: Technology, Society and Beyond.* Springer Nature Switzerland AG. doi:10.1007/978-3-030-48230-5

Duverger, P., & Steffes, E. (2012). Using YouTube Videos as a Primer to Affect Academic Content Retention. *Instutute of Education Science*, *23*(2), 51–66.

Dworkin, S. L. (2012). Sample size policy for qualitative studies using in-depth interviews. *Archives of Sexual Behavior*, *41*(6), 1319–1320. doi:10.100710508-012-0016-6 PMID:22968493

Ebehard, B., Podio, M., Alonso, A. P., Radovica, E., Avotina, L., Peiseniece, L., Sendon, M. C., Lozano, A. G., & Sole-Pla, J. (2017). Smart work: The transformation of the labour market due to the fourth industrial revolution (I4.0). *International Journal of Business and Economic Sciences Applied Research*, *10*(3), 47–66.

Ekanem, J. E. (2015). Problems and Prospects of Education Resource Centers in Nigeria. *Journal of Education and Practice*, *6*(30), 66–70.

Faraj, M., & Al-Abdallat, S. (2018). The Impact of YouTube and Facebook on the Achievement of Jordan University Students in English Language Course. *The Arab Journal for Quality Assurance in Higher Education, 11*(34), 3–24.

González, M. R., Gasco, J., & Llopis, J. (2016). Facebook and academic performance: A positive outcome. *Anthropologist*, *23*(1–2), 59–67. doi:10.1080/09720073.2016.11891924

Greenhow, C., & Robelia, B. (2009). Informal learning and identity formation in online social networks. *Learning, Media and Technology*, *34*(2), 119–140. doi:10.1080/17439880902923580

Kamaruddin, H., Rosilah, H., Norasmah, O., Wan, M. D., Zaki, A. W., & Md Sum, S. (2016). Meeting the Needs of Fourth Industrial Revolution (4IR) in Entrepreneurial Education in Malaysia: The Government's Role. *Intech, Tourism*, 13. https://www.intechopen.com/books/advanced-biometric-technologies/liveness-detection-in-biometrics

Kirschner, P. A., & Karpinski, A. C. (2010). Facebook® and academic performance. *Computers in Human Behavior*, *26*(6), 1237–1245. doi:10.1016/j.chb.2010.03.024

Kolhar, M., Kazi, R. N. A., & Alameen, A. (2021). Effect of social media use on learning, social interactions, and sleep duration among university students. *Saudi Journal of Biological Sciences*, *28*(4), 2216–2222. doi:10.1016/j.sjbs.2021.01.010 PMID:33911938

Krajcovic, P., & Cabyova, L. (2020). Use of Social Media for Marketing Communication of Socially Responsible Business Activities in Slovakia. In C. Karpacitis (Ed.), *Proceedings of the 7th European Conference on Social Media* (pp. 135–143). ACPI, UK.

Kulidtod, R., & Pasagui, N. (2017). Effects of Social Networking Media to the Academic Performance of the Students. *2nd International Conference on Educational Management and Administration (CoEMA)*, *45*(CoEMA), 59–64. 10.2991/coema-17.2017.11

Lamas, H. (2015). School Performance Review. *Journal of Educational Psychology, 3*(1), 313–386. http://revistas.usil.edu.pe/index.php/pyr

Lambić, D. (2016). Correlation between Facebook use for educational purposes and academic performance of students. *Computers in Human Behavior, 61*, 313–320. doi:10.1016/j.chb.2016.03.052

Lavuri, R., Navulla, D., & Yamini, P. (2019). Effect of social media networks on academic performance of Indian students. *Journal of Critical Reviews, 6*(4), 71–78. doi:10.22159/jcr.06.04.13

Leal-Rodrígueza, A. L., & Albort-Morant, G. (2017). Promoting innovative experiential learning practices to improve academic performance: Empirical evidence from a Spanish Business School. *Journal of Innovation and Kno, 4*, 50728.

Masino, S., & Niño-Zarazúa, M. (2016). What works to improve the quality of student learning in developing countries? *International Journal of Educational Development, 48*, 53–65. doi:10.1016/j.ijedudev.2015.11.012

Mim, F. N., Islam, M. A., & Paul, G. K. (2018). Impact of the use of social media on students ' academic performance and behavior change. *International Journal of Statistics and Applied Mathematics, 3*(1), 299–302.

Mishra, L., Gupta, T., & Shree, A. (2020). Online teaching-learning in higher education during lockdown period of COVID-19 pandemic. *International Journal of Educational Research, 1*.

Moghavvemi, S., Sulaiman, A., Jaafar, N. I., & Kasem, N. (2018). Social media as a complementary learning tool for teaching and learning: The case of youtube. *International Journal of Management Education, 16*(1), 37–42. doi:10.1016/j.ijme.2017.12.001

Mustafa, A. G., Taha, N. R., Alshboul, O. A., Alsalem, M., & Malki, M. I. (2020). Using YouTube to Learn Anatomy: Perspectives of Jordanian Medical Students. *BioMed Research International, 2020*, 1–8. Advance online publication. doi:10.1155/2020/6861416 PMID:32337267

Nizami, G. N., Naeem, Z., Arzoo, K., & Ismail, S. (2017). Impact of Facebook Addiction on Academic Performance Among Undergraduate Students. *Pakistan Journal of Rehabilitation, 6*(1), 45–50. doi:10.36283/pjr.zu.6.1/007

Olasina, G. (2017). An evaluation of educational values of YouTube videos for academic writing. *The African Journal of Information Systems, 9*(4), 2.

Osharive, P. (2015). *Social Media and Academic Performance of Students in University of Lagos*. doi:10.1007/978-3-319-67256-4_1

Pratama, Y., Hartanto, R., & Kusumawardani, S. S. (2018). Validating YouTube Factors Affecting Learning Performance. *IOP Conference Series. Materials Science and Engineering, 325*(1), 012003. Advance online publication. doi:10.1088/1757-899X/325/1/012003

Rouis, S., Limayem, M., & Salehi-Sangari, E. (2011). Impact of Facebook usage on students' academic achievement: Role of self-regulation and trust. *Electronic Journal of Research in Educational Psychology, 9*(3), 961–994. doi:10.25115/ejrep.v9i25.1465

Sheng, E. L., & Basaruddin, F. (2014). Ways of improving students performances in assessments from instructors and students perspectives. *International Journal of Asian Social Science*, *4*(2), 301–306.

Sima, V., Gheorghe, I. G., Subić, J., & Nancu, D. (2020). Influences of the industry 4.0 revolution on the human capital development and consumer behavior: A systematic review. *Sustainability (Switzerland)*, *12*(10), 4035. Advance online publication. doi:10.3390u12104035

Talaue, G., AlSaad, A., AlRushaidan, N., AlHugail, A., & AlFahhad, S. (2018). The Impact of Social Media on Academic Performance of Selected College Students. *International Journal of Advanced Information Technology*, *8*(4/5), 27–35. doi:10.5121/ijait.2018.8503

Tsai, M. J., & Tsai, C. C. (2003). Information searching strategies in Web-based science learning: The role of Internet self-efficacy. *Innovations in Education and Teaching International*, *40*(1), 43–50. doi:10.1080/1355800032000038822

Umar, T. I., & Idris, M. (2018). Influence of Social Media on Psychosocial Behaviour and Academic Performance of Secondary School Students. *Journal of Education and Entrepreneurship*, *5*(2), 36–46. doi:10.26762/jee.2018.40000013

UNICEF. (2019). *Investing in a safe, healthy and productive transition from childhood to adulthood is critical.* https://data.unicef.org/topic/adolescents/overview/#:~:text=Adolescents

KEY TERMS AND DEFINITIONS

Academic Performance: Is the attainment of goals in an education set-up.

Electronic Learning Resources: Are virtual or online learning repository.

Facebook: Is a social media platform that allows people to connect and share with friends and families online.

Fourth Industrial Revolution (4IR): Is the use of emerging technologies to redefine business processes through automation.

Freshmen: Are students in the first academic year.

Industry: Is a group of institutions or organisations focusing the supply chain through production and other related business activities.

Learning: The process of acquisition of knowledge.

Social Media: The networking of people using ICTs for communication, collaboration, socializing, and entertainment.

Technical Skills: Are hard skills or knowledge used to perform practical tasks.

Universities: Are highest level academic institutions.

YouTube: Is a social media platform for seamless sharing of video hosted by Google.

Chapter 6
Media and Parental Communication:
Effects on Millennials' Value Formation

Melanie B. Richards
https://orcid.org/0000-0001-8452-2153
East Tennessee State University, USA

ABSTRACT

Both popular media and scholarship have attributed adolescents' antisocial values and behaviors in part to media usage. Nevertheless, many scholars posit that media usage can be positive in certain contexts and that parent-child communication may abate negative media effects. An analysis of the educational longitudinal study data from 2002 to 2006 examines the effects both parent-child communication and various forms of media had on millennial teenagers' values development during a time of tremendous digital evolution. Results support that not all media was negative in effect on millennial value development and that parent-child communication may counteract some negative effects of media.

INTRODUCTION

The Millennial generation has often been stereotyped as lacking altruistic social values, as compared to previous generations (Ng, Schweitzer, & Lyons, 2010; Storn, 2005), a shift frequently blamed on Millennials' media diets in adolescence and beyond. Likewise, theorists across numerous academic disciplines believe today's media—encompassing diverse forms including television, video games, social media, and other online experiences—fosters an egoistic society (Gamson et al., 1992; Grindstaff & Turow, 2006). Media effects are important to essential cognitive orientations such as the development of agentic values (Bandura, 2006), and such effects are, of course, important in the lives of adolescents (Anderson et al., 2003; Greitemeyer, 2009). Distinctive media forms that dramatically increased or emerged during Millennials' adolescence hold particular study interest when discussing values adoption, including: violent video games, "reality" television shows, and a plethora of digital media interactions, including social networking (Coyne et al 2018; Torr, 2001).

DOI: 10.4018/978-1-7998-8318-0.ch006

Alongside the power of media, however, diverse competing forces shape teenagers' development of self and values systems during adolescence (Fine, 2004), especially the influence of parents and peers. Parents often feel that their influence is small, perceiving teens' egoistic values and poor behavior choices the result of "bad influences" from media and peers (Nathanson, 2001; Ragelienė & Grønhøj, 2020). Parental restrictions may even trigger rebellion from the teen (Nathanson, 2002). Nevertheless, numerous studies demonstrate the ways parents and peers can shape adolescents' development of values to encompass prosocial attitudes (Clark, 2011; Nathanson, 2002; Shin & Huh, 2011).

Rather than accede uncritically to the stereotype of media-addled, egoistic Millennials, then, it is crucial to examine how countervailing influences, especially the guidance of parents, shaped Millennials' values in adolescence. Values evolve slightly over a lifetime, but adolescence is a foundational milestone when young people begin to "construct their personal, institutional, and communal personas" (Fine, 2004). In this process, some values may be more central to self-definition than others. Accordingly, adolescents' perception of the importance of certain values, along with their experiences of displaying them, influences values hierarchy (Rosenberg, 1979). Likewise, values make up part of self-concept and are integral to Mead's socially developed "self" (Burke, 2004; Mead, 1934). Therefore, as comprehension of their cultural environment increases, adolescents' values evolve (Beech & Schoeppe, 1974), and hierarchical systems develop to support decision making (Rokeach, 1968).

To delineate and measure the differing power of media and parent-child communication to shape adolescents' development of values, this study empirically investigates the relationships among parent-child communication, various types of media consumption (including digital media), and structural factors in the development of teenagers' altruistic and egoistic values and the altruistic behavior of volunteerism, drawing on Education Longitudinal Study data from 2002-2006. More precisely, the following research question is examined: How did media operate in relation to Millennial adolescents' developing values and behaviors in comparison to parent-child communication and other influences, particularly at a time when teenagers were first fully immersed in the digital age?

LITERATURE REVIEW

Altruism, Prosocial Behaviors, and Parental Influence

Altruism is an intention to benefit another devoid of the expectation of reward (Wilson, 2008, para. 2), and motivated by selflessness and empathy (Eisenberg et al., 1991). Prosocial behaviors driven by altruism are "healthy, ethical" (Siu et al., 2013, p. 19), and conducive to a harmonious society. Although adolescence is often rife with competition around social capital, status, and academic performance (Nauert, 2015), adolescents "also cooperate with each other in altruistic or prosocial behaviors," providing emotional care that offsets competition and renders mutual support (Williams, 1987, p. 157).

Development of altruistic values within peer groups and the family can be enhanced or deterred by parental mediation (Clark, 2011; Nathanson, 2002; Shin & Huh, 2011), family communication patterns (Carlo et al., 2007), or peer mediation (Brechwald & Prinstein, 2011). In addition, family communication patterns and parental behaviors also contribute to development of empathy, which affects altruism and behavioral outcomes (Yoo et al., 2013; Kaufman, 2011). Some parental actions, such as parental solicitation and psychological control, negatively impact adolescents' prosocial behavior. However, actions such as sharing, care-giving, and mentoring by parents foster adolescents' development of empathy, altruism, and prosocial behaviors (Yoo et al., 2013).

In accordance with social cognitive theory, the expectation of rewards for altruism influences teenagers' future altruistic behavior (Hardy et al., 2010). Altruism is described frequently as a selfless, prosocial value, but it is not entirely devoid of reward expectancy (Krebs & Hesteren, 1994; Smith, 1981). Despite the significance of reward expectancy, however, much altruistic behavior retains prosocial effects and a selfless character. Padilla-Walker and Christensen (2011) demonstrate that "a close, supportive parent-child relationship is an important foundation for the development of prosocial behaviors" such as volunteerism (p. 545), in "which time is given freely to benefit another person, group or cause" (Wilson, 2000, p. 215) without being "mandated or coerced by others" (Haski-Leventhal, 2009, p. 272). Media certainly can introduce negative values and perceptions into the developing selfhood of adolescents (Anderson et al, 2003; Bar-on et al., 2001; Greitemeyer, 2011; Greitemeyer et al., 2010). Nevertheless, parental mediation and healthy parent-child communication and relationships can simultaneously contribute to development of altruistic behaviors.

Media, Parent-Child Communication, and Values Development

Across numerous disciplines, researchers have long affirmed media's capacity to shape selfhood (Bandura, 1986, 1994; Gamson et al., 1992; Gerbner & Gross, 1976; Grindstaff & Turow, 2006; Real, 1996; Van Gorp, 2007), while others have emphasized the power of individual agency to shape the experience of media (Blumler & Katz, 1974; Krosnick et al., 2003; Tan et al., 1997). A third, synthetic approach—drawing upon formulations of the "generalized other" in social psychology (Blumer, 2012; Mead, 1934)—emphasizes an individual's agency to challenge media effects (Jensen, 1991; Kellner, 1995). Such researchers chart the intersections of structural media influence and individual agency, considering media's representation of the generalized other as a structural force influencing values. In this context, effective parent-child communication may present a counterweight to the pressure of the generalized other in media, buttressing children's positive choices as they develop their own values.

Copious research into media influence on self-concept and identity has generated a number of insights essential to an analysis of the role of parent-child communication in adolescent value development. Adolescence is the crucial period in which identity formation is concentrated, and teenagers use media to help define self-concepts, identities, and beliefs (Blumler & Katz, 1974; Kellner, 1995, 2; Lloyd, 2002). Research has found, for example, that televised values resonate more highly with those who watch more TV (Potter, 1990). Additionally, media interaction can constitute ritual, and thus, effects may be strengthened as adolescents "co-author" media experiences (Real, 1996), a phenomenon important to the dynamics of adolescent peer groups. Indeed, interaction with peers may also enhance media effects, so teens experience the reinforcement of mediated values at multiple levels (Nathanson, 2001). However, just as parents can mediate negative effects of peers, they are also expected to counter negative media effects on values and prosocial norms (Siu et al., 2013; Warr, 1993). Such an abatement is observed when parents use some forms of media restriction. It is also expected that strong parental communication will lessen adoption of antisocial mediated values (Meeus et al., 2018; Padilla-Walker et al., 2016)

While such parental interventions can certainly counter antisocial effects of media, at the same time, prosocial values may emerge from media interactions themselves. Media is not fully negative in its effects, and there has been less study regarding prosocial behavior fostered by media, particularly when adolescents consume prosocial media, actively supported by parents (Gentile et al., 2012; Gentile et al., 2009; Meeus et al., 2018; Strasburger et al., 2012; Wilson, 2008). Research has demonstrated that exposure to content with prosocial behavior through video games (Gentile et al., 2009), musical lyrics (Greitemeyer, 2011; Greitemeyer, 2009), and/or television (Mares & Woodard, 2005) stimulates adolescents' prosocial values and associated behaviors. Moreover, the internet is more user-directed than media forms such as television, music, and video gaming, as it invites greater user generated content and facilitates active interaction, both with media and other users. Though some studies have found that internet use is more strongly associated with materialistic values (Rai et al, 2020)., it also holds great potential for prosocial utility.

Past research in this area has many limitations. In many media influence studies, researchers have focused on correlations in cross-sectional data and have failed to look at longitudinal samples. Researchers have also failed to analyze representative samples across the United States in the majority of studies, instead focusing on samples limited by region or school type. Typical structural and demographic variables that could help explain some of media's effect on values and behaviors have also not been thoroughly examined through multivariate analysis. Finally, and highly relevant to this study, parent-child communication has not been frequently introduced as a control variable.

METHODS

First, this study investigates how parent-child communication during Millennials' adolescence affected value importance in relation to media consumption. The research also examines whether concepts such as gender, race, and socioeconomic class help to explain potential effects. Second, the study asks how media consumption and parent-child communication affect the future altruistic behavior of volunteerism, with stated values and other variables as controls. This research effort concentrates on three media types: internet, television/DVDs, and video games.

Research Expectations

In the analysis of media effects on values in this study, it is hypothesized that this research will demonstrate that various forms of media function as representations of the generalized other and reinforce selected values. As supported in previous research, it is expected that heavy users of any media will have more egoistic values compared to those who are light users of that media (Carragee & Roefs, 2004; Gamson et al., 1992; Grindstaff & Turow, 2006; Potter, 1990). Egoistic values may not be innately harmful, but they do not directly improve society (e.g., valuing leisure time over community service). Egoistic values are also expected to decrease likelihood of future altruistic behaviors (i.e., volunteerism). These expected findings would support previous media effects research from a structural perspective.

However, also based on previous literature, exogenous concepts (e.g., socioeconomic class and gender) are expected to explain much of the negative effect of increased media consumption on values and behaviors (Andreoni & Vesterlund, 2001; Kamas et al., 2008; Piff et. al, 2010) and the addition of parent-child communication (as a control variable) is then expected to reduce the negative effect of media consumption to non-significance (Meeus et al., 2018; Padilla-Walker et al., 2016). This hypothesis is based on both previous findings from an agentic perspective that allows for co-authorship of the media experience and previous research supporting that parents have the ability to mediate antisocial effects of media. To determine if teenagers are adopting egoistic values they view in the media (versus a correlational effect), longitudinal multivariate analysis was applied.

Data Source/Preparation for Analysis

The Education Longitudinal Study data from 2002-2006 (ELS,2002), collected by the National Center for Education Statistics, was analyzed in this study. This data set operationalizes the concepts discussed and is a nationally representative, two stage stratified probability sample of over 15,000 10th-graders in 752 schools. The first stage of sample selection was schools, selected randomly with selection probability proportional to size/percentage of students represented. NCES' Common Core of Data (CCD) 1999-2000 and the Private School Survey (PSS) 1999-2000 were used as the sampling frame. Schools were deemed study eligible if they included a 10th grade. The final sample comprised 752 schools for an overall participation rate of approximately 62 percent (unweighted) (National Center for Education Statistics, 2005). The second stage of selection was students. Of 17,591 selected eligible students, 15,362 participated, for an overall response rate of approximately 87 percent (unweighted). Weighting of the ELS student questionnaire data (i.e., second follow-up cross-sectional weight F2QWT) was completed to compensate for unequal likelihood of selection of schools and students into the sample and to ensure data could be analyzed longitudinally.

These respondents were studied in the spring of 2002 and then again in the spring of 2004 and 2006 (National Center for Education Statistics, 2009). Several independent and dependent variables were created from the available data through techniques including aggregation and factor analysis in order to properly operationalize the desired concepts (i.e., Hours of TV/Video/DVD viewed per week, Hours of Video Games played per week, Frequency of Internet Use, Parent-Child Communication, Altruism, and Egoism).

Hours of TV/Video/DVD viewed per weekday was multiplied by five and then added to two times the Hours of TV/Video/DVD viewed per weekend day to create Hours of TV/Video/DVD viewed per week. The same process was completed to create Hours of Video Games played per week. The variable Frequency of Computer Use for Fun was limited to include only households in which the teenager had both a computer and internet access. The question analyzed read, "Whether at home, school, or some other place, how often do you use a computer for fun, such as talking to friends or relatives through E-mail, playing games, surfing the Internet, or listening to music?" (National Center for Education Statistics, 2005). This created variable is thus understood to be a proxy measure for Frequency of Internet Use and will be referred to as such.

Four components emerged in factor analysis for parent-child communication. Principal Components Analysis with Varimax Rotation summarized the data into factors that each measured something unique and were most discrete from one another (Abdi, 2003). Factors emerged along lines of the individual who displayed initiative for the interaction. The selected component, which can be viewed in Table 1, best represented parent-child communication as it covered a multitude of interactions with parents and best specified the theoretical intent of the variable. Means substitution was completed as long as the respondent answered at least half of the variables included before the composite variable was created.

Five components emerged in factor analysis regarding values. Two of these components aligned with the expectations for composite variables outlined previously (altruism and egoism). Principal Components Analysis with Varimax Rotation was again used to develop composite variables for values. Two other values variables that emerged in factor analysis were also used throughout as control variables:

importance of family/friends and importance of work. Data was only included in the composite variables if a value was present for all factors included in that variable. Additional details on both the primary variables used in the analysis and the factor analyses completed may be found in Tables 1 and 2.

Table 1. Factor analysis

Composite Variable	Variables Included	Factor Loading	Eigen Value
Parental Communication	How often discussed school courses with parents	.748	5.142
	How often discussed school activities with parents	.744	
	How often discussed things studied in class with parents	.737	
	How often discussed grades with parents	.630	
	How often discussed preparation for ACT/SAT with parents	.651	
	How often discussed going to college with parents	.714	
	How often discussed current events with parents	.667	
	How often discussed troubling things with parents	.571	
Altruism	Importance of helping others in community	.659	3.607
	Importance of working to correct inequalities	.717	
	Importance of being an active/informed citizen	.716	
	Importance of supporting environmental causes	.777	
Egoism	Importance of having lots of money	.614	1.212
	Importance of getting away from this area	.636	
	Importance of having leisure time	.591	

Analytic Strategy

Ordinary Least Squares (OLS) regression was completed in order to investigate how media consumption affects value formation and how this may be mediated by parent-child communication and other factors. Quantity of various types of media consumption in 2002 and parent-child communication were regressed on various value statements in 2004 and on various behaviors in 2006. Change in media consumption between 2002 and 2004 was also controlled for.

Hypotheses for this investigation, guided by previous theory and research literature, included:

Hypothesis 1: All initially significant effects of media on values (both altruism and egoism) will be reduced to non-significance upon introduction of the control variable, parent-child communication in 2002.

Hypothesis 2: Any negative impact of media consumption on the prosocial behavior of volunteerism will be reduced to non-significance upon introduction of parent-child communication (in 2002) and values (in 2004).

Table 2. Primary study variables

Variable	Scale	Mean
Hours/week spent watching TV/videos/DVDs in 2002	0-42 hours per week (continuous variable)	22.84
Hours/week played video/computer games in 2002	0-42 hours per week (continuous variable)	8.95
How often used computer for fun in 2002, for families who have a computer with access to the internet	1 Never 2 Rarely 3 Less than once a week 4 Once or twice a week 5 Everyday or almost everyday	N/A
Change in Hours/week spent watching TV/videos/DVDs from 2002-2004	-35 to +42 hours per week (continuous variable)	3.83
Change in Hours/week played video/computer games from 2002-2004	-35 to +42 hours per week (continuous variable)	5.23
Change in how often used computer for fun for families who have a computer with access to the internet from 2002-2004	-4 to +3 units (continuous variable)	-1.16
Altruism	4-12 (continuous variable)	8.74
Egoism	3-9 (continuous variable)	6.59
Frequency of volunteer service in past 2 years	1 Less than once a month 2 At least once a month, but not weekly 3 At least once a week	N/A
Parent-Child Communication (2002)	-13.04 to +10.23 (continuous variable)	-.100
Lives with Parents (2006)	1 No 2 Yes	N/A

RESULTS

Parental Communication Effects on Values

According to Hypothesis 1, increased parent-child communication was expected to negate media's influence on development of values. This hypothesis was partially supported. Regarding impact on media usage effects, parent-child communication decreased the effect of media consumption somewhat when it came to television and video games (by 14.3% and 26.1%, respectively). However, the inclusion of parent-child communication had an additive effect on internet usage and increased the impact of internet from non significance back to significance at b= -.038 (Table 3, Model 3). These findings suggested the possibility of interaction effects between parental communication and media consumption, which was also investigated and will be discussed later in this chapter.

Regarding values, increased parental communication had a positive effect on altruism (Table 3, Model 3) and also a small but statistically significant negative effect on egoism (Table 4, Model 3). Though these results did not fully support Hypothesis 1, parental communication is still a valuable addition in predicting altruism. In support, it should be noted that the standardized coefficient of parental communication compared to the standard coefficients of the various media variables is, in most cases, substantially higher. Therefore, parental communication is a much stronger positive influence on adoption of the value of altruism compared to the negative influence of the media.

Table 3. Unstandardized and standardized coefficients from the ordinary least squares linear regression of importance of altruism (2004) on hours/day of tv/ dvd viewing (2002), hours/day of video/computer playing (2002), internet use (2002), and other selected independent variables

	Composite Value: Importance of Altruism 2004			
	Model 1	***Model 2***	***Model 3***	***Model 4***
Hours per week watching TV/DVD 2002	-.007*** (.000) (-.044)	-.014*** (.000) (-.087)	-.012*** (.000) (-.071)	-.012*** (.000) (-.072)
Change in hours per week watching TV/DVD 2002-2004	-.002*** (.000) (-.013)	-.008*** (000) (-.046)	-.006*** (000) (-.034)	-.006*** (000) (-.034)
Hours per week playing Video Games 2002	-.038*** (.000) (-.215)	-.023*** (.000) (-.128)	-.017*** (.000) (-.098)	-.017*** (.000) (-.098)
Change in Hours per week playing Video Games 2002-2004	-.032*** (.000) (-.157)	-.021*** (.000) (-.104)	-.018*** (.000) (-.090)	-.018*** (.000) (-.091)
Frequency of Computer Use for Fun (for Individuals with Home Internet Access) 2002	-.025*** (.002) (-.012)	.003 (.002) (.001)	-.038*** (.002) (-.018)	-.038*** (.002) (-.018)
Change in Frequency of Computer Use for Fun (for Individuals with Home Internet Access) 2002-2004	-.005* (.002) (-.003)	-.009*** (.002) (-.005)	-.027*** (002) (-.016)	-.027*** (002) (-.016)
Female *(omitted category=Male)*		.459*** (.004)	.390*** (.004)	.389*** (.004)
Black *(omitted category=White)*		.752*** (.007)	.647*** (.007)	.646*** (.007)
Hispanic *(omitted category=White)*		.355*** (.006)	.383*** (.006)	.385*** (.006)
Asian *(omitted category=White)*		.270*** (.009)	.351*** (.009)	.351*** (.009)
"Traditional" Parental Structure *(omitted category= all other structures other than 2 parent mother/ father)*		.046*** (.004)	.043*** (.004)	.043*** (.004)
Number of Siblings		.001 (.001) (.000)	.010*** (.001) (.007)	.010*** (.001) (.007)
SES2		.147*** (.003) (.053)	.037*** (.003) (.013)	.038*** (.003) (.013)
Urban *(Omitted category= Rural)*		.273*** (.005)	.247*** (.005)	.247*** (.005)
Suburban *(Omitted category= Rural)*		.089*** (.005)	.084*** (.004)	.084*** (.004)
Northeast *(Omitted category= South)*		-.142*** (.005)	-.091*** (.005)	-.091*** (.005)
Midwest *(Omitted category= South)*		-.146*** (.005)	-.076*** (.005)	-.076*** (.005)
West *(Omitted category= South)*		.034*** (.005)	.034*** (.005)	.033*** (.005)
Parental Communication Composite Variable (2002)			.080*** (000) (.224)	.083*** (002) (.232)
Parental Communication x Hours per week watching TV/DVD 2002				.000*** (000) (.011)
Parental Communication x Hours per week playing Video Games 2002				.000*** (000) (-.004)
Parental Communication x Frequency of Computer Use for Fun (for Individuals with Home Internet Access) 2002				-.001*** (000) (-.016)
Intercept (Constant)	9.251*** (.011)	8.635*** (.013)	8.633*** (.012)	8.638*** (.012)
R2	.031	.063	.109	.110

Educational Longitudinal Study (2002-2004) (N=5,270).
Key: B*** (S.E) (Beta)
* p < .05 ** p < .01 *** p < .001

Table 4. Unstandardized and standardized coefficients from the ordinary least squares linear regression of importance of egoism (2004) on hours/day of TV/ DVD viewing (2002), hours/day of video/computer playing (2002), internet use (2002), and other selected independent variable

	Composite Value: Importance of Egoism 2004			
	Model 1	***Model 2***	***Model 3***	***Model 4***
Hours per week watching TV/DVD 2002	.010*** (.000) (.102)	.008*** (.000) (.078)	.008*** (.000) (.077)	.008*** (.000) (.077)
Change in hours per week watching TV/DVD 2002-2004	.008*** (.000) (.079)	.006*** (.000) (.060)	.006*** (.000) (.059)	.006*** (.000) (.058)
Hours per week playing Video Games 2002	.010*** (.000) (.094)	.003*** (.000) (.031)	.003*** (.000) (.029)	.003*** (.000) (.030)
Change in Hours per week playing Video Games 2002-2004	.003*** (.000) (.022)	-.001*** (.000) (-.009)	-.001*** (.000) (-.010)	-.001*** (.000) (-.009)
Frequency of Computer Use for Fun (for Individuals with Home Internet Access) 2002	.064*** (.001) (.050)	.082*** (.001) (.065)	.084*** (.001) (.066)	.083*** (.001) (.065)
Change in Frequency of Computer Use for Fun (for Individuals with Home Internet Access) 2002-2004	.017*** (001) (.016)	.019*** (001) (.018)	.019*** (001) (.018)	.018*** (001) (.017)
Female *(omitted category=Male)*		-.211*** (.003)	-.209*** (.003)	-.206*** (.003)
Black *(omitted category=White)*		.458*** (.004)	.461*** (.004)	.462*** (.004)
Hispanic *(omitted category=White)*		.129*** (.004)	.128*** (.004)	.122*** (.004)
Asian *(omitted category=White)*		.175*** (.006)	.172*** (.006)	.173*** (.006)
"Traditional" Parental Structure *(omitted category= all other structures other than 2 parent mother/father)*		-.085*** (.002)	-.085*** (.002)	-.084*** (.002)
Number of Siblings		-.047*** (.001) (-.054)	-.047*** (.001) (-.054)	-.047*** (.001) (-.054)
SES2		-.010*** (.002) (-.006)	-.006** (.002) (-.003)	-.007** (.002) (-.004)
Urban *(Omitted category= Rural)*		-.023*** (.003)	-.022*** (.003)	-.021*** (.003)
Suburban *(Omitted category= Rural)*		-.033*** (.003)	-.033*** (.003)	-.033*** (.003)
Northeast *(Omitted category= South)*		.051*** (.003)	.049*** (.003)	.049*** (.003)
Midwest *(Omitted category= South)*		-.002 (.003)	-.004 (.003)	-.003 (.003)
West *(Omitted category= South)*		-.043*** (.003)	-.043*** (.003)	-.042*** (.003)
Parental Communication Composite Variable (2002)			-.003*** (.000) (-.012)	-.020*** (.001) (-.093)
Parental Communication x Hours per week watching TV/ DVD 2002				.000*** (.000) (-.029)
Parental Communication x Hours per week playing Video Games 2002				.000*** (.000) (.016)
Parental Communication x Frequency of Computer Use for Fun (for Individuals with Home Internet Access) 2002				.005*** (.000) (.099)
Intercept (Constant)	5.988*** (.007)	6.286*** (.008)	6.286*** (.008)	6.280*** (.008)
R2	.017	.037	.037	.038

Educational Longitudinal Study (2002-2004) (N=5,270).
Key: B*** (S.E) (Beta)
* p < .05 ** p < .01 *** p < .001

Parental communication also significantly explained some of the variance in adolescents' levels of media consumption. The R-squared values in Table 3, Model 3 (predicting altruism) also increased significantly from .063 to .109 at the 99.9% confidence interval (according to an applied F-test) upon inclusion of this variable. Parental communication had no significant impact regarding media's influence on the value of egoism.

Interaction Effects on Values: Media and Parental Communication

There are various ways parents may approach children's media usage, and more involved parents are also more likely to employ mediation strategies (Warren et al., 2002). In the current study, it also appeared that parental communication might be interacting with media consumption to create different effects on teenagers' values. Therefore, those who consumed high amounts of media, but also had high levels of parental communication, would experience different effects from the media versus those with low parental communication.

To determine if this interaction was having an effect on values, three variables were created to represent interactions between the three types of media and parental communication. The effects of these three variables were then tested in Model 4 for both altruism and egoism. Although interaction effects were statistically significant for all media types, the media type that showed the strongest substantive interaction effects with parent-child communication was internet use. In Table 3, Model 4, adolescents with higher parental communication experienced a more negative influence of internet use on altruism. Conversely, in Table 4, Model 4, adolescents with higher parental communication also experienced a more positive influence of internet use on adoption of egoism.

Media use of both television and video games was multiplied by ten, twenty, forty and one hundred hours. This new media consumption quantity was then multiplied by parental involvement as a new independent variable and regressions on altruism and egoism were rerun. Each time, the unstandardized coefficient for the interaction effect still equaled zero. Because a number would not present itself, but the interaction effect was statistically significant, this effect must only be substantive in truly extreme cases.

Parental Communication Effects on Volunteerism

According to Hypothesis 2, increased parental communication was expected to negate media's influence on volunteerism. This hypothesis was generally not supported. Parental communication had no significant impact regarding media's influence on volunteerism except in the case of internet use. In this case, parental communication partially controlled for the positive effect of internet use on volunteerism. Overall however, for all behaviors, parental communication did little to control for media effects (Table 5, Model 4). Though these findings did not support Hypothesis 2, parental communication is again a valuable addition to all models due to its explanatory power on behaviors. The decision to include parental communication was also validated by an increase in the R-squared value in Model 4.

Table 5. Unstandardized and standardized coefficients from the ordinary least squares linear regression of frequency of volunteer service (2006) on media consumption and other selected independent variables

	Frequency of Volunteer Service (2006)				
	Model 1	*Model 2*	*Model 3*	*Model 4*	*Model 5*
Hours per week watching TV/DVD 2002	-.008*** (.000) (-.130)	-.009*** (.000) (-.138)	-.007*** (.000) (-.117)	-.007*** (.000) (-.107)	-.007*** (.000) (-.107)
Change in hours per week watching TV/DVD 2002-2004	-.005*** (.000) (-.083)	-.006*** (.000) (-.095)	-.005*** (.000) (-.082)	-.005*** (.000) (-.075)	-.005*** (.000) (-.075)
Hours per week playing Video Games 2002	-.009*** (.000) (-.139)	-.004*** (.000) (-.063)	-.002*** (.000) (-.033)	-.001*** (.000) (-.014)	-.001*** (.000) (-.012)
Change in Hours per week playing Video Games 2002-2004	-.006*** (.000) (-.077)	-.003*** (.000) (-.033)	-.001*** (.000) (-.013)	.000*** (.000) (-.005)	.000*** (.000) (-.005)
Frequency of Computer Use for Fun (for Individuals with Home Internet Access) 2002	.025*** (.001) (.032)	.012*** (.001) (.015)	.017*** (.001) (.022)	.005*** (.001) (.006)	.003*** (.001) (.004)
Change in Frequency of Computer Use for Fun (for Individuals with Home Internet Access) 2002-2004	.012*** (.001) (.018)	.002* (.001) (.002)	.004*** (.001) (.006)	-.003*** (.001) (-.004)	-.003*** (.001) (-.005)
Female *(omitted category=Male)*		.135*** (.002)	.089*** (.002)	.078*** (.002)	.080*** (.002)
Black *(omitted category=White)*		.083*** (.003)	.062*** (.003)	.041*** (.003)	.042*** (.003)
Hispanic *(omitted category=White)*		.094*** (.003)	.081*** (.002)	.100*** (.002)	.096*** (.002)
Asian *(omitted category=White)*		.142*** (.004)	.135*** (.004)	.161*** (.004)	.163*** (.004)
"Traditional" Parental Structure *(omitted category= all other structures other than 2 parent mother/father)*		.033*** (.002) (.021)	.025*** (.002) (.016)	.026*** (.002) (.017)	.028*** (.002) (.017)
Number of Siblings		.012*** (.001) (.022)	.009*** (.001) (.017)	.012*** (.001) (.022)	.012*** (.001) (.023)
SES2		.079*** (.001) (.074)	.069*** (.001) (.064)	.035*** (.001) (.033)	.035*** (.001) (.033)
Urban (Omitted category= Rural)		.090*** (.002)	.068*** (.002)	.063*** (.002)	.064*** (.002)
Suburban (Omitted category= Rural)		.022*** (.002)	.013*** (.002)	.013*** (.002)	.014*** (.002)
Northeast *(Omitted category= South)*		.099*** (.002)	.112*** (.002)	.124*** (.002)	.124*** (.002)
Midwest *(Omitted category= South)*		.005** (.002)	.015*** (.002)	.032*** (.002)	.034*** (.002)
West *(Omitted category= South)*		-.059*** (.002)	-.062*** (.002)	-.065*** (.002)	-.065*** (.002)
Composite Value 1: Altruism			.070*** (.000) (.182)	.058*** (.000) (.152)	.058*** (.000) (.152)
Composite Value 2: Egoism			-.057*** (.001) (-.091)	-.052 *** (.001) (-.083)	-.053 *** (.001) (-.085)
Composite Value 3: Family/Friends			.011*** (.001) (.018)	.004*** (.001) (.006)	.004*** (.001) (.006)
Composite Value 4: Work			.002*** (.001) (.004)	-.006*** (.001) (-.010)	-.005*** (.001) (-.008)
Parental Communication Composite Variable (2002)				.022*** (.000) (.164)	.001** (.001) (.011)
Lives with Parents (2006)				-.029*** (.001)	-.030*** (.001)
Parental Communication x Hours per week watching TV/ DVD 2002					.000*** (.000) (-.042)
Parental Communication x Hours per week playing Video Games 2002					.000*** (.000) (-.005)
Parental Communication x Frequency of Computer Use for Fun 2002					.006*** (.000) (.196)
Intercept (Constant)	1.696*** (.004)	1.489*** (.005)	1.110*** (.010)	1.371*** (.010)	1.358*** (.010)
R2	.023	.042	.081	.104	.107

Educational Longitudinal Study (2002-2006) (N=4,811),
Key: B*** (S.E.) (Beta)
* p < .05 ** p < .01 *** p < .001

It was also investigated whether interactions could be occurring between media and parental communication with regard to volunteerism (Warren et al., 2002). The only measure in which this interaction was substantive was internet use. Adolescents experiencing greater parent-child communication received a more positive return on internet use's effect on volunteerism. Therefore, those with more parent-child communication and higher internet usage were more likely to volunteer more frequently. Though interaction effects seemed substantively small (i.e., unstandardized coefficient of .006), these effects add up quickly when multiplied. For the highest levels of parental communication and internet use this coefficient would be multiplied approximately 50 times, resulting in an increase of .3 on the volunteerism frequency scale. This substantial effect at extreme levels of consumption is also expected due to the large standardized coefficient of .196.

DISCUSSION

This research provides new insights for how family communication and media affect cultural value development during adolescence. All forms of media have generally been considered to a) constitute a major influence on adolescents and b) produce a general detrimental effect on development of prosocial values and behaviors (Scharrer 2008; Harrison 2006; Sinton & Birch 2006; Tickel et al 2006; Sargent et al 2005; Huessman et al., 2003; Slater et al., 2003; Hofschire & Greenberg, 2002; Cope-Farrar & Kunkel 2002; Bushman & Anderson 2001; Johnson et al., 2002; Levine & Smolak, 1998; Real, 1996). The findings of this study do not fully support these assumptions. Many influences were found to affect values development in adolescence, and parent-child communication consistently proved more influential than media.

Media did have some negative social impact, as hypothesized, but it was not all powerful or uniformly negative in influence. Internet also operated differently than television and video games with regard to eventual behaviors. Thus internet, as a more agentic form of media, can also be used to achieve prosocial goals, particularly when parents are helping to direct prosocial usage.

Initially, findings regarding the interaction effect between parent-child communication and internet usage on the behavior of volunteerism appeared somewhat contradictory to the findings regarding how parental communication interacted with media consumption to affect values of altruism. A potential explanation of this perceived difference is that these two variables were measured in different waves of the study (i.e., altruism in 2004 and volunteerism in 2006) and teenagers may have used the internet differently as they aged, moving from entertainment to goal achievement. Additionally, functional offerings within digital media rapidly expanded and the internet increasingly became a location to find opportunities for volunteerism during this time period. Concurrently, volunteerism originating through online sources steadily increased in the early 2000s (www.volunteermatch.org, 2009; Amichai-Hamburger, 2008).

According to these findings, parent-child communication can also help to counterbalance media effects. Parental communication can help develop altruistic beliefs and discourage egoistic ones. Though parental communication does not fully control for media effects, it can help offset some of these effects. In other words, it appears that when parents are highly involved, but also allow high levels of online media consumption, teens are less likely to adopt prosocial cultural values. These findings support that co-usage and instructive mediation of the internet alone (without restrictions) may not be as effective as also employing restrictive mediation. That said, influences on altruism and egoism are small in a real-world application unless extremely high parental communication and internet use are concurrently taking place.

Another important finding was that R-squared values of all final models were, at the highest, .110 (in Table 3, Model 4). Consequently, it appears that both parents and policy makers have attributed inordinate power to the mass media that may not be deserved. It also appears that there are many influences on the development of values that are not controlled for in this analysis, likely including effects of peers, the effect of media *content* (outside of the current measurement of quantity), religious beliefs, political leanings, and other influences. Although this is a limitation of the data available, it remains a crucial finding that parents are in actuality more influential on values development than the media consumed during this time.

LIMITATIONS/FUTURE RESEARCH DIRECTIONS

It is difficult to disentangle the effects of the medium and the message in these results (McLuhan in Durham & Kellner, 2006). Content information was not available in the ELS:2002 data set, which is a limitation. It is also unknown what parental mediation strategies and other communications were in play. It is likely that both medium and message interacted with parent-child communication and caused varied effects. For example, we can know that an adolescent had higher usage of an internet-connected computer, but we do not know which sites he or she was visiting, or if parents were encouraging or restricting certain site usage.

Future work in this field would ideally ask about media content and parental mediation strategies and other communication in addition to quantity of media usage. Prospective research should also examine internet use and other new media use in a more direct manner, as internet usage was measured through proxy measures in this study (The Nielsen Company, 2009). This will be a critical lens going forward, considering how digital media has both increased and evolved and is now integrated into the other media forms examined in this study (Twenge et al, 2019). Digital media effects on the next generational of digital natives, Gen Z, would also be a worthy area of exploration to build upon this work.

In another limitation, the ELS 2002 data set does not provide ''why'' explanations, resulting in reliance on previous research and theory to guide interpretation. It is also possible that lack of additional time (resource scarcity) associated with high media usage discourages altruistic behaviors. Qualitative research could shed more light on the reasons behind different types of media use, parental communication and interventions employed, why different values are of varying importance, and why behaviors are completed. Because this data is not available, interpretations have been made based on the best available information.

CONCLUSION

These findings give greater insight into Millennials' value development and some guidance to parents today with regard to adolescents' media usage. Parents should be supportive of adolescents' prosocial, goal-oriented media use, but may wish to monitor and/or potentially restrict antisocial media use. Co-viewing, instructive mediation, and restrictive mediation are three popular mediation strategies parents may use (Warren et al., 2002), but restrictive mediation is likely the most impactful when it comes to encouraging prosocial values and behaviors (Meeus et al., 2018, Padilla-Walker., 2016). Nevertheless, restriction must also occur while communicating with adolescents more often (in ways that encourage prosocial values and behaviors) to help counterbalance media's potentially negative effects. When parents communicate effectively with their children and reinforce positive values and behaviors, they can positively impact children's prosocial attitudes and eventual behaviors.

REFERENCES

Abdi, H. (2003). Factor Rotations in Factor Analysis. In M. Lewis-Beck, A. Bryman, & T. Futing (Eds.), *Encyclopedia of Social Sciences Research Methods*. Sage.

Amichai-Hamburger, Y. (2008). Potential and promise of online volunteering. *Computers in Human Behavior*, *24*(2), 544–562. doi:10.1016/j.chb.2007.02.004

Anderson, C. A., Berkowitz, L., Donnerstein, E., Huesmann, L. R., Johnson, J. D., Linz, D., Malamuth, N. M., & Wartella, E. (2003). The Influence of Media Violence on Youth. *Psychological Science in the Public Interest*, *4*(3), 81–110. doi:10.1111/j.1529-1006.2003.pspi_1433.x PMID:26151870

Andreoni, J., & Vesterlund, L. (2001). Which is the Fair Sex? Gender Differences in Altruism. *The Quarterly Journal of Economics*, *116*(1), 293–312. doi:10.1162/003355301556419

Bandura, A. (1986). *Social foundations of thought and action: A social cognitive theory*. Prentice Hall.

Bandura, A. (1994). Social cognitive theory of mass communication. In J. Bryant & D. Zillmann (Eds.), *Media effects: Advances in theory and research* (pp. 61–90). Lawrence Erlbaum.

Bandura, A. (2006). Adolescent Development from an Agentic Perspective. In F. Pajares & T. Urdan (Eds.), *Self-Efficacy Beliefs of Adolescents* (pp. 1–43). Information Age.

Bar-On, M. E., Broughton, D. D., Buttross, S., Corrigan, S., Gedissman, A., González De Rivas, M. R., Rich, M., Shifrin, D. L., Brody, M., Wilcox, B., Hogan, M., Holroyd, H. J., Reid, L., Sherry, S. N., Strasburger, V., & Stone, J. (2001). Children, adolescents, and television. *Pediatrics*, *107*(2), 423–426. doi:10.1542/peds.107.2.423 PMID:11158483

Beech, R., & Schoeppe, A. (1974). Development of Value Systems in Adolescents. *Developmental Psychology*, *10*(5), 644–656. doi:10.1037/h0037016

Blumer, H. (2012). Symbolic interactionism. *Contemporary Sociological Theory, 62*.

Blumler, J. G., & Katz, E. (1974). The uses of mass communications: Current perspectives on gratifications research. *Sage (Atlanta, Ga.)*.

Brechwald, W. A., & Prinstein, M. J. (2011). Beyond homophily: A decade of advances in understanding peer influence processes. *Journal of Research on Adolescence*, *21*(1), 166–179. doi:10.1111/j.1532-7795.2010.00721.x PMID:23730122

Burke, P. (2004). Identities and Social Structure: The 2003 Coley-Mead Award Address. *Social Psychology Quarterly*, *67*(1), 5–15. doi:10.1177/019027250406700103

Bushman, B. J., & Anderson, C. A. (2001). Media Violence and the American Public: Scientific Facts Versus Media Misinformation. *The American Psychologist, 56*(6-7), 477–489. doi:10.1037/0003-066X.56.6-7.477 PMID:11413871

Carlo, G., McGinley, M., Hayes, R., Batenhorst, C., & Wilkinson, J. (2007). Parenting styles or practices? Parenting, sympathy, and prosocial behaviors among adolescents. *The Journal of Genetic Psychology, 168*(2), 147–176. doi:10.3200/GNTP.168.2.147-176 PMID:17936970

Clark, L. S. (2011). Parental mediation theory for the digital age. *Communication Theory, 21*(4), 323–343. doi:10.1111/j.1468-2885.2011.01391.x

Coyne, S. M., Warburton, W. A., Essig, L. W., & Stockdale, L. A. (2018). Violent video games, externalizing behavior, and prosocial behavior: A five-year longitudinal study during adolescence. *Developmental Psychology, 54*(10), 1868–1880. doi:10.1037/dev0000574 PMID:30234338

Eisenberg, N., Miller, P. A., Shell, R., McNalley, S., & Shea, C. (1991). Prosocial development in adolescence: A longitudinal study. *Developmental Psychology, 27*(5), 849–857. doi:10.1037/0012-1649.27.5.849

Fine, G. A. (2004). Adolescence as Cultural Toolkit: High School Debate and the Repertoires of Childhood and Adulthood. *The Sociological Quarterly, 45*(1), 1–20. doi:10.1111/j.1533-8525.2004.tb02395.x

Gamson, W. A., Croteau, D., Hoynes, W., & Sasson, T. (1992). Media Images and the Social Construction of Reality. *Annual Review of Sociology, 18*(1), 373–393. doi:10.1146/annurev.so.18.080192.002105

Gentile, D. A., Anderson, C. A., Yukawa, S., Ihori, N., Saleem, M., Ming, L. K., Shibuya, A., Liau, A. K., Khoo, A., Bushman, B. J., Rowell Huesmann, L., & Sakamoto, A. (2009). The effects of prosocial video games on prosocial behaviors: International evidence from correlational, longitudinal, and experimental studies. *Personality and Social Psychology Bulletin, 35*(6), 752–763. doi:10.1177/0146167209333045 PMID:19321812

Gentile, D. A., Nathanson, A. I., Rasmussen, E. E., Reimer, R. A., & Walsh, D. A. (2012). Do you see what I see? Parent and child reports of parental monitoring of media. *Family Relations, 61*(3), 470–487. doi:10.1111/j.1741-3729.2012.00709.x

Gerbner, G., & Gross, L. (1976). Living with television: The violence profile. *Journal of Communication, 26*(76), 173–199. doi:10.1111/j.1460-2466.1976.tb01397.x PMID:932235

Greitemeyer, T. (2009). Effects of songs with prosocial lyrics on prosocial thoughts, affect, and behavior. *Journal of Experimental Social Psychology, 45*(1), 186–190. doi:10.1016/j.jesp.2008.08.003

Greitemeyer, T. (2011). Effects of prosocial media on social behavior when and why does media exposure affect helping and aggression? *Current Directions in Psychological Science, 20*(4), 251–255. doi:10.1177/0963721411415229

Greitemeyer, T., Osswald, S., & Brauer, M. (2010). Playing prosocial video games increases empathy and decreases schadenfreude. *Emotion (Washington, D.C.), 10*(6), 796–802. doi:10.1037/a0020194 PMID:21171755

Grindstaff, L., & Turow, J. (2006). Video Cultures: Television Sociology in the "New TV" Age. *Annual Review of Sociology, 32*(1), 103–125. doi:10.1146/annurev.soc.32.061604.143122

Hardy, S. A., Carlo, G., & Roesch, S. C. (2010). Links between adolescents' expected parental reactions and prosocial behavioral tendencies: The mediating role of prosocial values. *Journal of Youth and Adolescence, 39*(1), 84–95. doi:10.100710964-008-9383-7 PMID:20091219

Harrison, K. (2006). Scope of Self: Toward a Model of Television's Effects on Self-Complexity in Adolescence. *Communication Theory, 16*(2), 251–279. doi:10.1111/j.1468-2885.2006.00270.x

Haski-Leventhal, D. (2009). Altruism and volunteerism: The perceptions of altruism in four disciplines and their impact on the study of volunteerism. *Journal for the Theory of Social Behaviour, 39*(3), 271–299. doi:10.1111/j.1468-5914.2009.00405.x

Hofschire, L. J., & Greenberg, B. S. (2002). Media's impact on adolescents' body dissatisfaction. In J. D. Brown, J. R. Steele, & K. Walsh-Childers (Eds.), *Sexual teens, sexual media: Investigating media's influence on adolescent sexuality* (pp. 125–149). Lawrence Erlbaum Associates.

Jensen, K. B. (1991). Media audiences Reception analysis: Mass Communication as the Social Production of Meaning. In K. B. Jensen & N. W. Jankowski (Eds.), *Handbook of Qualitative Methodologies for Mass Communication Research*. Routledge.

Johnson, J. G., Cohen, P., Smailes, E. M., Kasen, S., & Brook, J. S. (2002). Television viewing and aggressive behavior during adolescence and adulthood. *Science, 295*(0), 2468–2471. doi:10.1126cience.1062929 PMID:11923542

Kamas, L., Preston, A., & Baum, S. (2008). Altruism in individual and joint-giving decisions: What's gender got to do with it? *Feminist Economics, 14*(3), 23–50. doi:10.1080/13545700801986571

Kaufman, M. (2011). *How families facilitate the development of empathy in children: A family systems theory perspective*. https://www.semanticscholar.org/paper/How-families-facilitate-the-development-of-empathy-Kaufman/e7d280f10a3ed2cec5005379ff10766b1f11831d

Kellner, D. (1995). *Media Culture: Cultural Studies, Identity and Politics between the Modern and the Postmodern*. Routledge.

Krebs, D. L., & Van Hesteren, F. (1994). The development of altruism: Toward an integrative model. *Developmental Review, 14*(2), 103–158. doi:10.1006/drev.1994.1006

Krosnick, J. A., Anand, S. N., & Hartl, S. P. (2003). Psychosocial Predictors of Heavy Television Viewing Among Preadolescents and Adolescents. *Basic and Applied Psychology, 25*(2), 87–110. doi:10.1207/S15324834BASP2502_1

Levine, M. P., & Smolak, L. (1998). The mass media and disordered eating: Implications for primary prevention. In The prevention of eating disorders. Athlone.

Mares, M. L., & Woodard, E. (2005). Positive effects of television on children's social interactions: A meta-analysis. *Media Psychology, 7*(3), 301–322. doi:10.1207/S1532785XMEP0703_4

McLuhan, M. (2006). The Medium is the Message. In Media and Cultural Studies: Keyworks in Cultural Studies. Blackwell Publishing.

Mead, G. H. (1934). *Mind, Self, and Society*. University of Chicago Press.

Meeus, A., Beyens, I., Geusens, F., Sodermans, A. K., & Beullens, K. (2018). Managing positive and negative media effects among adolescents: Parental mediation matters—but not always. *Journal of Family Communication*, *18*(4), 270–285. doi:10.1080/15267431.2018.1487443

Nathanson, A. I. (2001). Parents versus Peers - Exploring the Significance of Peer Mediation of Antisocial Television. *Communication Research*, *28*(3), 251–274. doi:10.1177/009365001028003001

Nathanson, A. I. (2002). The Unintended Effects of Parental Mediation of Television on Adolescents. *Media Psychology*, *4*(3), 207–230. doi:10.1207/S1532785XMEP0403_01

National Center for Education Statistics. (2009). https://nces.ed.gov/surveys/ELS2002/

National Center for Education Statistics. (2005). *Educational Longitudinal Study of 2002*. https://nces.ed.gov/surveys/els2002/dec2005.asp

Nauert, R. (2015). Teenage Competition: Helpful or Harmful? *Psych Central*. Retrieved November 16, 2016 from https://psychcentral.com/news/2010/07/07/teenage-competition-helpful-or-harmful/15379.html

The Nielsen Company. (2009). *How Teens Use Media*. Author.

Ng, E. S. W., Schweitzer, L., & Lyons, S. T. (2010). New Generation, Great Expectations: A Field Study of the Millennial Generation. *Journal of Business and Psychology*, *25*(2), 281–292. doi:10.100710869-010-9159-4

Padilla-Walker, L. M., & Christensen, K. J. (2011). Empathy and self-regulation as mediators between parenting and adolescents' prosocial behavior toward strangers, friends, and family. *Journal of Research on Adolescence*, *21*(3), 545–551. doi:10.1111/j.1532-7795.2010.00695.x

Padilla-Walker, L., Coyne, S., & Collier, K. (2016). Longitudinal relations between parental media monitoring and adolescent aggression, prosocial behavior, and externalizing problems. *Journal of Adolescence*, *46*, 86–97. doi:10.1016/j.adolescence.2015.11.002 PMID:26641307

Piff, P. K., Kraus, M. W., Côté, S., Cheng, B. H., & Keltner, D. (2010). Having less, giving more: The influence of social class on prosocial behavior. *Journal of Personality and Social Psychology*, *99*(5), 771–784. doi:10.1037/a0020092 PMID:20649364

Potter, W. J. (1990). Adolescents' perceptions of the primary values of television programming. *The Journalism Quarterly*, *67*(4), 843–851. doi:10.1177/107769909006700439

Rai, R., Chauhan, C. & Cheng, MI. (2020). Materialistic values, brand knowledge and the mass media: Hours spent on the internet predicts materialistic values and brand knowledge. *Current Psychology*, *39*, 2140–2148. doi:10.1007/s12144-018-9900-0

Rageliené, T., & Grønhøj, A. (2020). The influence of peers′ and siblings′ on children's and adolescents′ healthy eating behavior. A systematic literature review. *Appetite*, *148*(104592). Advance online publication. doi:10.1016/j.appet.2020.104592 PMID:31927070

Real, M. R. (1996). *Exploring Media Culture: A Guide*. Sage Publications. doi:10.4135/9781483327600

Rokeach, M. (1968). A theory of organization and change within value-attitude systems. *The Journal of Social Issues*, *24*(1), 13–33. doi:10.1111/j.1540-4560.1968.tb01466.x

Rosenberg, M. (1979). "Which Significant Other?" and "Beyond Self-Esteem." In Conceiving the Self (pp. 83-98). Basic Books.

Sargent, J. D., Beach, M. L., Adachi-Mejia, A. M., Gibson, J. J., Titus-Ernstoff, L. T., Carusi, C. P., Swain, S. D., Heatherton, T. F., & Dalton, M. A. (2005). Exposure to movie smoking: Its relation to smoking initiation among US adolescents. *Pediatrics*, *116*(5), 1183–1191. doi:10.1542/peds.2005-0714 PMID:16264007

Scharrer, E. (2008). Media Exposure and Sensitivity to Violence in News Reports: Evidence of Desensitization? *Journalism & Mass Communication Quarterly*, *85*(2), 291–310. doi:10.1177/107769900808500205

Shin, W., & Huh, J. (2011). Parental mediation of teenagers' video game playing: Antecedents and consequences. *New Media & Society*, *13*(6), 945–962. doi:10.1177/1461444810388025

Siu, A. M. H., Daniel, T. L., & Ben, L. (2013). Prosocial norms as a positive youth development construct: A conceptual review. *Journal of Alternative Medical Research*, *5*(1), 19–27. doi:10.1100/2012/832026 PMID:22666157

Slater, M. D., Henry, K. L., Swaim, R. C., & Anderson, L. L. (2003). Violent Media Content and Aggressiveness in Adolescents: A Downward Spiral Model. *Communication Research*, *30*(6), 713–736. doi:10.1177/0093650203258281

Smith, D. H. (1981). Altruism, volunteers, and volunteerism. *Journal of Voluntary Action Research*, *10*(1), 21–36. doi:10.1177/089976408101000105

Strasburger, V. C., Jordan, A. B., & Donnerstein, E. (2012). Children, Adolescents, and the Media: Health Effects. *Pediatric Clinics of North America*, *59*(3), 533–587. doi:10.1016/j.pcl.2012.03.025 PMID:22643165

Stern, S. R. (2005). Self-Absorbed, Dangerous, and Disengaged: What Popular Films Tell Us About Teenagers. *Mass Communication & Society*, *8*(1), 23–38. doi:10.120715327825mcs0801_3

Tan, A., Nelson, L., Dong, Q., & Tan, G. (1997). Value Acceptance in Adolescent Socialization" A Test of a Cognitive-Functional Theory of Television Effects. *Communication Monographs*, *64*(1), 82–97. doi:10.1080/03637759709376406

Torr, J. D. (Ed.). (2001). *Is Media Violence a Problem?* Greenhaven Press.

Twenge, J. M., Martin, G. N., & Spitzberg, B. H. (2019). Trends in U.S. Adolescents' media use, 1976–2016: The rise of digital media, the decline of TV, and the (near) demise of print. *Psychology of Popular Media Culture*, *8*(4), 329–345. doi:10.1037/ppm0000203

Van Gorp, B. (2007). The constructionist approach to framing: Bringing culture back in. *Journal of Communication*, *57*(1), 60–78. doi:10.1111/j.0021-9916.2007.00329.x

Warr, M. (1993). Parents, Peers and Delinquency. *Social Forces*, *72*(1), 247–265. doi:10.1093f/72.1.247

Warren, R., Gerke, P., & Kelly, M. A. (2002). Is there enough Time on the Clock? Parental Involvement and Mediation of Children's Television Viewing. *Journal of Broadcasting & Electronic Media*, *46*(1), 87–111. doi:10.120715506878jobem4601_6

Williams, R. S. (2012). *Adolescence: An ethological perspective*. Springer Science & Business Media.

Wilson, J. (2000). Volunteering. *Annual Review of Sociology*, 215-240. doi:10.1146/annurev.soc.26.1.215

Wilson, B. J. (2008). Media and children's aggression, fear, and altruism. *The Future of Children*, *18*(1), 87–118. doi:10.1353/foc.0.0005 PMID:21338007

Yoo, H., Feng, X., & Day, R. D. (2013). Adolescents' empathy and prosocial behavior in the family context: A longitudinal study. *Journal of Youth and Adolescence*, *42*(12), 1858–1872. doi:10.100710964-012-9900-6

Chapter 7
Obesity Levels of Individuals With Intellectual Disabilities: Prediction for Intervention

Ebru Efeoglu
https://orcid.org/0000-0001-5444-6647
Kütahya Dumlupınar University, Turkey

Ayşe Tuna
Trakya University, Turkey

ABSTRACT

Individuals with intellectual disabilities (ID) have considerable health inequalities including higher levels of unmet health needs and a shorter life expectancy compared to the general population. The prevalence of obesity, a commonly accepted measure of health inequalities, is higher in people with ID than in the general population, and the factors leading to the increased prevalence among people with ID have not been well understood yet. This has become worse during the COVID-19 pandemic due to nationwide full and partial curfews. In this study, based on a dataset that comprises a set of parameters related to eating habits and physical conditions of a number individuals, the use of classification algorithms for predicting obesity levels of individuals with ID is proposed, and a performance analysis is made using well-known performance metrics. The results could be used by researchers and practitioners in this field to choose the best classifier for their mobile application solutions. Opportunities, research challenges, and future research directions in this topic are also presented.

INTRODUCTION

Intellectual Disabilities (ID) is an equivalent term for mental retardation onset before 18 years of age (American Psychiatric Association, 2013), and comprises additional support needs in at least two of the following domains: functional academic skills, communication, social and interpersonal skills, self-direction, self-care, home living, safety, health, work, leisure and use of community resources. The level of

DOI: 10.4018/978-1-7998-8318-0.ch007

ID can be categorized into four classes as profound, severe, moderate and mild. People with ID suffer from considerable health inequalities compared to the general population and have significant levels of unmet health needs (Lennox, Diggens, & Ugoni, 1997; Royal College of General Practitioners Working Party, 1990), a shorter life expectancy (Patja *et al.*, 2000; Bittles *et al.*, 2002) and higher mortality rates (McGuigan, Hollins, & Attard, 1995; Decouflé & Autry, 2002). Obesity and being overweight are independent risk factors for chronic diseases, which could result in reduced life expectancy (Hubert *et al.*, 1983).

As it has been shown, obesity is one of the internationally agreed health indicators for people with ID (Walsh, Kerr, & Van Schrojenstein Lantman-de Valk, 2003; Melville *et al.*, 2007) and one of the reasons of the reduced life expectancy and higher health needs of people with ID (Janicki *et al.*, 2002). Therefore, the International Association for the Scientific Study of Intellectual Disabilities clearly emphasized the impact and management of obesity on health and quality of life of people with ID (Lennox *et al.*, 2002). Body Mass Index (BMI) is a measure of body fat and applies to adults. It is calculated using Equation (1). Obesity is classified based on BMI value as listed in Table 1.

$$Body\,mass\,index = \frac{Weight}{Height * Height} \tag{1}$$

Table 1. Obesity classification

BMI (kg/m²)	Weight Status	Risk of Co-morbidities
<18.5	Underweight	Low
18.5-24.9	Normal	Average
25.0-29.9	Pre-obese	Increased
30.0-34.9	Obese I	Moderate
35.0-39.9	Obese II	Severe
≥40	Obese III	Very severe

(adopted from (World Health Organization, 1998))

The early onset of obesity requires the implementation of early intervention or secondary prevention to start as early as possible (Wen, Rissel, & He, 2017). Therefore, prediction of obesity plays a key role for further steps. In this study, a dataset comprised of a set of parameters related to eating habits and physical conditions of individuals between the ages 14 to 61 in Mexico, Peru and Colombia is used to predict obesity levels. Sequential Minimal Optimization (SMO) and K-Nearest Neighbors (KNN) algorithms have been used to predict the obesity levels of individuals and then the results obtained have been compared. For performance evaluation, confusion matrix, accuracy rate and error values have been used.

BACKGROUND AND RELATED WORK

After World Health Organization (1998) introduced BMI, the Centers for Disease Control and Prevention (CDC) introduced the clinical use of BMI in growth charts for young males and females (Guo, Wu,

Chumlea, & Roche, 2002). Based on a dataset that comprises 181 females and 166 males in the Fels Longitudinal Study, Guo *et al.* (2002) employed logistic models to predict adult overweight or obesity. They showed that a child's or adolescent's BMI value is an indicator of being overweight or obese in adulthood.

Hsieh, Rimmer, and Heller (2014) analyzed the prevalence of obesity in people with ID based on a dataset from a 4-year Longitudinal Health and Intellectual Disabilities Study and showed the factors associated with weight management status and obesity in this community. In their study, they focused on measures involving BMI, demographics, health behaviors, various health parameters, residential type and location, level of ID and related diagnoses. They concluded that people with ID are at the high risk of developing obesity, and particularly women with ID are at the high risk of developing morbid obesity.

Although most of the industrialized countries are aware of the rise in obesity in their countries, still there is a need to understand the health behaviors and health characteristics of people with ID (Bertapelli, Pitetti, Agiovlasitis, & Guerra-Junior, 2016). As shown in the literature (Rimmer & Yamaki, 2006), the prevalence of obesity among people with ID is higher compared to the general population. In case of the US, compared with other countries, adults with ID generally reside in smaller and less supervised settings and they have a substantially higher rate of obesity. Although obesity is a significant threat to the livelihood of people with ID and could have effects on their healthy years of life, independent living and community participation, there is still a lack of understanding and research on effective weight management and reduction strategies (Bertapelli, Pitetti, Agiovlasitis, & Guerra-Junior, 2016).

As well as adults, obesity is a serious problem among children. Childhood obesity is a substantial risk factor for vascular disease and cardiovascular morbidity; therefore, developing a methodology plays a key role in the success of large-scale genomic studies and validation of rare monogenic variants cause severe early-onset obesity. Based on the datasets obtained from Boston Children's Hospital (BCH) and Cincinnati Children's Hospital and Medical Center (CCHMC) Lingren *et al.* (2016) developed a rule-based algorithm to identify children with severe early-onset childhood obesity. They showed that the rule-based algorithm obtained an accuracy of 89.5% (CCHMC) and an accuracy of 77% (BCH) and proved that the best feature set was Unified Medical Language System concept unique identifiers, ICD-9 codes, and RxNorm codes.

Dugan, Mukhopadhyay, Carroll, and Downs (2015) used a decision support system to collect data prior to the second birthday of a group of children to create a dataset and then based on the dataset they conducted a study to predict childhood obesity after age two for those children. In their study, the authors used Naive Bayes, ID3, J48, Random Tree and Random Forest algorithms and proved that the ID3 model trained on the dataset obtained an accuracy of 85% and sensitivity of 89%. Their study is a first step towards building an automated system that relies on data from a clinical decision support system to predict obesity in children after age two.

PREDICTING OBESITY LEVELS USING CLASSIFICATION ALGORITHMS

The prevalence of obesity has been raising at a worrying rate in many countries. When the association of obesity with different comorbidities is taken into consideration, this is a significant threat to both the current and future health systems. Therefore, understanding the origins of obesity and developing predictive models for it and its related outcomes is critical in designing appropriate strategies to prevent it (Colmenarejo, 2020). Machine learning models that employ classification algorithms are useful tools for prediction studies since they can model nonlinear and complex relationships between variables and are suitable for being used with high-dimensional data. If they are provided with reliable datasets that contain many instances and predictor variables, which are nowadays available in repositories of electronic health records, machine learning models can generate highly accurate results (Colmenarejo, 2020). In the following subsections, first, performance metrics used in classification studies are introduced. Second, the dataset, results of a set of classification experiments and discussion on the results are presented.

Performance Metrics

Accuracy is the proportion of the correctly classified instances and all the instances and is calculated using Equation (2). In this equation, *TP* represents true positives, *TN* represents true negatives, *FP* represents false positives, and finally *FN* represents false negatives. True Positive Rate (TPR) is calculated using Equation (3) and False Positive Rate (FPR) is calculated using Equation (4). Compared with accuracy and other performance metrics, confusion matrix better visualizes the overall performance of a classification algorithm.

$$Accuracy = \frac{TP + TN}{TP + TN + FP + FN} \tag{2}$$

$$TPR = \frac{TP}{TP + FN} \tag{3}$$

$$FPR = \frac{FP}{FP + TN} \tag{4}$$

Relative Absolute Error (RAE) is another metric to measure the performance of predictive models. While good predictive models produce a RAE value close to zero, poor ones produce a RAE value greater than one (Cichosz, 2015).

Given that a classification algorithm predicted an instance as positive, what is the probability of the istance being indeed positive? This measure is called precision and is calculated using Equation (5). On the other hand, given a positive instance, what is the probability that the classification algorithm will properly identify it as positive? This measure is called recall and is calculated using Equation (6). F-measure, also called as F_1 score, is a measure that better suits to imbalanced classification problems and is calculated using Equation (7). Kappa coefficient is used to measure inter-rater reliability for qualitative items.

$$Precision = \frac{TP}{TP + FP} \tag{5}$$

$$Recall = \frac{TP}{TP + FN} \tag{6}$$

$$F_1 = 2 \times \frac{Precision \times Recall}{Precision + Recall} \tag{7}$$

Cross-validation is a technique used to assess how good a predictive model's performance in reality (Stone, 1974). In prediction problems, a predictive model is first given a dataset of known data to train itself, called training set. Then, it is given a set of unknown data against to test itself, called validation or testing set. The purpose of implementing cross-validation is to evaluate the ability of the predictive model to predict new data that was not used in estimating it so that overfitting and selection bias problems can be avoided (Cawley & Talbot, 2010). Each round of cross-validation first partitions a sample of dataset into a traning and testing sets, then performs the analysis on the training set, and finally validates the analysis on the testing set. By employing different partitions, cross-validation are realized several rounds, and then the validation results are averaged over those rounds (Grossman *et al.*, 2010).

Results and Discussion

Support Vector Machine (SVM) is a supervised learning model for regression analysis and classification. However, traning an SVM requires the solution of a very large quadratic programming (QP) optimization problem (Cortes & Vapnik, 1995). Therefore, Platt (1998) proposed a new algorithm for training SVMs, called SMO. SMO breaks the very large QP problem into a set of smallest QP problems so that they can be analytically solved and time-consuming QP optimizations are avoided (Platt, 1998).

KNN is a supervised machine learning algorithm which can be used for both regression and classification problems. It is a nonparametric learning algorithm because it does not assume anything about the underlying data (Altman, 1992). Another interesting feature of KNN is that it uses all the data for training. If KNN is being implemented, first, data points are transformed into their mathematical values. Then, it finds the distance between the mathematical values of these data points. Although there are others, Euclidean distance is the most common way of finding this distance. Based on the Euclidean distance, KNN computes the distance between each data point and the test data and then finds the probability of these points being similar to the test data and classifies it based on which points share the highest probabilities.

In this study, a dataset used in (Palechor & de la Hoz Manotas, 2019) has been used. The dataset consists of 2111 records that are based a questionnaire and total number of attributes is 17, which can be into three groups: eating habits (first one), physical condition (second one), and other attributes (third one). While preparing the dataset, which was also used in (Palechor & de la Hoz Manotas, 2019), after calculating BMI for each individual, the results were compared with the data provided by World Health Organization (1998) and the Mexican Normative (NOM-008-SSA3-2017). The first group of the attributes consists of features related to eating habits, frequent consumption of high-calorie foods, vegetable consumption frequency, number of main meals, inter-meal food consumption, daily water and alcohol consumption. The second group consists of physical state characteristics, calorie consumption monitoring, physical activity frequency, time used to use technology devices and transportation methods used. The third group consists of gender, age, height and weight. Different from Table 1, since pre-obese has been divided into two categories in this study, individuals have been divided into 7 classes: (a) Underweight, (b) Normal Weight, (c) Overweight Level I, (d) Overweight Level II, (e) Obesity Type I, (f)

Obesity Type II and (g) Obesity Type III. Two different classification algorithms have been used and their performances have been compared.

SMO and KNN algorithms have been used to predict the obesity levels of individuals and then the results obtained have been compared. In the performance evaluation, confusion matrix, accuracy rate and error values have been used. These performance metrics have been obtained both after testing on the same set (the first classification experiment) and after applying 10-fold cross validation (the second classification experiment). The confusion matrices of SMO and KNN algorithms are given in Figure 1 and Figure 2, respectively. Total number of TPs for each of the obesity category is listed in the diagonal (in red) in Figure 1 and Figure 2. When the figures are compared, the better performance of KNN algorithm can easily be seen. SMO algorithm has incorrectly predicted the obesity level of 271 individuals in the first classification experiment. After cross-validation, the algorithm has incorrectly predicted the obesity level of 330 individuals. On the other hand, KNN algorithm has correctly predicted the obesity level of all individuals in the first classification experiment, after cross-validation it has incorrectly predicted the obesity level of 176 individuals. The accuracy rates and error values of SMO and KNN algorithms are given in Figure 3 and Figure 4, respectively. KNN algorithm has achieved an accuracy of 100% in the first classification experiment and an accuracy of 82% when cross validated. In addition, it has had 0.32% error rate in the first classification experiment and 21% error rate when cross validated. On the other hand, SMO has achieved an accuracy of 86% in the first classification experiment and an accuracy of 84% when cross validated. In addition, it has had 84% error rate both in the first classification experiment and when cross validated.

Figure 1. Confusion matrices of SMO algorithm, a) without cross-validation, b) after 10-fold cross-validation

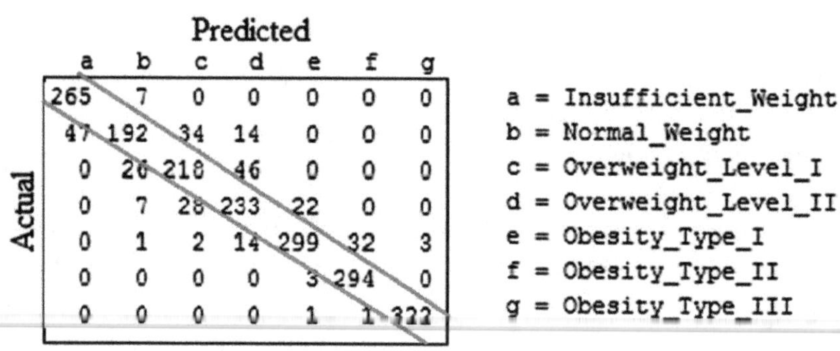

Actual \ Predicted	a	b	c	d	e	f	g
a	265	7	0	0	0	0	0
b	47	192	34	14	0	0	0
c	0	26	218	46	0	0	0
d	0	7	28	233	22	0	0
e	0	1	2	14	299	32	3
f	0	0	0	0	3	294	0
g	0	0	0	0	1	1	322

a = Insufficient_Weight
b = Normal_Weight
c = Overweight_Level_I
d = Overweight_Level_II
e = Obesity_Type_I
f = Obesity_Type_II
g = Obesity_Type_III

a)

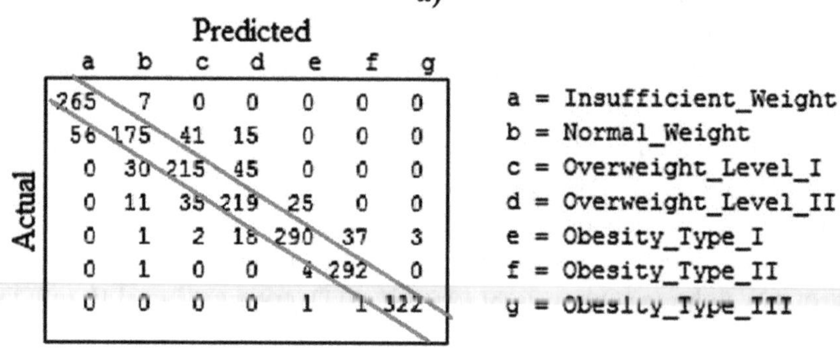

Actual \ Predicted	a	b	c	d	e	f	g
a	265	7	0	0	0	0	0
b	56	175	41	15	0	0	0
c	0	30	215	45	0	0	0
d	0	11	35	219	25	0	0
e	0	1	2	18	290	37	3
f	0	1	0	0	4	292	0
g	0	0	0	0	1	1	322

a = Insufficient_Weight
b = Normal_Weight
c = Overweight_Level_I
d = Overweight_Level_II
e = Obesity_Type_I
f = Obesity_Type_II
g = Obesity_Type_III

b)

Figure 2. Confusion matrices of KNN algorithm, a) without cross-validation, b) after 10-fold cross-validation

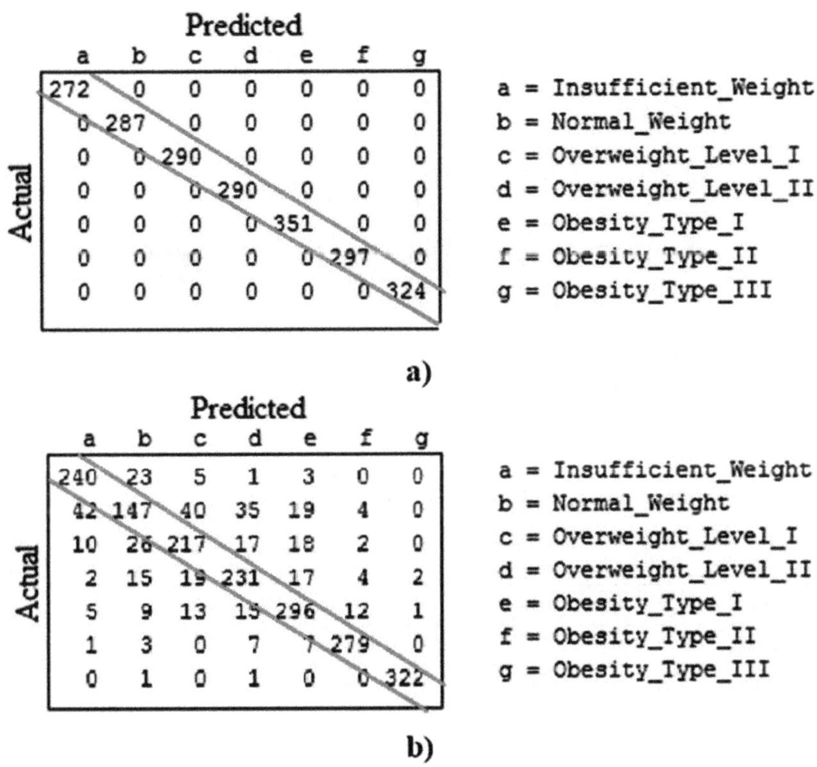

Figure 3. Accuracy rates of the algorithms, a) without cross-validation, b) after 10-fold cross-validation

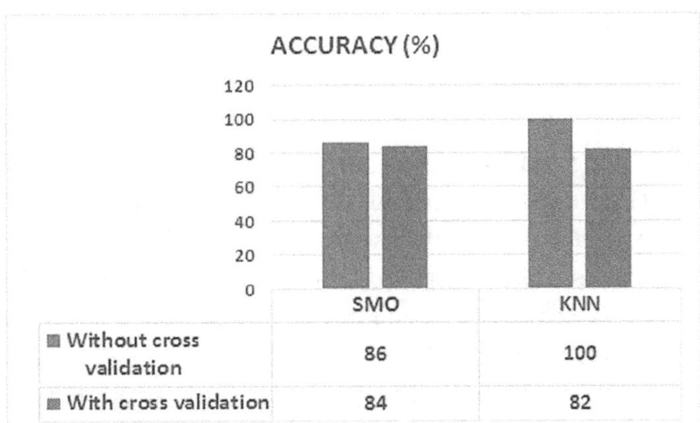

Other performance metrics of SMO and KNN algorithms are given in Table 2. While KNN algorithm provides a perfect harmony with the Precision, Recall, F-measure and Kappa statistical values reaching 1 in the first classification experiment, a small amount of decrease has been observed in these values when cross validated. On the other hand, SMO algorithm has achieved almost similar performance in both of the classification experiments.

Figure 4. Relative absolute errors (%) of the algorithms, a) without cross-validation, b) after 10-fold cross-validation

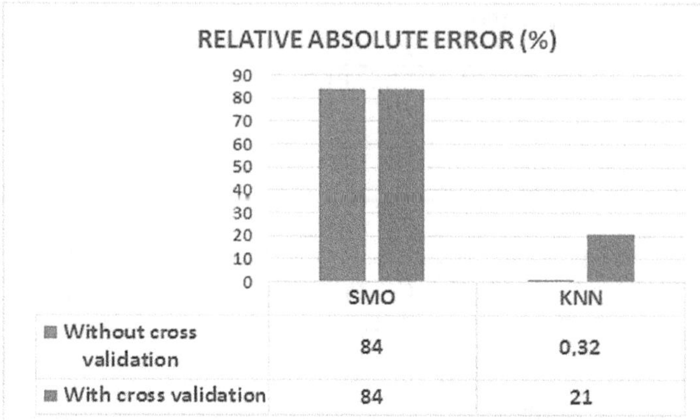

Table 2. Other performance metrics of SMO and KNN algorithms

		SMO	KNN
Without cross-validation	Precision	0.863	1
	Recall	0.864	1
	F-Measure	0.861	1
	Kappa	0.840	1
After 10-fold cross-validation	Precision	0.841	0.816
	Recall	0.842	0.820
	F-Measure	0.839	0.817
	Kappa	0.815	0.790

In order to objectively evaluate the success of this study, the results obtained in (Zheng & Ruggiero, 2017) have been reviewed. Zheng and Ruggiero (2017) employed Binary Logistic Regression (BLR), Improved Decision Tree (IDT), weighted KNN and Artificial Neural Network (ANN) to predict obesity in high school students. They used a dataset obtained from the 2015 Youth Risk Behavior Surveillance System (Kann *et al.*, 2016). The results in their study showed that BNR achieved an accuracy of 56.02% and a specificity of 54.77%, IDT achieved an accuracy of 80.23% and a specifity of 90.74%, weighted KNN achieved achieved an accuracy of 88.82% and a specifity of 93.44%, and finally ANN achieved an accuracy of 84.22% and a specifity of 99.46%. The results obtained by the weighted KNN are similar to the ones obtained in this study. Cervantes and Palacio (2020) carried out a similar study by employing Decision Tree, K-Means and SVM algorithms and compared the results they obtained through different performance metrics. They achieved slightly better results compared to Zheng and Ruggiero (2017), but the datasets employed in those studies were different.

OPPORTUNITIES AND CHALLENGES

In recent years, big data has been popular in public health and medicine and has been shown that major components of big data applications are greater volume, coverage, computational complexity and speed of availability. Timmins *et al.* (2018) conducted a study to show the contribution of 'found' data to obesity research and to describe the benefits and challenges in this field. Various sources and data types including geospatial, transport, retail sales, social media, weight management data, smart phones and wearable technologies were used in the study. The authors highlighted the variety of data uses in the literature and presented the examples that demonstrate four main ways in which 'found' data could complement conventional 'made' data. They also showed that 'found' data come with legal and ethical questions around ownership and access, validity, lack of control over data acquisition, commercial sensitivities, finding appropriate comparators, representativeness, costs, and complexities of data management, linkage and processing.

The prevalence of obesity increases everywhere in the world, therefore, nationwide surveys are implemented to measure the percentage of obese population (Dunstan *et al.*, 2020). On the other hand, since country-level food sales information is regularly available through different suppliers for each country, it can be used to predict the prevalence of obesity. Dunstan *et al.* (2020) used purchase and obesity prevalence data from 79 countries and implemented random forests, SVM and extreme gradient boosting to find out the most-relevant food category to predict obesity. They found that in the prediction of obesity, baked goods and flours was the most-relevant food category. Cheese and carbonated drinks was found as the second most-relevant food category.

Childhood obesity is a serious public health challenge, but research priorities for it have not been established yet (Hennessy *et al.*, 2019). There is a need to identify research priorities, reveal knowledge gaps, and investigate facilitators and barriers to knowledge translation in childhood obesity prevention (Hennessy *et al.*, 2019). Policymakers, researchers, and practitioners came together at a conference on obesity over 2 days in May 2017. During the conference, research priorities were identified and some key themes related to funding and resources and coproduction of research were determined with a focus on both social determinants and implementation research for preventing childhood obesity (Hennessy *et al.*, 2019). Because, detecting, managing and intervening severe early childhood obesity is critical to prevent the long-term risk of developing comorbidities related to obesity (Lingren *et al.*, 2016).

Weight management is a highly effective strategy for maintaining physical health and controlling chronic disease. Mobile health applications have the potential of managing weight and facilitating weight loss through strategies such as self-monitoring. Considering the growing interest on this topic, Vandevijvere and Kraak (2019) carried out a review to identify the effectiveness of mobile health technology in promoting healthy food consumption and physical activity. They analyzed a total of 39 articles in terms of treatment efficacy, impact on health-related behaviors, effective technology components and additional treatments in addition to participant characteristics. At the end of the research, they pointed out the need for following international standards both when carrying out and reporting research. Mobile health applications are generally considered as easy-to-use and helpful by patients and the use of mobile health interventions on obesity management and diabetes is quite promising; nevertheless, the results obtained in the efficacy of this approach varies a lot (Dounavi & Tsoumani, 2019). Therefore, Wang *et al.* (2020) analyzed 17 reviews on the use of mobile health interventions and included 6 of them in their meta-analysis. Although the effectiveness of mobile health interventions varied considerably by study, all of them showed that mobile health was a feasible option and was useful in improving patient health, especially for glycemic control and weight reduction.

The prevalence of adolescent obesity has been increasing and adolescents with ID may be at higher risk for obesity because of the psychosocial, behavioral and physical complications (Strahan & Elder, 2015). Strahan and Elder (2015) carried out a single subject design based case study to find out the effects of playing video games on an adolescent with Autism Spectrum Disorder (ASD) and obesity. On the Wii console, first, the participant played inactive video games for 6 weeks and then played active video games for the remaining 6 weeks. Physiological data of the participant were evaluated weekly at home. Results showed that active video game playing reduced BMI, stress and anxiety of the participant.

FUTURE RESEARCH DIRECTIONS

Starting and sticking to a diet plan is an effective method of weight loss. Similarly, modification of their existing diets is an effective method of weight loss for children with ASD. Nevertheless, because of the distinct characteristics of ASD it is rather difficult. Children with ASD might have an obsession for some foods due to sensory stimulation of texture, smell, or taste (Coury *et al.*, 2012). Efforts are needed to investigate the effects of dietary interventions for children with ASD (Healy, Pacanowski, & Williams, 2019).

Kelly (2019) analyzed the effects of an autonomy-supportive yoga intervention program on the self-determination and anxiety of people with developmental and ID. The findings in her research indicated that taking part in an autonomy-supportive yoga intervention could contribute to higher self-determination and decreased symptoms of anxiety in people with intellectual and developmental disabilities. In addition, it could promote autonomy and lead to feelings of relatedness among the participants. Moreover, several participants of her research reported having achieved weight loss and increased muscular strength (Kelly, 2019). The results validate the need for futher research to clearly identify the effects of yoga on people with ID.

As shown in genetic studies, small missing pieces of chromosome and changes to the lettering of genes can cause obesity or intellectual disability (Blanchet *et al.*, 2017). In 9 children with ID and obesity, mutations in myelin transcription factor 1-like gene, important for brain development, were shown (Blanchet *et al.*, 2017). Blanchet *et al.* (2017) reduced the levels of myelin transcription factor 1-like in a zebrafish and showed that this causes the reduction of oxytocin, a hormone involved in thought processes, and a problem with the development of the hypothalamus. This experiment is important to show the role of genes in ID and obesity. Being a multifactorial disease, obesity results from the interaction between lifestyle and genetic factors. As shown in the literature, strong effects of rare genetic variations were identified on obesity and personalized therapies for early-onset or syndromic obesity were designed (Ng & Bowden, 2013). Nevertheless, although common variants were identified in recently-conducted genome-wide association studies, their clinical values are limited. The genes encoding leptin, leptin receptor, melanocortin-4 receptor, prohormone convertase 1, proopiomelanocortin, single-minded homolog 1, and brain-derived neurotrophic factor are associated with early-onset severe obesity are associated with developing severe obesity at an early age (Ng & Bowden, 2013). In addition, carriers of mutations in some of these genes are associated with ID and developmental delays. This confirms that there is an interplay between the hypothalamic functions of energy homeostasis and body-weight regulation and neuro-development (Ng & Bowden, 2013). However, the common genetic variants currently known are only a small proportion of the genetic contribution to obesity, and these variants have differential effects in different populations (Ng & Bowden, 2013). Therefore, compared to the traditional clinical predictors, the value of these risk predictors is limited.

In parallel with the increasing rate of the prevalence of obesity, both the personal and personal burden of obesity-related complications are becoming more and more important. Although lifestyle-related changes and dietary measures are the key tools to counter obesity, sometimes pharmacological, even surgical, interventions are required (Williams, Nawaz, & Evans, 2020). However, the benefits of these interventions are questionable when surgical risks, side effects, weight loss efficacy and obesity relapse are taken into consideration. For the treatment of obesity there are approved drug therapies including the use of glucagon-like peptide-1 analogues as monotherapy and unimolecular dual or triple agonists that rely on the glucagon-like peptide-1 receptor and/or the gastric inhibitory peptide receptor and/or the glucagon receptor (Williams, Nawaz, & Evans, 2020). Other drugs with encouraging preclinical trial results also exist; however, further evidence is needed for their clinical use. It is desired that an anti-obesity drug produces sustained weight loss with no or minimal side effects. However, the mechanisms responsible for regulating energy balance have considerable built-in redundancy, overlap significantly with other physiological functions, and are affected by psychological, hedonic and social factors limiting the success of pharmacological interventions (Rodgers, Tschöp, & Wilding, 2012). Therefore, in most of the attempts for anti-obesity drugs, discovery programmes have not been successful due to failures in clinical development and withdrawals resulting from adverse effects. Thanks to recent improvements that drive to understand peptidergic signaling of hunger and satiety from the gastrointestinal tract, and of leptin-related homeostatic mechanisms and leptin's upstream pathways in the hypothalamus, new possibilities have emerged (Rodgers, Tschöp, & Wilding, 2012). Because of their body weight-lowering effects, Glucagon-like peptide-1 receptor agonists are successfully used in diabetes treatment. Therefore, other agents should be focused on for obesity treatment. A few combination approaches that use peptides and small molecules reached clinical trials (Rodgers, Tschöp, & Wilding, 2012).

CONCLUSION

Compared with the general population, individuals with ID, both adults and children, experience health inequalities in terms of life expectancy and unmet health needs. As it has been shown in many studies, obesity is more common among people with ID and this is one of the factors contributing to the shorter life expectancy of people with ID. Inactivity is one of the reasons for weight gain and has become more common among people during the COVID-19 pandemic because of nationwide full and partial curfews. In this chapter the use of computer-assisted tools for predicting obesity levels of individuals with ID has been proposed and the performance of different classifiers in predicting obesity levels has been compared using well-known performance metrics. When the performances of SMO and KNN algorithms have been compared, it has been seen that KNN algorithm is more successful in predicting obesity levels. The results obtained in this study could be useful in choosing the best algorithm for smart phone-based health monitoring applications.

In recent years, thanks to advancements in genetic technologies, genetic testing of people with early-onset or syndromic forms of obesity and their families have started to allow early diagnosis and personalized, appropriate intervention. Therefore, when patients with ID are involved, clinical geneticists and physicians should work together to explain the risk of obesity of their patients and to monitor their health.

If additional genetic variants are identified and how they interact with lifestyle are found out, the clinical applicability of these variants for the prediction of risks and implementation of personalized therapy will be possible. On the other hand, whatever the genetic profile of an individual is lifestyle modifications such as healthy diets and physical activities will remain the key of success in weight control. Although the evidence about their effectiveness in weight reduction and glycemic control is limited, mobile health interventions can play a key role in this regard if they are supported by evidence-based practices and based on valid measures.

ACKNOWLEDGMENT

This research received no specific grant from any funding agency in the public, commercial, or not-for-profit sectors.

REFERENCES

Altman, N. S. (1992). An introduction to kernel and nearest-neighbor nonparametric regression. *The American Statistician*, *46*(3), 175–185.

American Psychiatric Association, DSM-5 Task Force. (2013). *Diagnostic and statistical manual of mental disorders: DSM-5* (5th ed.). American Psychiatric Publishing, Inc. doi:10.1176/appi.books.9780890425596

Bertapelli, F., Pitetti, K., Agiovlasitis, S., & Guerra-Junior, G. (2016). Overweight and obesity in children and adolescents with Down syndrome-prevalence, determinants, consequences, and interventions: A literature review. *Research in Developmental Disabilities*, *57*, 181–192. doi:10.1016/j.ridd.2016.06.018 PMID:27448331

Bittles, A. H., Petterson, B. A., Sullivan, S. G., Hussain, R., Glasson, E. J., & Montgomery, P. D. (2002). The influence of intellectual disability on life expectancy. *The Journals of Gerontology. Series A, Biological Sciences and Medical Sciences*, *57*(7), M470–M472. doi:10.1093/gerona/57.7.M470 PMID:12084811

Blanchet, P., Bebin, M., Bruet, S., Cooper, G. M., Thompson, M. L., Duban-Bedu, B., Gerard, B., Piton, A., Suckno, S., Deshpande, C., Clowes, V., Vogt, J., Turnpenny, P., Williamson, M. P., & Alembik, Y. (2017). Clinical Sequencing Exploratory Research Study Consortium; Deciphering Developmental Disorders Consortium, Glasgow E, McNeill A. MYT1L mutations cause intellectual disability and variable obesity by dysregulating gene expression and development of the neuroendocrine hypothalamus. *PLOS Genetics*, *13*(8), e1006957. doi:10.1371/journal.pgen.1006957 PMID:28859103

Cawley, G. C., & Talbot, N. L. C. (2010). On Over-fitting in Model Selection and Subsequent Selection Bias in Performance Evaluation. *Journal of Machine Learning Research*, *11*(70), 2079–2107.

Cervantes, R. C., & Palacio, U. M. (2020). Estimation of obesity levels based on computational intelligence. *Informatics in Medicine Unlocked*, *21*, 100472. doi:10.1016/j.imu.2020.100472

Cichosz, P. (2015). *Data Mining Algorithms: Explained Using R*. Wiley. doi:10.1002/9781118950951

Colmenarejo, G. (2020). Machine Learning Models to Predict Childhood and Adolescent Obesity: A Review. *Nutrients, 12*(8), 2466. doi:10.3390/nu12082466 PMID:32824342

Cortes, C., & Vapnik, V. (1995). Support-Vector Networks. *Machine Learning, 20*(3), 273–297. doi:10.1007/BF00994018

Coury, D. L., Anagnostou, E., Manning-Courtney, P., Reynolds, A., Cole, L., McCoy, R., Whitaker, A., & Perrin, J. M. (2012). Use of psychotropic medication in children and adolescents with autism spectrum disorders. *Pediatrics, 130*(Suppl 2), S69–S76. doi:10.1542/peds.2012-0900D PMID:23118256

Decouflé, P., & Autry, A. (2002). Increased mortality in children and adolescents with developmental disabilities. *Paediatric and Perinatal Epidemiology, 16*(4), 375–382. doi:10.1046/j.1365-3016.2002.00430.x PMID:12445156

Dounavi, K., & Tsoumani, O. (2019). Mobile Health Applications in Weight Management: A Systematic Literature Review. *American Journal of Preventive Medicine, 56*(6), 894–903. doi:10.1016/j.amepre.2018.12.005 PMID:31003801

Dugan, T. M., Mukhopadhyay, S., Carroll, A., & Downs, S. (2015). Machine Learning Techniques for Prediction of Early Childhood Obesity. *Applied Clinical Informatics, 6*(3), 506–520. doi:10.4338/ACI-2015-03-RA-0036 PMID:26448795

Dunstan, J., Aguirre, M., Bastías, M., Nau, C., Glass, T. A., & Tobar, F. (2020). Predicting nationwide obesity from food sales using machine learning. *Health Informatics Journal, 26*(1), 652–663. doi:10.1177/1460458219845959 PMID:31106648

Grossman, R., Giovanni, S., John, E., Nitin, A., & Huan, L. (2010). *Ensemble Methods in Data Mining: Improving Accuracy Through Combining Predictions*. Synthesis Lectures on Data Mining and Knowledge Discovery. *Morgan & Claypool., 2*(1), 1–126. doi:10.2200/S00240ED1V01Y200912DMK002

Guo, S. S., Wu, W., Chumlea, W. C., & Roche, A. F. (2002). Predicting overweight and obesity in adulthood from body mass index values in childhood and adolescence. *The American Journal of Clinical Nutrition, 76*(3), 653–658. doi:10.1093/ajcn/76.3.653 PMID:12198014

Healy, S., Pacanowski, C. R., & Williams, E. (2019). Weight management interventions for youth with autism spectrum disorder: A systematic review. *International Journal of Obesity, 43*(1), 1–12. doi:10.103841366-018-0233-8 PMID:30305689

Hennessy, M., Byrne, M., Laws, R., Mc Sharry, J., O'Malley, G., & Heary, C. (2019). Childhood obesity prevention: Priority areas for future research and barriers and facilitators to knowledge translation, coproduced using the nominal group technique. *Translational Behavioral Medicine, 9*(4), 759–767. doi:10.1093/tbm/iby074 PMID:30011024

Hsieh, K., Rimmer, J. H., & Heller, T. (2014). Obesity and associated factors in adults with intellectual disability. *Journal of Intellectual Disability Research, 58*(9), 851–863. doi:10.1111/jir.12100 PMID:24256455

Hubert, H. B., Feinleib, M., McNamara, P. M., & Castelli, W. P. (1983). Obesity as an independent risk factor for cardiovascular disease: A 26-year follow-up of participants in the Framingham Heart Study. *Circulation*, *67*(5), 968–977. doi:10.1161/01.CIR.67.5.968 PMID:6219830

Janicki, M. P., Davidson, P. W., Henderson, C. M., McCallion, P., Taets, J. D., Force, L. T., Sulkes, S. B., Frangenberg, E., & Ladrigan, P. M. (2002). Health characteristics and health services utilization in older adults with intellectual disability living in community residences. *Journal of Intellectual Disability Research*, *46*(4), 287–298. doi:10.1046/j.1365-2788.2002.00385.x PMID:12000580

Kann, L., McManus, T., Harris, W. A., Shanklin, S. L., Flint, K. H., Hawkins, J., Queen, B., Lowry, R., Olsen, E. O., Chyen, D., Whittle, L., Thornton, J., Lim, C., Yamakawa, Y., Brener, N., & Zaza, S. (2016). Youth Risk Behavior Surveillance - United States, 2015. *MMWR. Surveillance Summaries*, *65*(6), 1–174. doi:10.15585/mmwr.ss6506a1 PMID:27280474

Kelly, C. T. (2019). *The Impact of an Autonomy-Supportive Yoga Intervention on the Self-Determination and Anxiety of Individuals with Intellectual and Developmental Disabilities: A Multi-Method Approach* (MSc Thesis). Clemson University. Retrieved on January 29, from https://tigerprints.clemson.edu/all_theses/3083

Lennox, N., Beange, H., Parmenter, T., Santos-Teachout, R., Evenhuis, R., Kerr, M., McElduff, A., Fraser, W., Böhmer, C. J., Davis, R., Turner, G., & Stewart, L. (2002). *Health Guidelines for Adults with an Intellectual Disability*. Retrieved on February 15 from http://www.iassid.org/pdf/healthguideline-2002.pdf

Lennox, N. G., Diggens, J. N., & Ugoni, A. M. (1997). The general practice care of people with intellectual disability: Barriers and solutions. *Journal of Intellectual Disability Research*, *41*(5), 380–390. doi:10.1111/j.1365-2788.1997.tb00725.x PMID:9373818

Lingren, T., Thaker, V., Brady, C., Namjou, B., Kennebeck, S., Bickel, J., Patibandla, N., Ni, Y., Van Driest, S. L., Chen, L., Roach, A., Cobb, B., Kirby, J., Denny, J., Bailey-Davis, L., Williams, M. S., Marsolo, K., Solti, I., Holm, I. A., ... Crimmins, N. (2016). Developing an Algorithm to Detect Early Childhood Obesity in Two Tertiary Pediatric Medical Centers. *Applied Clinical Informatics*, *7*(3), 693–706. doi:10.4338/ACI-2016-01-RA-0015 PMID:27452794

McGuigan, S. M., Hollins, S., & Attard, M. (1995). Age-specific standardized mortality rates in people with learning disability. *Journal of Intellectual Disability Research*, *39*(6), 527–531. doi:10.1111/j.1365-2788.1995.tb00573.x PMID:8746740

Melville, C. A., Hamilton, S., Hankey, C. R., Miller, S., & Boyle, S. (2007). The prevalence and determinants of obesity in adults with intellectual disabilities. *Obesity Reviews*, *8*(3), 223–230. doi:10.1111/j.1467-789X.2006.00296.x PMID:17444964

Ng, M. C. Y., & Bowden, D. W. (2013). Is genetic testing of value in predicting and treating obesity? *North Carolina Medical Journal*, *74*(6), 530–533. doi:10.18043/ncm.74.6.530 PMID:24316784

NORMA Oficial Mexicana NOM-008-SSA3-2017. (n.d.). *Para el tratamiento integral del sobrepeso y la obesidad*. Retrieved on February 16, 2021 from https://dof.gob.mx/nota_detalle.php?codigo=5523105&fecha=18/05/2018

Palechor, F. M., & de la Hoz Manotas, A. (2019). Dataset for estimation of obesity levels based on eating habits and physical condition in individuals from Colombia, Peru and Mexico. *Data in Brief, 25*, 104344. doi:10.1016/j.dib.2019.104344 PMID:31467953

Patja, K., Iivanainen, M., Vesala, H., Oksanen, H., & Ruoppila, I. (2000). Life expectancy of people with intellectual disability: A 35-year follow-up study. *Journal of Intellectual Disability Research, 44*(5), 591–599. doi:10.1046/j.1365-2788.2000.00280.x PMID:11079356

Platt, J. (1998). *Sequential Minimal Optimization: A Fast Algorithm for Training Support Vector Machines*. Microsoft Research Technical Report.

Prasher, V. P., & Janicki, M. P. (2002). *The Physical Health of Adults with Intellectual Disabilities*. Blackwell Publishing Company. doi:10.1002/9780470776216

Rimmer, J. H., & Yamaki, K. (2006). Obesity and intellectual disability. *Mental Retardation and Developmental Disabilities Research Reviews, 12*(1), 22–27. doi:10.1002/mrdd.20091 PMID:16435329

Rodgers, R. J., Tschöp, M. H., & Wilding, J. P. (2012). Anti-obesity drugs: Past, present and future. *Disease Models & Mechanisms, 5*(5), 621–626. doi:10.1242/dmm.009621 PMID:22915024

Royal College of General Practitioners Working Party. (1990). *Primary Care for People with a Mental Handicap*. Royal College of General Practitioners.

Stone, M. (1974). Cross-Validatory Choice and Assessment of Statistical Predictions. *Journal of the Royal Statistical Society. Series B. Methodological, 36*(2), 111–133. doi:10.1111/j.2517-6161.1974.tb00994.x

Strahan, B. E., & Elder, J. H. (2015). Video Game Playing Effects on Obesity in an Adolescent with Autism Spectrum Disorder: A Case Study. *Autism Research and Treatment, 128365*, 1–7. Advance online publication. doi:10.1155/2015/128365 PMID:26783457

Timmins, K. A., Green, M. A., Radley, D., Morris, M. A., & Pearce, J. (2018). How has big data contributed to obesity research? A review of the literature. *International Journal of Obesity, 42*(12), 1951–1962. doi:10.103841366-018-0153-7 PMID:30022056

Vandevijvere, S., & Kraak, V. (2019). Future directions to prevent obesity within the context of the Global Syndemic. *Obesity Reviews, 20*(S2), 3–5. doi:10.1111/obr.12887 PMID:31743591

Walsh, P. N., Kerr, M., & Van Schrojenstein Lantman-de Valk, H. M. (2003). Health indicators for people with intellectual disabilities: A European perspective. *European Journal of Public Health, 13*(Supplement 1), 47–50. doi:10.1093/eurpub/13.suppl_1.47 PMID:14533748

Wang, Y., Min, J., Khuri, J., Xue, H., Xie, B., Kaminsky, L. A., & Cheskin, L. J. (2020). Effectiveness of Mobile Health Interventions on Diabetes and Obesity Treatment and Management: Systematic Review of Systematic Reviews. *JMIR mHealth and uHealth, 8*(4), e15400. doi:10.2196/15400 PMID:32343253

Wen, L. M., Rissel, C., & He, G. (2017). The Effect of Early Life Factors and Early Interventions on Childhood Overweight and Obesity 2016. *Journal of Obesity, 3642818*, 1–3. Advance online publication. doi:10.1155/2017/3642818 PMID:28261496

Williams, D. M., Nawaz, A., & Evans, M. (2020). Drug Therapy in Obesity: A Review of Current and Emerging Treatments. *Diabetes Therapy: Research, Treatment and Education of Diabetes and Related Disorders*, *11*(6), 1199–1216. doi:10.100713300-020-00816-y PMID:32297119

World Health Organization. (1998). *Obesity: Preventing and Managing the Global Epidemic*. WHO.

Zheng, Z., & Ruggiero, K. (2017). Using machine learning to predict obesity in high school students. *IEEE International Conference on Bioinformatics and Biomedicine*, 2132-2138. 10.1109/BIBM.2017.8217988

ADDITIONAL READING

Aggarwal, C. C. (2014). *Data Classification: Algorithms and Applications* (1st ed.). CRC Press. doi:10.1201/b17320

Bhaumik, S., Watson, J. M., Thorp, C. F., Tyrer, F., & McGrother, C. W. (2008). Body mass index in adults with intellectual disability: distribution, associations and service implications: a population-based prevalence study. *Journal of Intellectual Disability Research*, *52*(Pt 4), 287–298. doi:10.1111/j.1365-2788.2007.01018.x PMID:18339091

Demšar, J. (2006). Statistical Comparisons of Classifiers over Multiple Data Sets. *Journal of Machine Learning Research*, *7*(1), 1–30.

Ferri, C., Hernández-Orallo, J., & Modroiu, R. (2009). An experimental comparison of performance measures for classification. *Pattern Recognition Letters*, *30*(1), 27–38. doi:10.1016/j.patrec.2008.08.010

Hruby, A., & Hu, F. B. (2015). The Epidemiology of Obesity: A Big Picture. *PharmacoEconomics*, *33*(7), 673–689. doi:10.100740273-014-0243-x PMID:25471927

Hsieh, K., Rimmer, J. H., & Heller, T. (2014). Obesity and associated factors in adults with intellectual disability. *Journal of Intellectual Disability Research*, *58*(9), 851–863. doi:10.1111/jir.12100 PMID:24256455

Maïano, C. (2011). Prevalence and risk factors of overweight and obesity among children and adolescents with intellectual disabilities. *Obesity Reviews*, *12*(3), 189–197. doi:10.1111/j.1467-789X.2010.00744.x PMID:20406413

Ogden, C. L., Yanovski, S. Z., Carroll, M. D., & Flegal, K. M. (2007). The epidemiology of obesity. *Gastroenterology*, *13*(6), 2087–2102. doi:10.1053/j.gastro.2007.03.052 PMID:17498505

Seidell, J. C., & Flegal, K. M. (1997). Assessing obesity: Classification and epidemiology. *British Medical Bulletin*, *53*(2), 238–252. doi:10.1093/oxfordjournals.bmb.a011611 PMID:9246834

Stefan, N., Birkenfeld, A. L., & Schulze, M. B. (2021). Global pandemics interconnected—Obesity, impaired metabolic health and COVID-19. *Nature Reviews. Endocrinology*, *17*(3), 135–149. doi:10.103841574-020-00462-1 PMID:33479538

KEY TERMS AND DEFINITIONS

Classification Accuracy: The rate of correct classifications, either for an independent test set, or using cross-validation.

Classification Algorithm: It assigns a given input data to a specific category.

Electronic Health Records: Patient-centered records that can be securely and instantly obtained by authorized users.

Intellectual Disability: It is a disability characterized by impaired intellectual and adaptive functioning.

Obesity: A chronic condition involving an excess amount of body fat.

Performance Metric: It objectively quantifies the performance of a classifier.

Chapter 8
Complexity of Digital Media Crowning the Mental Health of Adolescents

Disha Sharma
https://orcid.org/0000-0003-2124-0970
Amity University, Raipur, India

Sumona Bhattacharya
Atal Bihari Vajpayee University, India

ABSTRACT

Digital media is working as a different planet showing the disparities between the fantasies of what everyone thought about their lives and the reality of how they are actually living. It is important to have hedonic and eudemonic happiness in the life of an adolescent which contributes to overall well-being and to flourish with achievements, but 75% of 12-22 years are on digital media and spend on average two hours a day there, and this issue needs to be addressed. The first section of the chapter deals with the disruptions created with the digital media in order the way adolescents compare their lives with everyone highlighted on media. The other section targets the direct impact of the same on adolescent lives and analyses the various recovery measures and stages required and various techniques the parents and peers can use to deal with such situations. The basic purpose of this is to add value in the world of economy of attention and how to outgrow it without hurting oneself and not turning micro moments into macro moments of digital media.

INTRODUCTION

Do you know? That about 25 percent of 5 years old use the phone, and over 90 percent of teens do so? About, three-quarters use the internet, and this number increases to 50 percent by the age of 9 and to at least 90 percent by the age of 18. Well, the word 'revolution' has an idea of change, and seeing them move in the media and the process of evolution has a revolutionary impact on our lives. The age

DOI: 10.4018/978-1-7998-8318-0.ch008

of print is passing and the assumptions and all the practice associated with it are now becoming media specific. In oxford digital media is defined as "An umbrella term often treated as synonymous with new media or computer-mediated communication." If you are using technology, it is expected that you are using Digital Media. Thus, Digital media is a technology and a content united, which is the blending of a product, game, app, or any interactive website (Labrecque, vor dem Esche, Mathwick, Novak, & Hofacker, 2013). Digital Media can take the form of any news, blog post, journal, photography, stories, art, and all of which people can take in. Contributed by a team of professional or a single entrepreneur, digital media deliver experiences. Digital Media results in animation, social media, video, augmented, and virtual reality, data, and reports.

This appears to be no limit as to the type and numbers of Digital Media. Hence, by narrowing our definition, we say that *'Digital Media could be news, video, audio, software, or other content created, assessed or stored in the Digital form'*. Another question could be what do we intend with the concept of Digital Media? Who produced it? Is the outcome desirable? These questions entice us to question the motives of its production.

At first, it was just to keep in touch with friends, finding and sharing information with a broader circle. Digital Media is a necessity but adolescents used it so much it actually got in the way of personal connection with people (Debell & Chapman, 2004).

Traditional Media like print, radio, television are mostly 'one way,' meaning you can't really do anything but listen, watch, read it. In today's network world Digital Media is interconnected and interactive. Digital Media is a part of an infinite world where you can connect to others as easily as they can connect to you. Finding community with common interests and beliefs; instant access to other people and information; and can tap knowledge anywhere. Anyone can post anything online; this raises a question on the authenticity. Thus there is always a dilemmic thought whether what digital media is doing to us is justified or not? In spite of the so many positive approaches, why the ultimate impact on a specific age group is undesirable?

The question is how adolescents are using social media? Data portal's 'Global Overview Report' says, Close to half a billion new social media users. 1.3 billion years spent using the internet. Trillions of dollars spent on e-commerce (Kemp, 2021). They claim that in the last 12 months digital media became an even more essential part of people's lives, with social media, e-commerce content, and gaming. As per Simon, 66.6% (5.22 Billion) of the world population are mobile phone users, out of which 89.27% are having access to the internet. They claim to have a growth of 13.2% (490 Million) in the active internet users since Jan 2020. Since, 2016 the 12.8 billion population is been an average social media user (*Fig 1*).

Filipinos are the world's largest consumer of social media leading with an average of 4 hours, whereas on the other end japan is spending less than an hour. However, India is the worldwide average user of social media with 2.25 hours *(Fig: 2),* anything this much time doing has lasting effects on us.

WHAT PEOPLE ARE DOING ONLINE?

Data from App Annie shows that average people spent more than 4 hours in a day using their phone, which equals 3.5 trillion cumulative hours using the phone over the past 12 months. 44% of these 4 hours are being spent on communication apps, 26% in entertainment and video apps, and 9% in gaming. As per thinkwithgoogle data, watch time on YouTube of episodes has more increased 50%, of highlights and scene compilation videos has quadrupled, watch time on entertainment video has doubled, and record growth of 90% seen in predictions video. The experts believe that the 'digital divide' that existed previously on the basis of demographic and socioeconomic lines is now abolished completely, which means there are no differences between the sexes in overall computer or Internet use rates (Qutteina, Hallez, Mennes, De Backer, & Smits, 2019).

Figure 1.

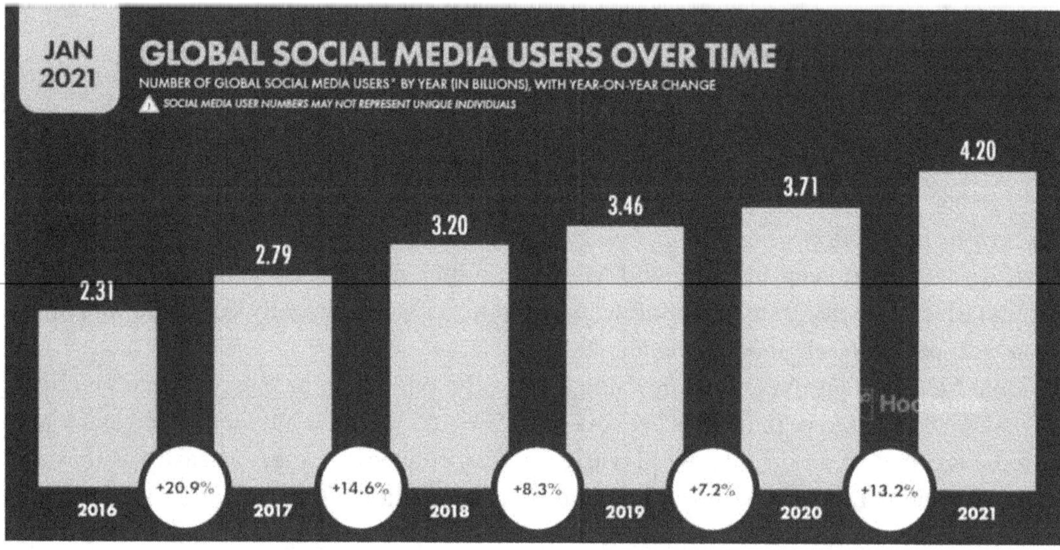

Figure 2.

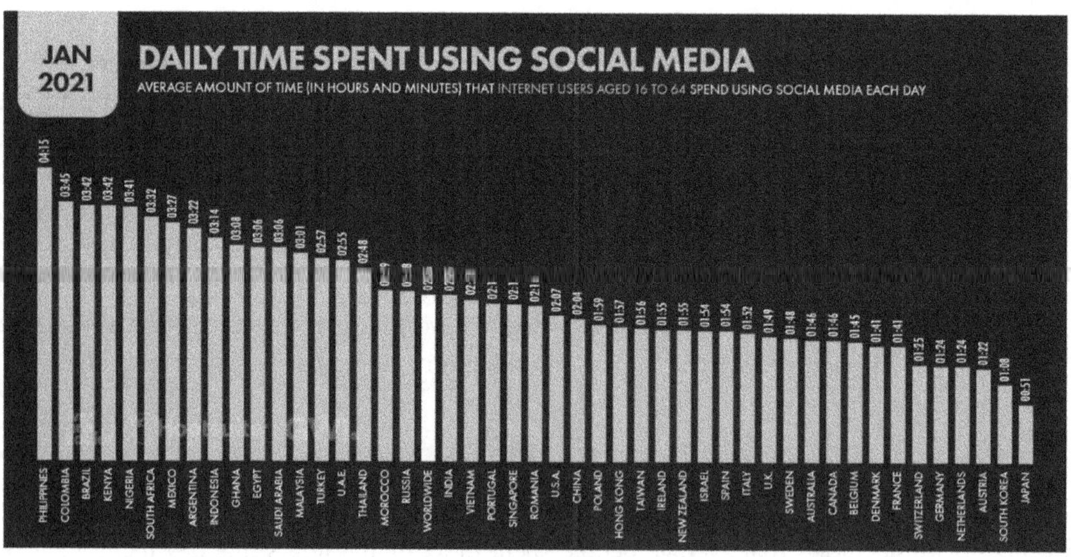

Let's break up the statistics and understand what actually adolescents are up to? According to Pew Research Center, growing share of internet say they use YouTube leads to current usage of 81%; Facebook continue to download the landscape with 69%. There are certain cites/apps used widely and have strong following among adolescents are Instagram, Snapchat and TikTok *(Fig: 3)*.

Figure 3.

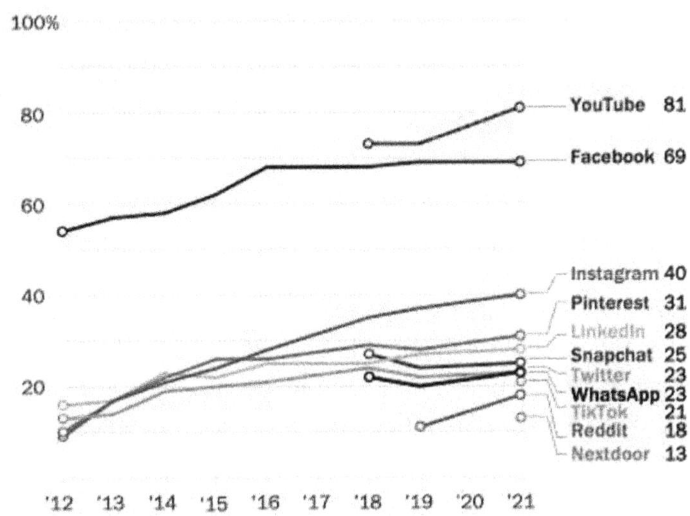

Keeping the same inference for the adolescents, maximum time spent by them is on YouTube followed by Instagram, Facebook, and Snapchat. The amazing fact is the parent group (assume 30-49) statistics are almost in the same sequence. Every seven out of ten say they visit the websites on daily basis. Among all the social media 'Twitter' has the least weightage also the user visits it less frequently. In India, adolescents are heavy users of the Internet, it has become a part of their daily routine.

Social Media do have a positive effect on adolescents, whether by teaching social skills, strengthening relationships, or just being fun. Though the persistent uses can lead to negative impacts, predominantly on the mental health and safety of the users. Based on this argument it can be inferred that the statistics are the situation is eviler than expected, as all the websites/apps are nowhere in relation to any productivity or study-based platforms. We are not wrong to call it the havoc of Digital Media. This raises the matter of concern for the guardians to measure the strength of havoc causing by Digital Media.

MEASURING THE STRENGTH OF DIGITAL MEDIA HAVOC

Adolescents have to learn to manage time when it comes to using media. Digital media is working as a different planet showing the disparities between the fantasies of what everyone thought about their lives, and the reality of how they are actually living. What is digital media doing to an adolescent? What is digital media doing to peers? As adolescents are the most contributing demographic on digital media (Debell & Chapman, 2004). The internet era came around one and a half-decade ago, then generations are young enough to grow up with digital media and old enough to have cognitive engagements, with it in a way that an adolescent probably couldn't (Zucco et al., 2018).

What causes all this havoc? Actually, all these all end up to the hazardous causes very similar to drugs, smoking, alcohol, and other harder drugs. The pleasure center of the brain has a stimulus to everything we do, and the particular hormone secreted by the brain is 'Dopamine'. So every time you check your phone, notification, likes, message, comments it truly triggers the same hormone. Experts considered this activity as a disorder. The havoc related to the above-related issues, if these go unchecked have potential to become full blown mental-health issues (Williams, Page, & Page, 2019). Let's discuss them below.

1. **NOMO Phobia:** It is a phobia out of cellular phone contact. ***'No-Mobile-Phone'*** Phobia. Doctors called this as syndrome of over-Connection with phone. The reason could be loss of mobile phone, mobile battery, missed calls, loss of internet etc. There are people who have their mobile phone switch on every time. The people keep on checking their phone every minute repeatedly. Researchers compare **NOMO Phobia** equivalent to the felling of wedding day jitters, having surgical procedure etc. Resultant people seen carrying charger or power bank. Studies shows that over use of mobile phones and social media is correlated to higher anxiety and lower performance. Though to present over selves complete and constantly updated is need of an hour, this is a kind of obligation.
2. **Ringxiety:** The other word given to it is 'Phantom vibration syndrome or Phantom ringing syndrome'. According to Rothberg (2010), the term is not a syndrome but is better considered as a tactile hallucination since the brain perceives a sensation that is not actually present. Research suggests that 9 out of 10 adolescents experienced phantom vibration. This generally develops after carrying a cell phone set to use a vibrating alert. Ringxiety is out there and it does not differentiate on demographic parameters. Psychiatrist says 25 percent of the patients visiting them can hear' virtual rings'. Doctors call it a disease of thought and imagination.
3. **Internet Addiction:** Do you play video games on the internet? Are you compulsively shopping online? Are you scrolling videos on Facebook? Do you use a computer excessively and it is interfering with your offline life? There is a disorder called internet addiction disorder (IAD) that causes neurological complications, psychological disturbances, and social problems (Cash, Rae, Steel, & Winkler, 2012). The symptoms observed in clinical practices show it overlapping the symptoms commonly associated with a behavioral addiction. As a human being an in-individual is a reward wilier, they always get back to what gives them pleasure or a sense of reward. It could be a picnic, a family gathering, meeting with a favorite person, or anything, coincidentally the Internet is compensating for these pleasures with Digital media. Where googling, opening numbers of tabs, cross-checking every news, comparing the college or course are addictive unlikely what we consider it.
4. **FOMO Phobia:** ***'Fear Of Missing Out'*** a little phrase, we've all thrown around. Well it is actual social anxiety from the fear that you are missing a potential event, opportunity or connection. FOMO keeps you connected with people, but it made you feel inadequate. Person sitting in dining, sipping coffee, watching tv, or may be writing an essay for school, a single beep of our device can drag all of our attention to that sound, eventually we all are struggling. FOMO not only distance us from people, but it keeps us from enjoying the present moment. Ultimately it impedes us from achieving our goals. FOMO becomes more interesting in context of modern research, according to one survey from Harvard, key to being happy at any phase of life is connection and accomplishment, but FOMO keeps us from it. Technology gives us the platform to instantaneously connect

and form around the world, but there is a limit on how much we can connect, and over connection leads us to stress.

5. **Highlight Reels:** Highlight reel is the collection of the best and the brightest moments Digital Media is our personal highlight reel. The place where we put up our wins or we look great, when we are out with friends and family, vacating at some exotic location. Steven Furtick (pastor), said that *'We struggle with insecurity because we compare our behind the scenes with everyone else's highlight reels'*. Digital media has vanished the line of 'Reel' vs. 'Real', we are constantly comparing ourselves to others. Comparable factor was available way before the intervention of digital media, but now it is happening all the time and it is directly linked to you.
6. **Social Currency Greed:** Currency is something we use to attribute value to a good or service. In Digital media, these likes, the comments, the shares, the subscribers, the followers have become the form of social currency by which we attribute value to something. Marketers also called it as 'Economy of attention', it's just everything is competing for your reaction, and when you give some kind of finite attention it becomes a recorded transaction attributing value. The problem is we are the product here and we are letting others attributing value to us. Adolescents take the photo/video down, just because they didn't get the desired number of likes. Seems like taking the product off the shelf just because it wasn't selling fast enough. This is alarming, this is changing our sense of identity. We are trying to validate our self-worth of what others think about us, and quantifying it for others to see. An adolescent will take 100 photos to make sure, they get the right one, then waiting for the perfect time to post it. This is a sheer obsession.
7. **Social Disconnection:** Digital media intervention has changed the way of social interaction, when we start interacting only on social media platforms the outcome leads to social disconnection. Experts say people who have access to their phones are less likely to talk to other people. A 14 years old girl says 'Who needs people, when you have a smartphone?', and the irony is she is no wrong here. If you going to any new place, google map will help; if you cooking a new dish, YouTube is there; need new drees for an upcoming event, you can visit online shopping sites; deadline of the homework assignment, just google it simply. This unfavorable effect of phones on social connectedness, in turn, should have downstream negative consequences for the mood of adolescents. Research suggests that for emotional well-being feeling social connection is very important. Although, the presence of smartphones won't let them feel disconnected as they always cultivate a sense of connection with the online world.
8. **Lack of Concentration:** Experts call it 'Attention Span' and yes Digital World is affecting it. When was the last time you actually read a book? When was the last time you watched a movie without any distraction? How many tabs are open on your smartphone? How often do you check your newsfeed? According to Associated Press, the average attention span our attention span is now only 8 seconds long, beating the 9-second attention span of goldfish. Do you see a problem here? Well, 'Attention Span' is not a concrete state of mind to be described in a blanket definition. As per the definition of Jonathan Schwabish 'Just because one might believe that we have a shorter attention span than some sort of fish doesn't mean that we are not able to hold attention. It's high time that adolescents need to make some strict discipline and abide by it. Learn to say 'no' and start focusing (Littlefield, 2020).
9. **Gaming/Gambling:** 'WHO' raises issues regarding video game addiction, called it a gaming disorder, and also registered it as one of the disorders in their list. Massive multiplayer online role-playing games rely on compulsion loop, which is played every day. Presently, the structure

of gaming has changed, on the ground level, the core addiction of gaming can be clearly observed in adolescents. These games are designed in such a way that they fetch an adolescent's attention and drag them into the addiction of these games. These games (ex. Pubg) play with an individual fantasy, make them choose the desired name, character, look, option to interact with random people around the globe, and can have access at any point in time. This leads to lots of preoccupation of people who play games (Ghasemi, Rabiei, Davoodi, & Rabiei, 2017). Even when they do not play they feel restless, moody, irritable, etc. Generally, a person plays games to change the mood, to relieve depressive feelings, but is it really happening?

10. **Financial Burden**: Is Social Media making you overspend? Adolescents keep on doing online shopping, gaming, and gambling, online. The pleasure will lead to loss of time, money, status, and the productive status of mind. A study from the Allianz Life Insurance Company added social media expenditure in the list of necessities along with rent, card bills, education loans, etc. As per Carter (2021), millennials says that social media tends to compare their life style and wealth with peers. Resultant a big chunk of teens ends up doing impulsive buying, careless expenditures, auto debit subscriptions, online gambling and gaming, etc. In spite of temptation to spend more a user must try to positioning themselves to be in good financial shape.

Figure 4.

0% 20 40 60 80 100

	YouTube	Facebook	Instagram	Pinterest	LinkedIn	Snapchat	Twitter	WhatsApp	TikTok	Reddit	Nextdoor
Total	81	69	40	31	28	25	23	23	21	18	13
Men	82	61	36	16	31	22	25	26	17	23	10
Women	80	77	44	46	26	28	22	21	24	12	16
White	79	67	35	34	29	23	22	16	18	17	15
Black	84	74	49	35	27	26	29	23	30	17	10
Hispanic	85	72	52	18	19	31	23	46	31	14	8
Ages 18-29	95	70	71	32	30	65	42	24	48	36	5
30-49	91	77	48	34	36	24	27	30	22	22	17
50-64	83	73	29	38	33	12	18	23	14	10	16
65+	49	50	13	18	11	3	7	10	4	3	8
<$30K	75	70	35	21	12	25	12	23	22	10	6
$30K-$49,999	83	76	45	33	21	27	29	20	29	17	11
$50K-$74,999	79	61	39	29	21	29	22	19	20	20	12
$75K+	90	70	47	40	50	28	34	29	20	26	20
HS or less	70	64	30	22	10	21	14	20	21	9	4
Some college	86	71	44	36	28	32	26	16	24	20	12
College+	89	73	49	37	51	23	33	33	19	26	24
Urban	84	70	45	30	30	28	27	28	24	18	17
Suburban	81	70	41	32	33	25	23	23	20	21	14
Rural	74	67	25	34	15	18	18	9	16	10	2

These changes in a person are neither spontaneous nor going to resolve quickly, as these habits are the result of continuous use of digital media over a period of time. We started believing that smartphones complete an individual, imagine a person who is doing these days in and out without smartphones, he would not feel complete, and this has become an extended part of the body. Edging the development leads to change in interest, social disconnection, not involving themselves with friends, less time given to study, lost relationships, financial crunches, etc. Above all their screen-time increases exponentially, leads to poor eyesight, and weight gain primarily. Many studies suggested that it can convert to anxiety, depression, and internalizing problems if it exceeds the average of 2 hours. Hyperactivity, inattention, academic problem, cognitive development, reduced sleeping hours are some challenges faced by the parents.

BEYOND TASTY TRIVIA: ANALYSIS OF ADVERSE EFFECT OF DIGITAL MEDIA

Do you know one in every six mobile phones is actually affected with a facal matter? Increased use of digital media is changing people's everyday lives. You're dealing with the addictive generation. What devastates you is, there are kids who are committing suicide due to social media. Digital media is ending up destroying relationships, costing productive time, costing money, cost your peace and it will make your life worse. Beyond demographic connectivity, caste, age, creed, status, individual life is persuadable, and it's all about how your mind works. Big cloud tech companies are hiring attention engineers, it is just like borrowing principles from casino gambling. They are there to make products more attractive more addictive.

You see, we all are vulnerable to social approval. And it's as if we've been infected, it's as if you are waiting for that validation searching feedbacks. This dopamine is making it more and more addictive, and what worse is that adolescents don't have the coping mechanism to deal with the adverse effect. Adolescents are suffering from anxiety, they do not have the skillset to ask for help, and this is turning them more towards social media, which offer them temporary relief. People look like they are having a much better life than they actually do. Here's a note 'Digital world isn't real'. Hence, it's high time we should stop using others' social media lives as benchmarks. They are injecting things, they know what causes depression, what leads to addiction. Inevitably our lives are becoming more transparent, we are bleeding emotions and it is just pulling us. Take a note of how it is impacting adolescents live in the below points *(Fig 5)*:

Physical Health

Among the two fold side effect, the hazard on physical health is more visible and dangerous. It amounts to crime to give smartphones or access to other screen media to children under 10 years old. Because it is really bad for a cognitive development. It has the potential to destroy your physical activity totally, it can keep you sitting at one place doing nothing just scrolling and eating or maybe not (Robinson & Smith, 2021). It can proportionally increase your weight, poor you eyesight, create unnecessary headache, affect your personal hygiene, and make you feel lethargic & tired all time. Let's be honest and discuss the adverse physical effect below:

1. **Eye problems:** By eye problems, the author means, Retina Damage; Blurred Vision; Light sensitivity; dry eyes; Itchy or watery eyes; Eyesight weakness. Today it is normal to bring your smartphones to your bed, while it is not believed to be good for the eyes. Direct and constant exposure

Figure 5.

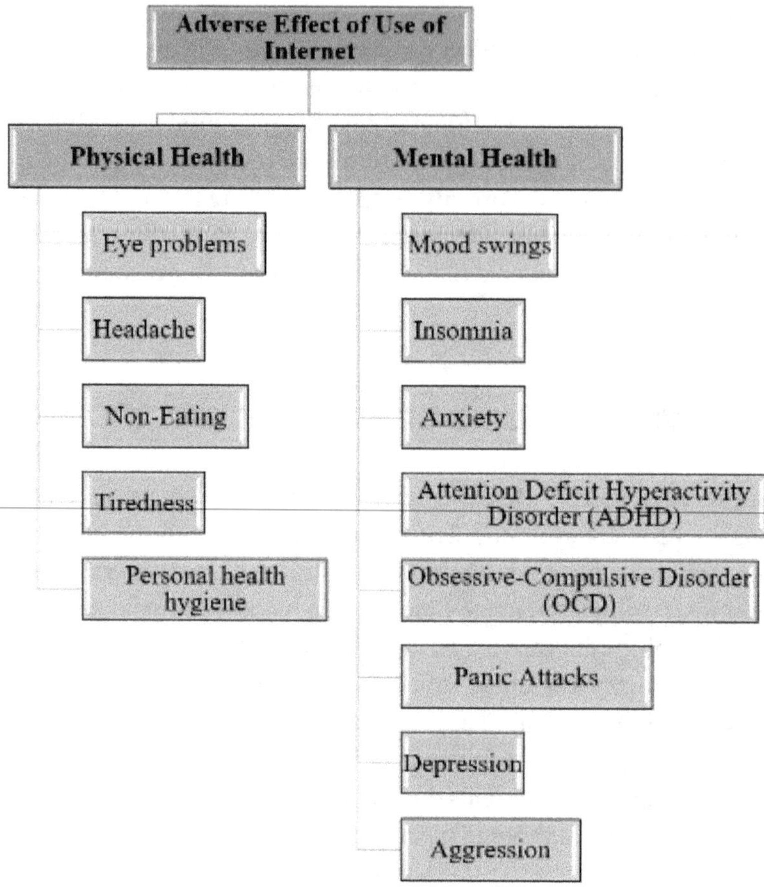

to the blue light transmitted by LED devices can end up hurting your retina. Exposure of eyes to a certain wavelength of blue light during nights reaches deeper into the eyes than UV rays and can develop AMD (Age-related macular degeneration) ("Eye Health: Technology's Effect on Eye's", 2021). AMD directly affects your central vision that is it affects your ability to see the object right in front of you.

2. **Headache**: No one can deny the potential of overusing the phone in trigging the headache. So what causes headaches? Long hours of watching TV, working on a laptop, using a smartphone, playing video games are the reasons behind this. This can lead to exhaustion, eyestrain, and lack of blood circulation. It is believed that eyestrain is the common reason for headaches. This is due to the connection between the eyes and the brain. Eyestrain could be because of the brightness of the screen, font size of the text, changing animations in gaming, etc. As all these factors lead to overworking of eyes and it is not able to focus, this can also lead to muscle spasms.
3. **Non-Eating Disorder**: Can we really establish a connection between non-eating and the use of the internet? According to the National Eating Disorder Association, an average exposure of two hours in social media has increased self-objectification and body image concerns. Social media expose the users to unrealistic ideals of beauty, the perfect body shape, thinspiration, weight loss posts, diet plans, or more. The nature of social media lends itself towards comparison, and we

often judge over selves with the virtual world of highlight reels. Another possible reason is being lethargic, staying at one place, and probably considering eating as a waste of time (Jarrett, 2021). Thus it is always advisable to be mindful when it comes to the diet and eating habits.

4. **Tiredness**: A study from Hong Kong suggests that every time a phone falls the heart rate increases by 15 Percent, this is even faster than in the case of stage phobia. At one place it is keeping you restless mentally and tired most of the time. Hours of playing video games, less exposure to outdoor activity, less walking, etc. have led to the accumulation of energy and can end up making us more lethargic, tired, and maybe fat. Social media is evolving every day, and we have to adapt.
5. **Personal health hygiene**: Hereby personal hygiene we meant to say about not taking bath daily, not shaving, not cutting your nails, changing clothes, even brushing of teeth. How worse is this? What if it can lead you to death? Well not taking bath or maintain personal hygiene can cause infection to our body. It's our immune system that keeps on fighting the bacteria and viruses trying to enter into our body, but the proportionate increase in bacteria and virus in our body can lead to the situation of immune compromise.

Mental Health

Does your adolescent become angry, irritable, anxious, or even violent when the phone is taken away? Does your adolescent skips or avoid social gatherings to use smart phone instead? Is your adolescent personal care, friendships, family relationships or school work negatively affected by smart phone use? Are there any major changes in mood that can't otherwise be explained? The younger you are, the more you suffer from digital media. If a child is constantly clicking on a featureless surface ever the same hand motion (Gao et al., 2020; O'Sullivan et al., 2021). Then the higher end cognitive center of brain, that we use for thinking and cognitive analysis, and it harm their sensory and motor skills (Rodriguez, 2021). Let's discuss the epidemic cause of Digital media towards our mental health:

1. **Mood Swings:** It clearly refers to rapid change of mood. These swings can be positive or negative just like stress is. 'Eustress' is considered to be a positive type of stress. In case of adolescents, they are very rapid responder to any type of situation, their ups and downs are easily noticeable and often remedial. They are young, adaptable and addictive as well. The construction of digital media is based on the concept of addiction. Hence, change in the pattern, change in the routine can cause mood swings (Magson et al., 2021).
2. **Insomnia:** Do you know smart phones causes sleeplessness? Sleeplessness is the biggest cause of diabetics, which can directly lead you to cardio vascular disorders like heart attack and strokes. Sleeplessness has potential to increase the risk of cause of deaths, and Sleeplessness is a gift from smartphone (Fobian, Avis, & Schwebel, 2016). Adolescents find hard time falling asleep in the usual time. 70 percent on adolescents are seeing falling asleep late at school days and post midnight on others. Digital media is affecting your sleep efficiency, creating sleeplessness and often ending up at being insomniac. Apart from that the messages and call tones are so vulnerable to the sleep onset, adolescents reported being woken up at least once they fall asleep.
3. **Anxiety:** Anxiety is contagious. Hutchins, (2021) Social media use and increased screen time can lead to an increased in depression and anxiety among teens and adolescents. The time spent by teems is directly proportional to the level of anxiety among teens. By anxiety we clearly mean becoming angry for no reason, often irritable, most of the time behaving anxiously, and sometimes

they end up being violent (Shafer, 2017). The addictive pattern framing of the internet resulting in robust anxiety among adolescents and it is becoming severe. The need of the hour is asking for a change in with the approach, it is asking to become calmer and less impulsive.

4. **Attention Deficit Hyperactivity Disorder (ADHD):** Anxiety is contagious. Hutchins, (2021) Social media use and increased screen time can lead to an increased in depression and anxiety among teens and adolescents. The time spent by teems is directly proportional to the level of anxiety among teens (Coyne et al., 2020). By anxiety we clearly mean becoming angry for no reason, often irritable, most of the time behaving anxiously, and sometimes they end up being violent. The addictive pattern framing of the internet resulting in robust anxiety among adolescents and it is becoming severe. The need of the hour is asking for a change in with the approach, it is asking to become calmer and less impulsive.
5. **Obsessive-Compulsive Disorder (OCD):** OCD in childhood is an impairing condition. OCD reflects some specific set of distressing symptoms incorporating intrusive and repetitive thoughts, and often time-consuming rituals. Doctors consider it a severe type of disorder, but also studies suggest proper treatment can lead to early recovery (Krebs & Heyman, 2015). The commonly recommended treatment for OCD is cognitive behavior therapy. Hence, WHO suggests children should be offered interventions it can relieve distress and impairment in children.
6. **Depression:** If you are using your smart phone a lot, then actually you are not interacting with physical social world. As per a study 10-15 years age group when spends more than three hours a day in social media, then they possess twice the risk to be depressed when they're 18 (Hutchins, 2021). Thy also identified the correlation between the use of internet and death rates. This is big cause of it and we have to be cautious to it. The popcorn brain[1] is more prone to depression. Internet being constantly providing media that hits dopamine is all leading to depression.
7. **Aggression:** The experts calls aggression the core adverse effect of Digital Media. Digital media is causing dementia in long run. The more time you spent in internet the more likely you to have elevated blood pressure. An experiments shows that if your smart phone rings it increases your heartrate speed and blood pressure goes up. We are doing a lot of activities that cause addiction, and developmental dysfunction. In a summaries way, more exposure to electronic media increases the potential of adolescents behave aggressively. Aggression can often lead to public health threat.

We have come to grips with a lot of technology and sometimes the solution is radical. You need morals to use the technology especially smartphones (Raju, Valsaraj, & Noronha, 2015). You don't want to give this exposure to your adolescents and let them go loose with this. This piece of technology is causing serious problems. How long does the harm of Digital Media continue to be noticeable in human beings? Maybe across lifespan. Smartphones are that risky, it has the potential to hinder your brain development, affect your health, and education, and we must handle it carefully.

KEEP THE DISTANCE: JUST SAY 'NO'!

Your Childhood Was Immune From Digital World, Your Adolescents Isn't...

If you leave the internet, you will be having all the time to accomplish the things that you desired. We have to understand that, the presence of technology is because of us and it can never be vice versa (Kushlev,

Proulx, & Dunn, 2017). This simply means that we can live without technology and exist wisely in this biosphere. What if you don't have social media account? What if you don't shop online? What if you switch from Netflix to Television? Are you going to die? Or the world is going to end? Well to all these questions the authors have a single answer that is 'You will be fine'. You still going to have friends, you will be knowing what is going around the world, you may be shopping from some exotic mall, and you can still collaborate and connect through the world.

But are you really 'OK' without Digit World? They say at some point one has to start, it will be hard initially, but you will be ending better off. There is a virtual line of the internet world, an edge line. Before that everything is controlled and composed, beyond that, the extracts are disastrous and chaotic. The mental state of adolescents is less matured to convince themselves for keeping the distance and start a journey towards a more sustainable life. The positive side of the Digital world cannot be ignored, but one who crosses the limit is liable to pay price. People can withdraw themselves and use this technique adequately. Do you the adolescents will also be able to do so? Adolescents are more likely to fall into this trap, deeper and harder. Hence, Abstinence is not an option, but you can 'practice safe Digital media'.

Adolescent age is vulnerable, you are exposed to things that may be the first time in life, an age of shaping whole life. And the most important factors at that time are peace of mind, focus, and utilization of time to the full extent. Adolescents can set boundaries, switch off the internet, put it in airplane mode, or just delete the app. The world is open to productive, rare, and valuable materials (Panagiotopoulos, Barnett, Bigdeli, & Sams, 2016). So if you do not carry a fancy Social Media Account which you consider to be vital for your success, or you just don't want to miss any chance of opportunity, then it's not justified. At the same time, you can invest in building a new skill set, be creative, and explore. But then again the withdrawals are always difficult, similar to a detox process, initially, it can be uncomfortable, can make you feel anxious. Thought if you are consistent and persistent, things will start settling down and you will adapt to the situation. The observable Benefits of disconnecting from social media are:

- Increased focus
- Peace of mind
- Less anxious
- Sleep efficient nights
- More productive hours
- More creative
- More social in physical
- Active body (Less lethargic)

Alarming Indication for Parents/Guardians

The flood of information and knowledge is an easy way to confuse a child in adolescent age. In the name of entertainment, all stuff of adultery is served to the brain and they get succumbed to the exposure of reality over the world of fantasy (Boitard, Pourazad, Nasiopoulos, & Slevinsky, 2015). After the conduct of few types of research, a noticeable change has been observed among the adolescent, which can be treated as alarming indications and there must be positive mind-sets of parents /guardians to take measures from overcoming the issues (Ess, 2017).

Behavioural Change

- It can be observed that the child becomes very possessive about his/her looks apparel, makeup, and hairstyles and gives full attention to the appearance of self among the public. In this aspect, it is often found that there arises a tremendous disagreement in the choices of the child and parents.
- The adolescent is found to be arrogantly demanding for expensive handsets, two-wheelers, and other commodities of luxury to build up his imaginary repute among peers without understanding its importance and without paying the value of hard-earned money.
- The adolescent is much inclined towards the brands of each product, irrespective of the sensibility towards the purchasing power of the family. They may become extremely arrogant and start threatening the parents in case of any refusal to them.
- The mind-set of the adolescent is "just enjoying the life" and the alarming condition arises when they are unable to select a goal, to be achieved.
- Peer group takes prime importance in their life and they lack sensitization towards the family responsibilities, lack emotions, and respect towards the elderly persons and other family members with absolute rejection of family values.
- Deterioration in grades obtaining in examinations and losing interest in all educational activities carelessly.
- At this stage of life, they have a high level of inquisitiveness towards all available materials of intoxication. The attempts taken for experimenting can become serious ailments in a later stage.
- The adolescents also undergo biological changes in their body so they are highly attracted to the opposite sex and may become attention seekers and go beyond boundaries to satisfy their curiosities.
- Mostly the adolescent argues with or without logic and tries to justify all his/her behaviour. They dislike being disciplined especially in food and sleeping habits.
- Most of the time they will say 'sorry' to close the chapter rather than any realization of committing mistakes and thus repeat their insensible behaviours.

The above-mentioned behavioral changes are alarming indications and the parents or guardians must be aware enough to take measures to divert them and make them more responsible towards themselves, family, and society.

Coping Mechanism

Rome is not built in a day, so the application of coping strategies will not act miraculously overnight. A systematic influencing and effective mechanism is required to be adopted to safeguard the interest of the adolescent and develop a balanced behavior among the cohort (Vidal, Lhaksampa, Miller, & Platt, 2020).

As per our research analysis, the following coping strategies can help out the adolescent to get rid of media impacts, social blows, and control anger-related emotions. The mechanism must be inclusive of expression of love, affection, family values, and most importantly learning to take responsibility.

- A well-structured routine must exist in the family related to eating and sleeping habits, celebrating festivals and participating in family functions, etc. without imposing the same is mandatory.

- Creating a joyful atmosphere in the family with silent prompting of the presence of all members and giving high importance to the adolescent.
- Encouraging the child for involvement in any field of art or sports in parallel to the basic course of education.
- Help the adolescent to fix short-term goals and achieving the same. This will help them to develop the habit of goal fixation.
- Monitor the media use but do not interfere in all affairs of the adolescent being parent or guardian.
- Give responsibilities and express your trust in them for fulfilling them.
- Allow their decision-making and appreciation for the same.
- Indicate their mistakes instead of blaming them.
- Provide support in problem-solving, being a good listener rather than a dictator.
- It is very important to have free talks over sensitive issues such as sexual orientation, all materials of intoxication, violence, religions, etc so that they have a balanced view and reactions on such subject matters (Caton & Chapman, 2017).
- Taking their suggestions on important decisions will make them more sensible and responsible in all affairs.

All of the strategies if adopted systematically by the parents and guardians, then they can help the adolescent to keep away from the ill effects of continuous mobile interaction, followed by, stress fear, and anger. This will also help the adolescent in capacity building and developing the skill variant.

CONCLUSION

The connection between mental health and digital media is inevitable. Since its inception, the existence of digital media is based on the grounds of increasing productivity of an individual and keeping them connected to the world. But the studies have always come up with one answer that is 'Social media is making us unsocial'. The world of the hashtag is not working much favourably for adolescents especially. The reason behind is overdoing of internet thing, having a smartphone, or maybe because your peers are using it. This can results in depression, anxiety, ADHD, or even death. The art of social media making us unsocial and disconnected from the physical world is needs to be discussed. The adolescents are exposed to it, they can see their parents using smartphones, friends uploading objects, or maybe their teacher teaching through smartphones. Digital media is designed to be addictive, and it is not ending up any soon. For adolescents the existence of digital media is perpetual, and literally, they cannot contemplate beyond it. They are getting addicted to it, they are becoming unsocial, their physical health is deteriorating, they are not in good mental health, and they are unaware of all these happening with them. The statistics on the use of social media are scary, and an adolescent is expected to consider it as a priority for resolution. Decoding the physical hazard caused by social media can include; Headache, eye problems, non-eating, tiredness, physical hygiene. Beyond Physical hazards, the mental health issues include; mood swings, insomnia (Sleeplessness), anxiety, ADHD, OCD, Panic attacks, depression, and aggression. Checking your phone 50-100 times, having ringxiety, Internet addiction, NOMO phobia, FOMO phobia, the greed of social currency, social disconnection, lack of concentration, and having gaming obsession is alarming and disturbing and something to be worried about. This makes you question your self-worth, the greed of validation generates, you start questioning your safety, and there are high probabilities of encountering online harassment. We as an individual are solely responsible for every situation we encounter as a consequence of not using internet appropriately.

Digital media was there, it is here, and it will going to exist, which means it's not going anywhere. So quitting social media and getting back into it after some time will not going to make any sense. Absence is not an option anymore. Therefore, you exist appropriately, by spending less time in the Digital world and practice 'safe social'. You wouldn't blame Television for showing a bad tv show, Instagram is not making you go on vacation, or say Twitter does not make people write a hateful post. Hence, timely precautions and keeping a safe distance from social media can save you.

As the level of maturity among adolescents is not experienced and developed. The time will expect the peers and parents to save and guide them. Thus, we need to have preventive strategies and coping mechanisms. To fix the issue parents can start by recognizing the problem because awareness is critical. They can keep a check on adolescents by auditing their digital media regime. Help your adolescents creating a better online experience, by reading blogs, make them follow some creative page or channel, this activity can help them explore the unexplored this will also help in modeling good behaviour in them.

To manage the see-saw of the digital world, one needs to develop the skill of balancing. Digital media is an imperative tool for each and every individual, but one cannot go long way without it. To sustain here and maintain the balance we all need to join hands and work towards a healthy social media interaction that has a holistic wellbeing approach. *'If not now, then when; If not you, then who?'*

REFERENCES

App Annie. (2021). *The App Analytics and App Data Industry Standard*. Retrieved 4 May 2021, from https://www.appannie.com/en/?sfdcid=7016F000002MS1c&utm_campaign=ww-logo-201910-1910-digital-2021-partnership&utm_content=report-&utm_medium=partnership&utm_source=digital-2021

Boitard, R., Pourazad, M. T., Nasiopoulos, P., & Slevinsky, J. (2015). Demystifying High-Dynamic-Range Technology: A new evolution in digital media. *IEEE Consumer Electronics Magazine*, *4*(4), 72–86. doi:10.1109/MCE.2015.2463294

Carter, S. (2018). *Social media may be making you overspend—and it's not just because of the ads* [Blog]. Retrieved from https://www.cnbc.com/2018/03/15/social-media-may-make-you-overspend-and-its-not-just-because-of-ads.html

Cash, H., Rae, C. D., Steel, A. H., & Winkler, A. (2012). Internet Addiction A Brief Summary of Research and Practice. *Current Psychiatry, 8*(8), 292–298. Retrieved from https://www.ingentaconnect.com/content/ben/cpsr/2012/00000008/00000004/art00010

Caton, S., & Chapman, M. (2017). *The Use of Social Media and People with Intellectual Disability*. Academic Press.

Coyne, S. M., Young, B., Rogers, A. A., Zurcher, J. D., Stockdale, L., Booth, M., ... Booth, M. (2020). Does time spent using social media impact mental health? An eight year longitudinal study. *Computers in Human Behavior*, *104*(1), 1–9. doi:10.1016/j.chb.2019.106160

Debell, M., & Chapman, C. (2004). *Computer and internet use by children and adolescents in 2001: Statistical analysis report*. National Center for Education Statistics. doi:10.1037/e609922011-001

Ess, C. (2017). Digital Media Ethics. Oxford Research Encyclopedia of Communication, 2017(Sep), 1–47. doi:10.1093/acrefore/9780190228613.013.508

Eye Health: Technology's Effect on Eye's. (2021). Retrieved 6 May 2021, from https://guardionhealth.com/technologys-effect-eyes/#:~:text=One%20of%20the%20most%20common,to%20a%20painful%20tension%20headache

Fobian, A. D., Avis, K., & Schwebel, D. C. (2016). Impact of media use on adolescent sleep efficiency. *Journal of Developmental and Behavioral Pediatrics*, *37*(1), 9–14. doi:10.1097/DBP.0000000000000239 PMID:26651090

Gao, J., Zheng, P., Jia, Y., Chen, H., Mao, Y., Chen, S., Wang, Y., Fu, H., & Dai, J. (2020). Mental health problems and social media exposure during COVID-19 outbreak. *PLoS One*, *15*(4), 1–10. doi:10.1371/journal.pone.0231924 PMID:32298385

Ghasemi, V., Rabiei, K., Davoodi, S., & Rabiei, H. (2017). Mobile Internet Usage among Adolescents and Young Adults in Iran: A Sociological Survey. *Tarih Kültür ve Sanat Arastirmalari Dergisi*, *6*(1), 863. doi:10.7596/taksad.v6i1.785

Hutchins, M. (2021). *Social Media Use Linked To Anxiety, Depression Among Teens, New Study Finds.* Retrieved 6 May 2021, from https://www.wbur.org/hereandnow/2020/01/09/social-media-anxiety-depression-teens

Jarrett, J. (2021). *Is There a Link Between Social Media and Eating Disorders?* Magnolia Creek. Retrieved 6 May 2021, from https://www.magnolia-creek.com/eating-disorder-recovery-blog/social-media-and-eating-disorders/

Kemp, S. (2021). *Digital 2021: Global Overview Report — DataReportal – Global Digital Insights.* Retrieved 4 May 2021, from https://datareportal.com/reports/digital-2021-global-overview-report

Krebs, G., & Heyman, I. (2015). Obsessive-compulsive disorder in children and adolescents. *Archives of Disease in Childhood*, *100*(5), 495–499. doi:10.1136/archdischild-2014-306934 PMID:25398447

Kushlev, K., Proulx, J. D. E., & Dunn, E. W. (2017). Digitally connected, socially disconnected: The effects of relying on technology rather than other people. In Computers in Human Behavior (Vol. 76). doi:10.1016/j.chb.2017.07.001

Labrecque, L. I., vor dem Esche, J., Mathwick, C., Novak, T. P., & Hofacker, C. F. (2013). Consumer power: Evolution in the digital age. *Journal of Interactive Marketing*, *27*(4), 257–269. doi:10.1016/j.intmar.2013.09.002

Littlefield, A. (2020). *No, You Don't Have the Attention Span of a Goldfish* [Blog]. Retrieved from https://www.ceros.com/inspire/originals/no-dont-attention-span-goldfish/

Magson, N. R., Freeman, J. Y. A., Rapee, R. M., Richardson, C. E., Oar, E. L., & Fardouly, J. (2021). Risk and Protective Factors for Prospective Changes in Adolescent Mental Health during the COVID-19 Pandemic. *Journal of Youth and Adolescence*, *50*(1), 44–57. doi:10.100710964-020-01332-9 PMID:33108542

O'Sullivan, K., Clark, S., McGrane, A., Rock, N., Burke, L., Boyle, N., Joksimovic, N., & Marshall, K. (2021). A qualitative study of child and adolescent mental health during the COVID-19 pandemic in Ireland. *International Journal of Environmental Research and Public Health, 18*(3), 1–15. doi:10.3390/ijerph18031062 PMID:33504101

Panagiotopoulos, P., Barnett, J., Bigdeli, A. Z., & Sams, S. (2016). Social Media in Emergency Management: Twitter as a Tool for. *Technological Forecasting and Social Change, 111*(October), 1–27. Retrieved from https://www.sciencedirect.com/science/article/pii/S0040162516301196%0Ahttp://dx.doi.org/10.1016/j.techfore.2016.06.010%0Ahttps://linkinghub.elsevier.com/retrieve/pii/S0040162516301196

Parekh, R. (2017). *What Is ADHD?* Retrieved 6 May 2021, from https://www.psychiatry.org/patients-families/adhd/what-is-adhd#:~:text=Attention%2Ddeficit%2Fhyperactivity%20disorder%20(,common%20mental%20disorders%20affecting%20children

Qutteina, Y., Hallez, L., Mennes, N., De Backer, C., & Smits, T. (2019). What Do Adolescents See on Social Media? A Diary Study of Food Marketing Images on Social Media. *Frontiers in Psychology, 10*(November), 1–12. doi:10.3389/fpsyg.2019.02637

Raju, N. J., Valsaraj, B. P., & Noronha, J. (2015). Online Social Networking: Usage in Adolescents. *Journal of Education and Practice, 6*(22), 80–84. http://search.proquest.com/docview/1773229886?accountid=14744

Robinson, L., & Smith, M. (2021). *Social Media and Mental Health - HelpGuide.org*. Retrieved 6 May 2021, from https://www.helpguide.org/articles/mental-health/social-media-and-mental-health.htm#:~:text=Excessive%20social%20media%20use%20can,negative%2C%20self%2Dperpetuating%20cycle%3A&text=Using%20social%20media%20more%20often,depression%2C%20anxiety%2C%20and%20stress

Rodriguez, A. (2021). *Social media fatigue: 10 facts you should know*. Retrieved 6 May 2021, from https://www.mycustomer.com/community/blogs/annarodriguez/social-media-fatigue-10-facts-you-should-know#:~:text=According%20to%20Techopedia%2C%20social%20media,spent%20online%20maintaining%20these%20connections.%E2%80%9D

Rothberg, M. B., Arora, A., Hermann, J., Kleppel, R., Marie, P. S., & Visintainer, P. (2010). Phantom vibration syndrome among medical staff: a cross sectional survey. *BMJ, 341*(2). doi:10.1136/bmj.c6914

Shafer, L. (2017). *Social Media and Teen Anxiety*. Retrieved 6 May 2021, from https://www.gse.harvard.edu/news/uk/17/12/social-media-and-teen-anxiety

Vidal, C., Lhaksampa, T., Miller, L., & Platt, R. (2020). Social media use and depression in adolescents: A scoping review. *International Review of Psychiatry (Abingdon, England), 32*(3), 235–253. doi:10.1080/09540261.2020.1720623 PMID:32065542

Williams, P. (2019). *To what extent is Social Media responsible for a rise in mental health issues in Students in America*. Academic Press.

Zucco, R., Lavano, F., Anfosso, R., Bianco, A., Pileggi, C., & Pavia, M. (2018). Internet and social media use for antibiotic-related information seeking: Findings from a survey among adult population in Italy. *International Journal of Medical Informatics*, *111*(December), 131–139. doi:10.1016/j.ijmedinf.2017.12.005 PMID:29425624

ENDNOTE

1 Dubbed novelty addiction.

Chapter 9
Stress and Anxiety in the Digital Age:
The Dark Side of Digital Technology

Humera Waseem Khan
University of Delhi, India

Arti Jain
Daulat Ram College, India

ABSTRACT

While digitalization undoubtedly has improved the access to information and communication with the world, increased cases of cyberbullying, harassment, cyber abuse, and suicide numbers have also surfaced among teens. The hype of the online world, the celebrity culture, and their high status seldom generate the fear of inadequacy, dissatisfaction, and isolation which ultimately aggravate the feeling of anxiety and depression. This chapter will underline the causes of the digital stress and technology-related anxiety among adolescents, their deteriorating psychophysical behavior, factors disintegrating their mental fitness and physical well-being, and most importantly, the answer to the problem. Various small steps can be taken to avoid serious problems such as talking with your family and friends. Some other approaches include cutting off from the negative digital surroundings, limiting the watch time, etc. With the growing online world, there is an urgent need to control the exposure one gives to the social world.

INTRODUCTION

We, the homo sapiens, being titled as social creatures, feel the urge to stay connected with family, friends and the world to thrive in life. Thus communication becomes integral for survival and growth. Although digitalization, undoubtedly has strengthened the process of public interaction, on the contrary the art of social exchange has become bleak and impartial. Being constantly bridged on social media has become spirit of today's generation. Youngsters use these digital media as a way of escape from the real world because the virtual world is shown as luxurious and aesthetically pleasing. The social network initially

DOI: 10.4018/978-1-7998-8318-0.ch009

created to expand the horizons of knowledge and information, it's irregular, extreme and unnecessary usage has lead to sorrow, disappointment, frustration and isolation amongst the users. Face to face interactions has decreased drastically which is ultimately responsible for depression and loneliness. There are numerous benefits that one encounter while surfing the web. Be it academics, sports, education, politics or entertainment, no field is untouched by social media platforms. This blessing of social world due to unnecessary usage has become a bane. The management of stressors and resources is specifically important in this increasingly digitized and digitalized Working world.

Objectives of the Chapter

1. Effects of digital media on social experiences and development of adolescents.
2. Effect on real life developmental proficiency due to extreme internet use.
3. Positive aspects of social network.
4. Deleterious outcomes of the internet community.
5. Defining sources of stressors.

BACKGROUND

Mobile media attaches us to people, content and various services alone from time and place restrictions but this benefit is also challenging individuals' autonomy (Vanden Abeele, De Wolf, & Ling, 2018, 5-14). The problem is more pronounced among adolescents and teens. Adolescence is the age when the aspirations for strong acceptance, enthusiastic hope and rational psychiatric, social and cultural changes develop (Youniss, 1987) The constant urge to get attention and satisfaction are the root cause why they find themselves cemented to the screen all the time. Initially developed internet world has innumerable benefits. The number of hours that adolescents spend on digital is speedily increasing. This in turn is concomitantly associated with the pervasiveness of mental health issues. (Brown, 2004,363-394).

Somc of thc positivc aspects of social media includes staying updated with family and friends, using online world as a platform for approaching new communities, for expanding business networks, it also serves as a channel for creativity and self expression and availability of easily accessible and reliable sources of knowledge and information.(Robinson, 2021)

Digital Stress

Digital stress has now become a global problem and has a direct influence on body and soul. Two major kinds of digital stressors are identified.

Type 1: Mean and intimidating attacks on individual's that leads to communal disgrace and humiliation.

1. Mean and intimidating attacks: Young generation use social media as place where they hang out with their virtual friends. Posting day to day activities, relationship intimacy, confidential information about close relatives make their private life more vulnerable. People who intend to bully and harass the users, take advantage of these personal details to choose their victims. Through social media platforms and text messaging, youth receive unwanted messages. These predators threatened

the victims and in order to maintain their so called social status, these victims often find them in an inescapable trap of online world.

2. Communal disgrace and humiliation: While private criticism is conducted through directly contacting, messaging, public shaming is usually done by broadcasting of information. Majority cases encounter circulation of compromised photographs or extremely personal details to the recipient not intended to share by the user.
3. Impersonation: It refers to the use of digital world to hide an individual's recognition and profess to be someone else. Impersonation is encountered through various ways. Two major ways are discussed here. First one includes hacking and logging into someone else's account either to mock, embarrass or trouble the person being targeted. The second kind accounts for impersonation through fake accounts or using false identities.

Type 2: feeling strangled and invasion without consent.

1. Feeling strangled: This stressor is also referred as feeling smothered. It is defined as the fear of suffocation that the user develop due to non required and uncontrolled access to someone close to their personal information. The unnecessary intervention becomes throttling after a certain time and the user begin detecting a sense of extreme engagement in their personal lives.
2. Breaking and invading: It implies the access to log into another person's personal account or checking their electronic gadgets without their permission. This either happens through checking the contact list or reading their personal messages or surfing their browser history. All these actions are viewed as invading digital privacy. Sometimes family members and friends act upon these in order to gain access to private communication of the user. The panic of being digitally unprotected surfaces into agitation and envy (Weinstein & Selman, 2014, 391-409).

Celebrity Culture

Film celebrities, sports personalities, social media stars, television icons have always been the blaze of public attention and focal point of attraction for their audience. Their personal lives, eating habits, daily routine all has turned into a couth which can be termed as the celebrity culture. It has become a trait of communion, specially in developed world. The definition of being a celebrity has also changed in recent years. Now it is more often applied to those who are popular for being popular. Having powerful crowd penetration and influence to buy, product marketers are aided by these celebrities as their main brand promoters and advertising platforms. Most people look up to celebrities for glamour, splendid way of living and personal nurturing. They usually provide an authority of validation to their audience. Celebrities have a profound impact on youngsters body image. Specifically for girls appearance is the top most concern. The influence become so strong that the thin line between real and reel life disappears.("How Celebrities Influence Teens and Why it Matters, Newport Academy", 2021) Decorated photographs, flawless bodies, alluring destinations all force young people to meet non realistic body standards. According to body image survey conducted by The today show and aol.com, 80 percent of teen girls compare themselves with these celebrities and models they see online. Dissatisfaction by one's own body, skin color, shape and size is connected with various mental health challenges. For instance, poor diet in order to obtain the perfect body can cause eating disorders which has the direct relation with mental fitness and health problems like depression and low self esteem.("Social media and celebrity culture 'harming young people', Body image, The Guardian", 2021)

Addictive Design

Millions have been spent on strategies to grab one's attention to various social media networking sites and applications. The features and tools are intentionally made addictive that indirectly prompt parts of brain and keeps the use glued to the display.

1. Endless scrolling: Not only architects and marketers, but specialized teams of psychologists, IT experts and designers try as hard as possible to get the job done. Initially created to refine ease of use and general using experience, this feature of social media applications has become so compelling that users end up wasting hours just sliding through their screen.
2. Likes, hearts and claps: Satisfaction, seeking attention and uplifting self respect are the key reasons that governs the exceptional indulgence of today's youth in the online world. They keep checking their social media accounts again and again to sense the feel of validation, applaud and contentment from peers. ("The 4 Features That Make Social Media Apps Addictive", 2021)

A much interesting fact revolves around the designing of these social media platforms. From main mascot to various internal properties all infographics are promisingly inviting to the users. Choice of colors and graphics make huge impact. Different colors struck the user's mind differently. Some striking social media applications' colors such as bright red, yellow, green, different hues of blue are used to arouse feelings of love, excitement, calmness, good luck and enlightenment. Techniques such as HEX color codes and RGB values form the basis for these online media logos. (Ocampo, 2021)

Cyberbullying on Another Level

The very popular Bluc Whalc gamc is an cxamplc of taking cyber harassment on a different level. The killer game is an online challenge that one cannot find in app stores and websites. It is shared among undercover groups through various social media platforms. The killer game forces you to enter a virtual world where the users are pressurized indirectly to complete a series of 50 instructions in 50 days while the last task is to commit suicide. ("Blue Whale: The truth behind an online 'suicide challenge' – BBC News", 2021). Each task is when after the completion of the previous task and a proof of completion, the players are instructed to upload photos of them undertaking the challenge. The game is named on a serene mammal because whales are often depicted as icon of solitude and isolation (Rossow, 2021).

Fear of Missing Out

A social media paradox that is becoming increasingly compulsive and troublesome is fear of missing out (Chai et al., 2019,85-92). Also termed as FOMO has been defined as a form of social anxiety emerging from the believe that others are having meaningful, luxurious and pleasurable life. It represents individual's high level of concern. Sense of deficiency, negative emotions and unnecessary and involuntary feel of jealousy, low self respect are some very dangerous consequences of this phenomenon (Przybylski et al., 2013,1841-1848).

CONCLUSION

Innumerable mobile technologies functioning through wireless system framework such as laptops, tablets has flooded our leisure time with information, entertainment and business opportunities. Users feel connected with the online world pervasively. The effect of online media on adolescents' can no longer be overlooked. Adolescence no wonder is the period of intense emotions, strong beliefs, internet, unrealistic expectations, desires and enthusiasm. When feeling of betrayal or dissatisfaction arises among individuals, there are plenty of Digital tools to provide them the satisfactory empathy and compensation. These tools are unknowingly damaging to brain as well as body. This generates a need for comprehensive reason for apprehension about the Digital world. Celebrity fashion, habit forming designing of various applications and games has the situation more problematic and serious.

REFERENCES

Blue Whale. (2021). *The truth behind an online 'suicide challenge'*. BBC News. Retrieved 3 October 2021, from https://www-bbc-com.cdn.ampproject.org/v/s/www.bbc.com/news/blogs-trending-46505722.amp?amp_js_v=a6&amp_gsa=1&usqp=mq331AQKKAFQArABIIACAw%3D%3D#aoh=16332251619975&referrer=https%3A%2F%2Fwww.google.com&amp_tf=From%20%251%24s&ampshare=https%3A%2F%2Fwww.bbc.com%2Fnews%2Fblogs-trending-46505722

Brown, B. B. (2004). Adolescents' relationships with peers. In R. M. Lerner & L. Steinberg (Eds.), Handbook of Adolescent Psychology (2nd ed., pp. 363–394). John Wiley & Sons.

Chai, H.-Y., Niu, G.-F., Lian, S.-L., Chu, X.-W., & Sun, X.-J. (2019). Why social network site use fails to promote well-being? The roles of social overload and fear of missing out. *Computers in Human Behavior, 100*, 85–92. doi:10.1016/j.chb.2019.05.005

How Celebrities Influence Teens and Why it Matters. (2021). *Newport Academy*. Retrieved 2 October 2021, from https://www.google.com/url?q=https://www.newportacademy.com/resources/empowering-teens/celebrities-influence/&usg=AOvVaw3t-vK9SiJBcWeUwduiNSfS

Ocampo, A. (2021). *How Colour Psychology can be Applied to Mobile Apps. Top mobile App Developers*. Retrieved 2 October 2021, from https://thisisglance.com/how-colour-psychology-can-be-applied-to-mobile-apps/

Przybylski, A. K., Murayama, K., DeHaan, C. R., & Gladwell, V. (2013). Motivational, emotional, and behavioral correlates of fear of missing out. *Computers in Human Behavior, 29*(4), 1841–1848. doi:10.1016/j.chb.2013.02.014

Robinson, L. (2021). *Social Media and Mental Health*. Retrieved 2 October 2021, from https://www.google.com/url?q=https://www.helpguide.org/articles/mental-health/social-media-and-mental-health.htm&usg=AOvVaw1FQOMnN_Iilaj6ssJP7yqQ

Rossow, A. (2021). Cyberbullying Taken To A Whole New Level: Enter The 'Blue Whale Challenge'. *Forbes*. Retrieved 3 October 2021, from https://www.forbes.com/sites/andrewrossow/2018/02/28/cyberbullying-taken-to-a-whole-new-level-enter-the-blue-whale-challenge/?sh=4e2037212673

Social media and celebrity culture 'harming young people'. (2021). *The Guardian.* Retrieved 2 October 2021, from https://amp-theguardian-com.cdn.ampproject.org/v/s/amp.theguardian.com/lifeandstyle/2018/jul/23/social-media-and-celebrity-culture-harming-young-people?amp_js_v=a6&amp_gsa=1&usqp=mq331AQKKAFQArABIIACAw%3D%3D#aoh=16331002636511&referrer=https%3A%2F%2Fwww.google.com&amp_tf=From%20%251%24s&ampshare=https%3A%2F%2Fwww.theguardian.com%2Flifeandstyle%2F2018%2Fjul%2F23%2Fsocial-media-and-celebrity-culture-harming-young-people

The 4 Features That Make Social Media Apps Addictive. (2021). *Medium.* Retrieved 2 October 2021, from https://bettermarketing.pub/the-4-features-that-make-social-media-apps-addictive-1ce7190f1702

Vanden Abeele, M. M. P., De Wolf, R., & Ling, R. (2018). Mobile media and social space: Howanytime, anyplace connectivity structures everyday life. *Media and Communication*, *6*(2), 5–14. doi:10.17645/mac.v6i2.1399

Weinstein, E., & Selman, R. (2014). Digital stress: Adolescents' personal accounts. *New Media & Society*, *18*(3), 391–409. doi:10.1177/1461444814543989

Youniss, J. (1987). *Adolescent Relations with Mothers, Fathers and Friends.* The University of Chicago Press.

Chapter 10
The Impact of Mindfulness Practices and the Implementation of Technology in Higher Education

Shaakira Sharif
The Chicago School of Professional Psychology, USA

Aubrey Statti
The Chicago School of Professional Psychology, USA

Kelly M. Torres
The Chicago School of Professional Psychology, USA

ABSTRACT

Maintaining efficient time management and learning how to balance different facets of life can be difficult for undergraduate students. Students can have difficulty with maintaining their schedules and academic responsibilities, which can disrupt daily living functionality. Mindfulness is an evidence-based practice that can help students establish a self-care routine. Additionally, establishing a mindfulness regimen can help to improve students' academic performance. With the inclusion of technology, mindfulness mobile applications have gained popularity over the past decade and have provided a convenient method for students to engage in guided meditations.

INTRODUCTION

Incorporating Mindfulness Technology Into the College Environment

Attending college indicates that young adults have reached at least partial independence from their caretakers and are responsible for making crucial life decisions. Establishing a balance between attending classes, work, and other personal obligations can be challenging, but that balance is needed to have a successful

DOI: 10.4018/978-1-7998-8318-0.ch010

college experience (Appleby, 2017). According to Appleby (2017), there are multiple components for succeeding in college, such as the soft skills of interpersonal and critical thinking competencies and a goal-orientated mind frame. Additionally, establishing a self-care routine is another helpful component that can help with student success. For example, Parmer (2017) mentioned that maintaining self-care practices can be difficult for college students, but it can be an outlet for stress reduction. Although there are multiple self-care routines for students to reduce stress and improve their academic performance, mindfulness is a technique that can be useful to improve mental clarity and help with the issues associated with stress and concentration levels.

CHALLENGES FOR COLLEGE STUDENTS

Examining the various challenges that college students face may help higher education administrators understand the need for wellness practices, resources, and support from faculty that can help students with their stress levels, academic achievement, and overall well-being (Acharya, 2018; Arora & Singh, 2017; Beauchemin, 2018; Epstein, 2015; Roy, 2018). Common challenges that college students typically deal with include mental health issues, academic achievement, high stress levels, and concentration problems (Acharya, 2018; Alderson Broaddus University (n.d); Francis and Horn, 2017). Therefore, helping students identify wellness practices that can work to improve their well-being and help with their academic progress may be beneficial. The Research and Planning Group for California Community Colleges (The RP Group) (n.d.), a non-profit organization that is focused on advocacy and legislative issues for college, stated that six factors that contribute to success for college students include being directed, focused, nurtured, engaged, connected, and valued. Students receive adequate direction from college faculty and mentors and work to obtain their goals, are focused on achieving various tasks related to college, are nurtured by their learning institution, engaged in their classes and extracurricular activities on campus, feel connected with their peers and learning institution, and feel valued when they are acknowledged for their abilities and achievement (Parmer, 2017; Ramos-Salazar & Hayward, 2018; The RP Group, n.d.). This chapter seeks to highlight these success factors and the various challenges that college students typically experience throughout their coursework. Particularly, this chapter provides suggestions on how to best assist students throughout their academic journey and also to provide information that can help college administrators and support staff with developing effective programming for students may improve educational outcomes (Holinka, 2015; Shelesky et al., 2016; Whitehead; 2017; Winerman, 2017).

MENTAL HEALTH FOR COLLEGE STUDENTS

The National Alliance on Mental Illness (NAMI) (2019), stated that one in five adolescents ages 13-18 years have a severe mental illness and one in five adolescents have a learning disability. Additionally, 11% of adolescents have a mood disorder, 10% have a conduct disorder, and 8% have an anxiety disorder (NAMI, 2019). Sixty-four percent of college students who are no longer attending college classes have cited mental health issues as the primary reason for not continuing their courses (NAMI, 2019). According to prior research, students who are diagnosed with a mental health disorder or experiencing emotional distress may require additional support within the college setting (Alderson Broaddus University, n.d.; Braider et al., 2019; Mitchell et al., 2019). The American College Health Association (ACHA) is a research-based organization that focuses on advocacy, research, and education for undergraduate and graduate students (ACHA, n.d). The American College Health Association National College Assessment, administer by The ACHA, accesses the following domains:

1. General Health of College Students,
2. Disease and Injury Prevention,
3. Academic Impacts,
4. Violence, Abusive Relationships and Personal Safety,
5. Tobacco, Alcohol, and Marijuana Use,
6. Sexual Behavior,
7. Nutrition and Exercise,
8. Mental Health,
9. Sleep.

Results from the latest assessment on mental health indicated that of the 54,497 undergraduate students, male and female, surveyed, from 98 different schools, 57% of respondents experienced feelings of hopelessness, being overwhelmed (88%), exhaustion (85%), and anxiety (66%) within the past 12 months (American College Health Association-National College Health Assessment II, 2019). Within the same survey, 44% of male students reported more than average stress, 12% of female students experienced tremendous stress, and nearly 18% of both female and male undergraduate students experienced feelings of suicidal ideation. The results of the study demonstrated that students are experiencing difficulty in areas involving mental clarity and wellness (ACHA, n.d.).

STRESS IN THE COLLEGE ENVIRONMENT

Another element that that adversely effects college students is high stress levels that cause anguish and can impact functioning. In a medical or biological context stress is a physical, mental, or emotional factor that causes bodily or mental tension. Stresses can be external (from the environment, psychological, or social situations) or internal (illness, or from a medical procedure). Stress can initiate the "fight or flight" response, a complex reaction of neurologic and endocrinologic systems. (pg. 2). Additionally, in a survey conducted by the American Psychological Association (2019), three out of four adults experienced physical and emotional stress symptoms, such as headaches, exhaustion, and significant changes in their sleeping habits. As originally reported by the Anxiety and Depression Association of America via the American Institute of Stress (2019), nearly 80% of college students sometimes or often experience stress. Additionally, factors that led to stress for students include lack of sleep, financial obligations, need to balance workload from classes, and adjustment to new surroundings and social groups (American Institute of Stress, 2019; American Psychological Association, 2019; EAB, 2020; Ingersoll, 2019). It is valuable to identify previous research that focuses on stress from college students. Acharya et al. (2018) explored life stressors for college students through the Student Stress Survey (SSS), which listed 41 stressors related to college. The eight most common stressors that were identified by students were: participating in social activities, working with new people, changing sleeping habits, changing eating habits, increasing class workload, receiving a lower grade than anticipated, placing oneself in an unfamiliar situation, and changing their living environment (Acharya et al., 2018).

Another component intertwined with stress is the difficulty for students to cope with life stressors and utilize healthy coping mechanisms. Identifying healthy coping mechanisms can be useful in reducing stress and enhance the quality of life for students (Acharya, 2018; Gnilka et al., 2015). For example, Beauchemin (2018) implemented wellness interventions in order to decrease stress levels for students. In Beachemin's study, researchers designed a seven-week brief-solution focused wellness intervention where research participants completed participated in group discussions on stress and stress reduction techniques and completed a strengths-based goal worksheet at the conclusion of each study (Beauchemin, 2018). Within the study, research participants were given the Perceived Stress Scale (PSS) to determine stressors and the Five Factor Wellness Evaluation of Lifestyle (5F-WEL) to identify wellness practices. Results indicated that students benefited positively from the interventions and implications noted that improving wellness can decrease stress levels (Beauchemin, 2018).

With there being a link between high stress levels, maladaptive behavior, health outcomes, it is important to examine factors surrounding these issues. Shaktin et al. (2016) investigated whether there are statistical differences between college students enrolled in the risk and resilience course (RR) (experimental group) and child and adolescent psychopathology course (control group). As part of the procedures, students in the RR received the skill-building intervention for one semester that focused on coping skills that help with reducing stress and helped reduce negative thoughts. Findings from the study showed that students in the RR experimental group had a reduction in stress levels and improvements in their negative thinking patterns. Further, it was noted that the intervention is a useful option for other college students and for future studies, it might be helpful to implement the intervention for students throughout their college careers (Shaktin et al., 2016). In connecting these studies, it shows the importance of managing stress and different methods that can be used to help college students (Acharya, 2018; Beauchemin, 2018; Gnilka et al., 2015; Shaktin et al., 2016).

COUNSELING SERVICES AND INTERVENTIONS FOR COLLEGE STUDENTS

With limited counseling and case management resources in college counseling centers many students need these services to help them with their life functioning and increase their academic outcomes (Epstein; 2015; Francis & Horn, 2017; Holinka, 2015; Shelesky et al., 2016). Shelesky et al. (2016) developed a case management model that included a behavioral component to help students to address their mental health and life situations needs. Conclusions from the article indicated that it is helpful for college counseling centers to have a full-time case manager that addresses the importance of providing case management services, referring students to community resources, and delivering behavioral interventions (Shelesky et al., 2016). Also, case managers can be resourceful and assist college students with a wide variety of needs. For example, since case managers can locate resources and provide behavioral interventions and it would be helpful to have a case manager on a full-time basis. Still, funding an additional salary may pose an issue for different colleges (Epstein, 2015; Shelesky et al., 2016).

Shefet (2018) also discussed the need of counseling resources in college counseling centers. As Shefet discussed, there is high percentage of students seeking counseling due to various environmental and social factors and despite the need for more counseling services Colleges are often not able to meet the demand of the student population due to budget restrictions and limited resources; such as support staff. To address the issue of limited counseling services, students may be put on a waitlist, provided with community referrals, offered group counseling and limited individual sessions, and have a walk-in clinic option (Shefet, 2018).

Based on the research from Shefet (2018), it is evident that there is a need for more counseling resources for students to address their mental health needs efficiently. Methods such as a waitlist and limited sessions are not always effective and can lead students to feel rejected and can be harmful to critical mental health issues such as suicidal and homicidal ideation. However, community referrals, group counseling, and a walk-in clinic are plausible options and provides more options for students (Shefet, 2018).

In detailing the issues that lead to college students experiencing difficulties associated with completing college, Bishop (2016) stated that it is imperative to focus on precollege factors and the most common factors include socioeconomic status, high school grade point average, and first-generation college status. Consequently, low retention rates are also an issue for college students and psychological factors, academic ability, and challenges with social adjustment are common risk factors within this realm. One support source for college students who are having difficulties is the inclusion of counseling services provided on-campus for students. Through archival data, Bishop (2016) classified students as either low-risk or high-risk based on the pre-college factors and it was noted in the data whether students engaged in counseling services. Key elements from the study noted that psychological factors were the top concerns for students beginning counseling services, but high risk-students were not always aware of counseling services on-campus (Bishop, 2016). Conclusions from the study noted that college counseling centers can increase their presence on college campuses, conduct outreach efforts to high-risk college students, and counselors can utilize efficient counseling techniques for working with this population of students (Bishop, 2016).

Lastly, Epstein (2015) provided a review of counseling resources on community college campuses. It was noted that community college counseling centers typically do not have enough resources to provide students with comprehensive mental health care (Epstein, 2015 Roy, 2018). Nevertheless, resources from the Affordable Care Act (ACA) provides additional counseling, academic, and wellness resources for higher learning institutions in some states, which can be helpful for community colleges. Even though researchers specifically focused on community colleges, the researcher raised concerns about how budgetary limitations can effect access to support services resources, which speaks to a larger issue about how monetary restrictions can effect and limit the access of counseling resources for students (Epstein, 2015).

WELLNESS AND TECHNOLOGY SUPPORT FOR COLLEGE STUDENTS

With the limited amount of counseling resources at college counseling centers, it is evident that alternative measures should be examined to provide counseling to students. The Mayo Clinic (2020) described telehealth as the usage of digital information to communicate and provide services to clients remotely. For example, college students who had eating disorders showed success in completing online therapy sessions coupled with biofeedback (Novotney, 2017). Considering the limited access to counseling services, college students can utilize teletherapy for counseling services, and it can increase school satisfaction (Meilman et al, 2016; Nobleza et al., 2018).

Increasing the visibility of college counseling centers and their resources is also essential to reaching students who have mental health disorders. Concerning this issue, it would be helpful for college counseling centers to adopt telehealth into their realm of services. In response to broadening the availability of counseling services for college students, Johnson & Kalkbrenner (2017) were focused on whether the

Mobile Health (mHealth) system was useful for college students. With the mHealth mobile application, it has been proven to reduce symptoms of depression, anxiety, and substance use, but currently there is no literature specifically tying these results to college students (Johnson & Kalkbrenner 2017). Based on previous literature reviews involving college students who have used the mHealth system, it was mentioned that it was helpful with increasing awareness of counseling services, students were willing to use a mobile platform for counseling, and the application was useful in promoting healthy mental health strategies (Haithcox-Dennis et al., 2015; Kazemi et al., 2019). Additionally, students who had access to the mHealth application were provided with a viable and convenient resource, but despite the usefulness of mHealth in various research studies, there is limited research on the effectives of mHealth and other types of technology for college student; which can be explored in future research (Johnson & Kalkbrenner 2017).

FACTORS THAT CONTRIBUTE TO COLLEGE STUDENT SATISFACTION

Regarding student satisfaction, it is important to examine the factors that contribute to students being satisfied with college and how it effects their life outside of the academic realm. For instance, Holinka (2015) found that stress was negatively correlated with stress and helping college students develop and implement coping skills can be utilized in counseling sessions and help students reduce stress and help with school and life satisfaction. Focusing on self-efficacy is another important factor when it comes to examining academic satisfaction for students. With self-efficacy, it focuses on the personal belief of being capable of performing tasks and college students levels of self-efficacy impacts their academic performance and environmental supports helps students with developing healthy self-efficacy throughout their college experience (Garriot et al., 2015; Ramos-Salazar & Hayward, 2018). Researchers within the research study also mentioned that college self-efficacy is a strong predictor of academic satisfaction, but it is not indicative of school performance. Another indicator of satisfaction among college students is their mental status and healthy coping skills and considering the fact that that mental status is an important indicator, it would be beneficial for college to help provide resources that can help students establish healthy coping mechanisms and provide additional resources and supports that can help students in order to enhance healthy life functioning (Gnilka, et al., 2015; Ramos-Salazar & Hayward, 2018; Shatkin, 2016). After reviewing the literature on the access of counseling services and prevalence of mental illness amongst college students, it is evident that more resources are needed on college campuses (Epstein, 2015; Shefet, 2018; Shelesky et al., 2016). With this in mind, it is apparent that college students are in need of counseling services that will help them cope with the demands of college and personal obligations outside of their academic careers.

Due to these circumstances, alternatives to traditional individual counseling were presented as a way of creating more opportunities for students to engage in counseling services. Other options include the following: group counseling, technology (mHealth platform), full-time case management services with a behavioral component, and community referrals (Johnson & Kalkbrenner 2017; Shefet, 2018). The effects of telehealth have also indicated a way of helping students increase their access to counseling services (Meilman & Weatherford, 2015; Nobleza, et al., 2018). Additionally, school and life satisfaction are also factors when examining whether counseling would be useful for students (Garriot et al., 2015; Gnilka, et al. m, 2015). With all the aforementioned services, there are still gaps in the literature on the usage of technological means for college students such as, telehealth for and the impact on student supports on mental clarity and academic achievement in the college setting.

BACKGROUND INFORMATION ON MINDFULNESS

The historical background of mindfulness reveals that the self-care technique is rooted in the Satipaṭṭhāna and Zen Buddhist practices and focuses on being present in the moment (Bhikkhu, 2018). Beginning with the Satipaṭṭhāna practice, it is one of the introductory stages of meditation practiced by Buddhist monks as a means of achieving enlightenment and the foundation of mindfulness (Bhikkhu, 2018; Encyclopedia Britannica, 2015). The concept of Zen is also centered around achieving enlightenment through the meditation and reaching a calm state of awareness (Nagatomo, 2019). Additionally, the title of the mindfulness concept is derived from the Seven Factors of Awakening, which is also a pillar of the Buddhist religion (Piyadassi, 2013). Key contributors to the expansion of mindfulness into Western culture include Thích Nhất Hạnh and Dr. John Kabat-Zinn (Nisbet, 2020).The concept of mindfulness has been used for mental health and medical diagnoses such as depression, anxiety, substance abuse, and chronic pain as a means of reducing stress level, enhancing mental clarity, and increasing cognitive functioning and some common mindfulness techniques which include deep breathing, mediation, yoga, and body scanning (Dean, 2014; Kirmayer, 2015).

EFFECTIVE MINDFULNESS INTERVENTIONS

When examining mindfulness in the college setting, it is important to consider effective interventions that help to improve daily functioning and mental clarity for students. Mindfulness-Based Stress Reduction (MBSR) is non-pharmaceutical intervention that is used to help reduce stress levels and implements mindfulness meditation practices (Alsubaie et al., 2017; Felver et al., 2018; Niazi & Niazi, 2011). This intervention was developed by Dr. John Kabat-Zinn at the University of Massachusetts. He conducted an 8-week intervention that was used as a stress management coping mechanism for stress related and mental health issues such as anxiety and depression and for chronic and debilitating diseases (Alsubaie et al., 2017; Niazi and Niazi, 2011). With the MBSR program, participants are also able to engage in techniques such as yoga and meditation, which help to establish a mindfulness mindset (Felver et al. 2018; Niazi & Niazi, 2011). In a comprehensive review on MBSR interventions in the home setting, it was noted that research participants engaged in mindfulness practices on an average of 30 minutes per day and reported positive outcomes throughout the process (Parson et al., 2017).

Specifically, with college students, MBSR has proven to reduce psychological distress in college students (Felver et al., 2018). In a research study by Felver et al., (2018), researchers were interested in investigating how mindfulness-based practices help reduce the feelings of stress for students as well as replicating a previous study that focused on the same methodology. For the non-experimental design of the research study, there was a total of 21 college students, 81% Female and 19% Male, with a racial makeup of 48% White, 38% Asian, 10% Hispanic, and 5% identified as Other. Students were recruited for the study through advertisements listed at the college counseling center and once students expressed interest in the group. Participant interviews were conducted to share the methodology of the study and expectations of students (Felver et al., 2018). Throughout the course of the study, students completed questionnaires focused on trait mindfulness and psychological distress, participated in mindfulness practices, and kept a journal and results from the study indicated that there was decrease in stress for students and increase in trait mindfulness (Felver et.al, 2018).

Additionally, in a meta-analysis of MBSR and its effects on psychological well-being conducted by Querstret et al. (2020), it was mentioned that this intervention is one of the strongest evidence-based interventions when implementing mindfulness practices. After analyzing 97 non-clinical studies that used the MBSR program intervention, the researchers concluded that the program was beneficial to in reducing depressive and anxiety symptoms, overthinking, and enhancing the quality of life for research participants. Although MBSR presented promising outcomes in the research studies, researchers recommended that in future research studies it would be imperative to examine whether the duration and mode of delivery of the MBSR interventions have an effect on the MBSR program outcomes. Additionally, reducing mindfulness practices to under 2.5 hour within a week has a negative impact on the effectiveness of the MBSR intervention (Querstret et al., 2020).

Another intervention that has been noted as effective when implementing mindfulness is Mindfulness-Based Cognitive Behavioral Therapy (MB-CBT). With MBCBT, it incorporates the concepts of cognitive behavioral therapy and includes practices from MBSR (Querstret et al., 2020; Sipe and Eisendrath, 2012). Like the MBSR, MB-CBT is also an 8-week mindfulness intervention and is helpful to alleviate stress and symptomology of mental health issues with techniques such as deep breathing, yoga, and meditation (Goldberg et al., 2019; Sipe & Eisendrath, 2012). Research on MBCT has been used with different populations and various mental health diagnoses such as depression, anxiety, obsessive compulsive disorder, and addiction (Key et al., 2017; Song & Lindquist, 2015).

Group interventions have also been successful in helping research participants engage and be consistent with mindfulness practices (Alsubaie et al., 2017; Parcover et al., 2018). Since there is a sense of community with groups, mindfulness is sometimes practiced in this format. For example, Juberg et al. (2019) focused on group interventions centered around undergraduate college students and studied how students were affected by a mindfulness-based intervention delivered on campus by mental health professionals. Researchers provided pre and post-test self-report quantitative measures on depression, anxiety stress, mindfulness, experiential avoidance, self- compassion, and engaged living. Students were prompted to complete mindfulness activities every day throughout the study and over the course of eight weeks, students were part of a closed mindfulness group with each session lasting 1.5 hours and the opportunity to process their mindfulness activities throughout the week within a group setting (Juberg et al., 2019). When comparing pre/post-test self-report measures, it was noted that at the conclusion of the study, depression, anxiety, stress, experiential avoidance, and engaged living were reduced (Juberg et al., 2019). Additionally, mindfulness and self-compassion increased. By examining the results of this study, mindfulness-based interventions can be a useful treatment option in the university setting and can be used for students to use with their busy schedules. However, one of the limitations of the study is that there were no follow-up measures at the conclusion of the research study (Juberg et al., 2019).

MINDFULNESS IN THE HIGHER EDUCATION SETTING

Detailing effective mindfulness practices that can be used to accommodate the needs and the complexities of college students is needed to help students enhance their academic progress and college experience (Fortin, 2018; Galante et al., 2018; Mahfouz et al., 2018). According to the American Addiction Centers (2019), students with the majors of education, medical, psychology, arts, science, social sciences, information technology, law and criminal justice, business finance, marketing and communication, and engineering are the most stressed. Some of the top issues that students stress over include exams, finances, academic performance, social relationships, and occupation (American Addiction Centers, 2019; EAB, 2020; Roy, 2018). One maladaptive coping mechanism that students tend to engage in is substance abuse. For example, students use study drugs, such as Ritalin and Adderall to enhance their mental sharpness and concentration while completing college assignments (American Addiction Centers, 2020). Seventy four percent of college students were able to obtain study drugs without a prescription from friends who had a prescription and 33% obtained study drugs from other resources, such as social groups on campus and through their primary care physician (American Addiction Centers, 2020). Additionally, binge drinking for coping and social acceptance were among other substance abuse issues for college students, which can significantly impact their academic performance (American Addiction Centers, 2020; Yagoda, 2016). Developing a healthy coping regiment, such as mindfulness to deal with stress, academic pressure, and other life challenges can provide a plethora of benefits for college students including enhancements in their daily functioning (Fortin, 2018).

For example, Galante et al. (2017) utilized a mindfulness-based intervention to help with resilience and stress levels of students in a researched study named the Mindful Student Study (MSS). Over the course of one semester, 309 out 616 students were randomly assigned to the experiment group and students identified as having critical history of mental illnesses. Regarding the procedures of the study, students in the experimental group practiced mindfulness as part of a specialized 8-week intervention and received mental health support as needed. With the Mindfulness Skills for Students intervention, students were a part of face-to-face group intervention and completed at home meditation exercises. Data was captured at different points within the study to continuously check the academic progress and well-being of students. For instance, when students missed a face-to-face session, the researchers would follow-up via email to check-in to see if the student missed a session due to an undesirable reaction to mindfulness practices and offered an alternative session they could attend. Data gathered from the study showed that students in the experimental and control groups had similar baseline measures, but after completing mindfulness practices the experimental group had better academic outcomes and scored higher on the well-being and resilience measures (Galante et al., 2017). One consideration from the study is to offer multiple options for students to engage in mindfulness, such as an online format.

Dvořáková et. al. (2017) tested the feasibility and effectiveness of mindfulness-based practices for undergraduate students. The sample of population of college students included; 109 freshman undergraduate students containing 72 female and 37 males with a racial makeup of the following backgrounds: Caucasian (50%), Asian (26%), African American (5%), Hispanic (8%), and mixed (10%) and comprised of 16% international students (Dvořáková et. al., 2017). Students were recruited via email and through flyers and those who qualified for the study were invited to an in-person interview in order to discuss the study in detail and to obtain informed consent. Baseline questionnaires were sent via email to students through a secure link to gather information on stressors. The intervention took place over the course of eight weeks and students completed sessions on mindfulness and then engaged in practices outside of the sessions. Outcomes from the data showed that students in the experimental group reported lower levels of depressive feelings and anxiety after engaging in mindfulness practices. Posttest measures were also sent via email to follow-up on stressors. Conclusions from the study indicated that mindfulness was useful in helping students establish healthy coping practices. Also, quantitative measures may not fully represent the thoughts of students towards the benefits of mindfulness. The limitations within the study showed that quantitative measures may not capture all benefits or perceptions of mindfulness.

Vidic and Cherup (2019) examined whether mindfulness-based practices strengthen students' perceptions on stress, resilience, and self-efficacy. Research participants were given four pre-test questionnaires on the basis of perceived stress, resilience, self-efficacy and perfectionism. During the study, students engaged in mindfulness practices at the start of the class and had time to reflect on their responses to meditation. Additionally, students were given follow-up information on the positive effects of mindfulness and researchers provided posttest measures on the same subject areas. Results of the study indicated that the experimental group had a significant decrease in stress levels and showed an increase in levels of resilience and self-efficacy in comparison to the control group. Researchers stated that expanding the methodology of the study, would be imperative in terms of lengthening the time of the intervention and using a larger research sample would be useful to generalize results (Vidic & Cherup, 2019).

Bamber and Schneider (2020) analyzed student's perceptions of mindfulness and best practices of mindfulness-based interventions for college. All research participants were undergraduate college students and the study focused on students from various ages, gender, ethnicities. Based on the literature review, it was mentioned that students saw a change in their stress levels due to mindfulness practices. Additionally, students reported that mindfulness was viewed as a coping mechanisms and alleviated feelings of stress and anxiety and it improved their concentration levels and overall sense of well-being (Bamber & Schneider, 2020). Findings from the study mentioned that mindfulness-based interventions should be tailored to meet the need of the research participants in order to maximize outcomes.

Similarly, Morgan (2017) studied undergraduate students using a quasi-experimental design using interventions such as, stress reduction practices such as mindfulness, yoga, and equine-assisted learning activities over the course of a three week period. During the course of the study, students in the experimental group continued to participate in one of the three activities and kept journals and daily logs throughout the study to document their experiences with mindfulness practices. After the study was completed, researchers compared the pre-test results of stress levels between the experimental and control groups and it was noted that the experimental group had a reduction in their stress levels and acknowledged that mindfulness practices were useful for stress reduction. Based on the participants' journals, the researchers noted that the psychosocial issues, financial concerns, and career planning were some of the stressors that students had during the course of the study.

Ramler et al. (2016) studied the impact of mindfulness-based stress reduction and how it helps students adjust to their first year of college. Results of the study stated that mindfulness was a helpful component for students in the experimental group versus the control group and males appeared to respond more to the mindfulness techniques versus the females. Additionally, mindfulness helped with stress reduction and adjustment to college. Based on the results from the research study, researchers stated that incorporating mindfulness interventions into the college setting should be strongly considered for students. Findings from this study indicated that mindfulness can have a strong impact on helping college students adjust to the college setting and learning about different components (stress and adjustment). These measures should be considered in future research studies on college students (Ramler et al., 2016).

Studying the impact of mindfulness on academic achievement is another research focus for college students. Mahfouz et al. (2018) focused on what effects mindfulness interventions have on stress and academic success and by using the convenience sampling method (recruited from a larger study) with a mixed methods methodology. Students who completed assessment measures that focused on levels of anxiety, life satisfaction, and depression engaged in mindfulness practices through the Just Breath program (Mahfouz et al., 2018). The Just Breath program is an 8-week mindfulness-based program that was adopted from Learning to Breath program. This program was adapted to meet the needs of college students (Mahfouz et al., 2018). Researchers concluded from the initial and 1-year follow-up interviews that the Just Breath mindfulness program had a positive impact on improved organization and time management, commitment to a healthy lifestyle, improved emotional awareness and relationships, and self-compassion and as part of the discussion on the research findings. The researchers mentioned that mindfulness programs can be useful to help students with their academic success and wellness (Mahfouz et al., 2018).

Additionally, Chenlin (2018) tested the efficacy of a mindfulness training intervention program in reducing academic stress amongst undergraduate students. Prior to the study, researchers sent detailed information and an invitation to administrators at one university in order to request permission for their students to participate in the research study. Researchers randomly selected experimental and control groups. Participants completed surveys on stressors and perceived reduction of stress activities. For approximately eight weeks, students in the experimental group engaged in a mindfulness-based group, tracked their progress with the intervention, and showed a significant decrease in their stress level versus the control group (Chenlin, 2018).

Lin and Mai (2018) examined the short-term and long-term effects of students who engaged in mindfulness practices on a weekly basis over a period of three months. Mindfulness practices were incorporated into the classroom activities for freshman college students who were selected for the experimental group. Students in the control group received minimal mindfulness interventions. To examine whether mindfulness had an effect on academic performance for students in the experimental group, formative assessments were provided throughout the research study. Summative assessments were given at the end of the study in order to track the final progress of students (Lin & Mai, 2018). The distinct difference in the mindfulness practices that the experimental group received versus the control group was that the experimental group engaged in mindfulness prior to beginning of the class through mediation and deep rhythmic breathing. Outcomes of the study showed that students in the experimental group had higher scores on the formative assessment after week two of the study and showed better outcomes academically on a short-term basis when comparing the scores of students in the control group (Lin and Mai, 2018). The researchers also mentioned that adding a qualitative component to the research study could help with learning more about how mindfulness has impact on academic performance.

Similarly, Cole et al. (2016) studied the impact of mindfulness and ego resilience in reducing academic stress levels of college students. Regarding ego resilience, it is defined as the adaptability and emotional response to stressful situations and environments (American Psychological Association, n.d.). Ghanaian students were amongst the research participants and were noted as having multiple barriers to accessing and completing college. Students enrolled in psychology courses were among the participants and given a number of assessments measures. Assessments used in the study were The Mindful Attention Awareness Scale, The Ego-Resiliency Scale, The State-Trait Anxiety Inventory (Form Y)–Trait Form, The Center for Epidemiological Studies–Depression, and The Academic Expectations Stress Inventory. After analyzing the responses from the survey, it was mentioned there were links between academic stress, depression, and stress level. It was also determined that mindfulness was helpful in reducing stress levels and ego resilience was tied to reducing depressive symptoms for students (Cole et al, 2016). As a final takeaway from the study, researchers debriefed students after the study and noted that it is important to consider and reflect on culture considerations as a factor for the results of the study.

As a follow-up study to MSS, which was one of the largest randomized control trials with mindfulness and academic achievement Bóo et al. (2020), used qualitative measures to gain insight on student's perceptions of mindfulness on their academic performance. Six students who were originally part of the MSS study, participated in interviews after one academic year for the purpose of investigate their engagement with mindfulness practices. After reaching students via email and conducting semi-structured interviews, researchers used a thematic analysis approach to evaluate data (Boo et al, 2020). Students had a range in academic performances throughout the following academic year. Key themes that were extracted from the interview data included mindfulness having a positive effect on academic performance being used as a coping mechanism to combat stress, enhancing concentration and memory, increasing self-awareness, and reducing procrastination behaviors (Boo et al., 2020). Alternatively, a few students shared that engaging in mindfulness can sometimes be overwhelming and become a burden. Implications of the study mentioned that the results of the interview can be helpful for mental health providers in the college setting to understand the challenges that students are facing and what resources are needed to help students throughout their college journey (Boo et al., 2020).

MINDFULNESS TECHNOLOGY AND MOBILE APPLICATIONS

Since the start of MBSR and MB-CBT programs, mindfulness practices have evolved over time and include the usage of mobile applications that are accessible for users through technological devices such as cellular phones and tablets. Mindfulness medications mobile applications were introduced in 2008 through the Apple Inc. application store (Apple, 2018). Additionally, with access to mindfulness through technology, students can engage in mindfulness practices in a convenient manner, which can be useful when dealing with daily stressors (Ackerman, 2020; Dean, 2014; Flett et al., 2020; van Emmerik, 2018).

Currently, there a number of different mindfulness mobile applications that offer a variety of options and include techniques such as deep breathing, guided meditation, and relaxation (Ackerman, 2020; Willgress, 2020). Mani et al. (2015) conducted a systematic review on Mindfulness applications that are available through the iTunes and Google Marketplace. At the time of the research study, there were approximately 563 readily available mobile applications that contained a combination of mindfulness training, daily reminders, and explanations of guided meditation practices. As part of the analyzing the applications, the Mobile Application Rating Scale (MARS) was used within the study. The MARS is an assessment developed by Queensland University of Technology that measures mobile health applications and is used to as a guide to assist health professionals and application developers when developing new mobile applications and recommending the to consumers (Ministry of Health – Manatū Hauora, 2017). Sections on the MARS assessment includes the application classification, affiliations with other social sights, age appropriateness, and the quality of the application (Ministry of Health – Manatū Hauora, 2017). When rating the mindfulness applications, Headspace, Smiling Mind, iMindfulness, and Mindfulness Daily were among the top applications for mindfulness due to high scores on the MARS subscales and were deemed as aesthetically pleasing for users (Ministry of Health – Manatū Hauora, 2017).

Mak et al. (2018) conducted a study on the efficacy of three mobile app–based program such as the mindfulness-based program, self-compassion program, and cognitive behavioral psychoeducation program in improving mental well-being. Participants were randomized between 3 conditions. Over the course of four weeks participants spent 10-15 min daily reviewing the course content and practicing various related exercises. Research participants completed pre/posttest measures and a three-month follow-up by completing web-based measures of mental well-being, psychological distress, mindful-awareness, and self-compassion. Students in all three groups showed improvements in mental clarity and life stressors. Researchers stated that gamification should be considered for future mobile mindfulness-based apps (Mak et al., 2018). Having an example of different mobile mindfulness apps can be useful in designing the research study. Also, the pre/posttest measures show that it is important to have a follow-up study at the conclusion of the research study.

In another research study on mindfulness and technology, the central research question from Loiacono et al. (2018) was concentrated on whether integrating technology can help college students engage in mindfulness practices. In tying the research question with the purpose of the study, researchers investigated the integration of technology into evidence-based mindfulness interventions and tailoring programs to the needs of college students. College students were recruited through online measures such as email, Twitter, and Facebook, and after responding to the recruitment attempts, students completed an online consent form prior to in-person interviews and completing questionnaires. Assessments measured the stress levels, depressive symptoms, and anxiety before and after the study to determine whether there was a correlation between utilizing mindfulness application and a reduction in these content areas (Loiacono et. al, 2018). With both the interviews and questionnaires, students were asked what type of mindfulness measures they were likely to engage in on an ongoing basis. Results from both the qualitative and quantitative measures noted that students were likely to engage in the Mindfulness app on the phone with guided meditation and eight weekly two-hour classes and one all-day class on a Saturday or Sunday with presentations versus teaching and practice of mindfulness and yoga practices. One conclusion from the study stated that mobile apps are the preferred method because it provides a convenient option for students with their hectic schedules.

Huberty et al. (2019) were interested in examining whether the "calm" mobile application can help reduce stress levels and enhance meditation practices for college students. Students were recruited via social media and email listservs and placed in randomized and waitlist control groups. Prior to the study, students completed questionnaires on perceived stress, mindfulness, and self-compassion prior to the research study. During the study, the research participants completed mindfulness practices through the "calm" application over the course of eight weeks. After the eight-week intervention and 12-week follow-up of the group, it was noted that the "calm" app was useful in helping students reduce their stress levels by engaging in mindfulness mediations. Additionally, some students continued to utilize mindfulness practices at the completion of the study, and student's levels of self-compassion and mindfulness were increased at the completion of the study. Overall, it was noted that the Calm mobile application is efficient in delivering mindfulness-based practices. Since there is limited literature on this topic, future research is needed to expand on the topic of mindfulness for college students (Huberty et al., 2019).

CONCLUSION AND RECOMMENDATIONS FOR FUTURE RESEARCH

Overall, there are various issues such as academic challenges, mental health issues, and inadequate sleep patterns that can affect college students and their academic performance and well-being throughout their academic careers. Mindfulness has a wide variety of techniques that can be implemented to improve concentration and create a sense of wellness for college students. This chapter discussed and analyzed the effects of mobile mindfulness practices and interventions for college students. Additionally, the different mindfulness interventions, such as MB-CBT and MBSR were noted as being effective for college students and was useful to enhance mental clarity and academic performance. Future research should include an exploration into the different types of aspects of technology that students find useful when completing mindfulness practices and determine what aspects of the mobile technology helps students stick to a routine.

REFERENCES

Acharya, L., Jin, L., & Collins, W. (2018). College life is stressful today – Emerging stressors and depressive symptoms in college students. *Journal of American College Health*, *66*(7), 655–664. doi:10.1080/07448481.2018.1451869 PMID:29565759

Ackerman, C. (2020, February 2). + *Reviews*. https://positivepsychology.com/mindfulness-apps/

Alderson Broaddus University. (n.d.). *Common Issues for College Students*. https://ab.edu/counseling-services/common-issues-for-college-students/

Alsubaie, M., Abbott, R., Dunn, B., Dickens, C., Keil, T. F., Henley, W., & Kuyken, W. (2017). Mechanisms of action in mindfulness-based cognitive therapy (mbct) and mindfulness-based stress reduction (mbsr) in people with physical and/or psychological conditions: A systematic review. *Clinical Psychology Review*, *55*, 74–91. doi:10.1016/j.cpr.2017.04.008 PMID:28501707

American Addiction Centers. (2019). *Coping with College Stress: Exploring perceived stress and coping methods in college*. https://americanaddictioncenters.org/learn/college-coping-mechanisms/

American Addiction Centers. (2020, June 1). *Guide to Substance Abuse and Mental Health in College*. https://americanaddictioncenters.org/rehab-guide/college

American College Health Association. (2010). *American College Health Association-National College Health Assessment II: Reference Group Data Report Spring 2010*. Linthicum, MD: American College Health Association. https://www.acha.org/HealthyCampus/HealthyCampus/Student_Objectives.aspx

American College Health Association. (2019). *American College Health Association-National College Health Assessment II: Undergraduate Student Executive Summary Spring 2019*. American College Health Association.

American College Health Association. (n.d.). *Who We Are*. https://www.acha.org/ACHA/About/About_ACHA/ACHA/About/About_ACHA.aspx?hkey=9deff142-0898-4f21-8a69-f5e0621eb59c

American Institute of Stress. (2019, September 9). *Stress: An Epidemic Among College Students*. https://www.stress.org/stress-an-epidemic-among-college-students

American Psychological Association. (2019). *Stress in America: Stress and Current Events. Stress in America™ Survey*. https://www.apa.org/news/press/releases/stress/2019/stress-america-2019.pdf

American Psychological Association. (n.d.). *Ego resilience*. https://dictionary.apa.org/ego-resiliency

Apple. (2018, July 5). *The App Store turns 10*. https://www.apple.com/newsroom/2018/07/app-store-turns-10/

Appleby, D. C. (2019). The soft skills college students need to succeed now and in the future. *Education Directorate, 5*(2). https://www.apa.org/ed/precollege/psn/2017/09/soft-skills

Arora, N., & Singh, N. (2017). Factors Affecting the Academic Performance of College Students. iManag's. *Journal of Educational Technology, 14*(1), 47–53. doi:10.26634/jet.14.1.13586

Bamber, M. D., & Schneider, J. K. (2020, January 7). College students' perceptions of mindfulness-based interventions: A narrative review of the qualitative research. *Current Psychology (New Brunswick, N.J.).* Advance online publication. doi:10.100712144-019-00592-4

Beauchemin, J. D. (2018). Solution-focused wellness: A randomized controlled trial of college students. *Health & Social Work, 43*(2), 94-100. http://dx.doi.org.tcsedsystem.idm.oclc.org/10.1093/hsw/hly007

Bhikkhu, T. (2018, March 5). *The Buddha's Original Teachings on Mindfulness*. https://tricycle.org/trikedaily/satipatthana-sutta-mindfulness/

Bishop, K. K. (2016). The relationship between retention and college counseling for high-risk students. *Journal of College Counseling, 19*(3), 205–217. doi:10.1002/jocc.12044

Bohon, L. M., Cotter, K. A., Kravitz, R. L., Cello, P. C. Jr, Fernandez, Y., & Garcia, E. (2016). The Theory of planned behavior as it predicts potential intention to seek mental health services for depression among college students. *Journal of American College Health, 64*(8), 593–603. doi:10.1080/07448481.2016.1207646 PMID:27386898

Bóo, S. J. M., Childs, F. J., Cooney, S., Datta, B., Dufour, G., Jones, P. B., & Galante, J. (2020). A follow-up study to a randomized control trial to investigate the perceived impact of mindfulness on academic performance in university students. *Counselling & Psychotherapy Research, 20*(2), 286–301. doi:10.1002/capr.12282

Braider, L., La Lima, C., Crimarco, N., Hollander, B., Reid-Russell, A., Kane, J., & Greenwald, B. (2019). Characterization of psychiatrically hospitalized college students. *Journal of American College Health, 67*(7), 654–660. doi:10.1080/07448481.2018.1500471 PMID:30365917

Burns, M. E., Houser, M. L., & Farris, K. L. B. (2018). Theory of planned behavior in the classroom: An examination of the instructor confirmation-interaction model. *Higher Education: The International Journal of Higher Education Research, 75*(6), 1091–1108. doi:10.100710734-017-0187-0

Carmack, H. J., & Heiss, S. N. (2018). Using the theory of planned behavior to predict college students' intent to use linkedin for job searches and professional networking. *Communication Studies, 69*(2), 145–160. doi:10.1080/10510974.2018.1424003

Chenlin, Y., ChangJiang, L., Jingzhou, H., & Jinfeng, W. (2018). Academic stress and evaluation of a mindfulness training intervention program. *NeuroQuantology: An Interdisciplinary Journal of Neuroscience and Quantum Physics, 16*(5).

Cole, N. N., Nonterah, C. W., Utsey, S. O., Hook, J. N., Hubbard, R. R., Opare-Henaku, A., & Fischer, N. L. (2015). Predictor and moderator effects of ego resilience and mindfulness on the relationship etween academic stress and psychological well-being in a sample of Ghanaian college students. *The Journal of Black Psychology, 41*(4), 340–357. doi:10.1177/0095798414537939

Cooper, G., Barkatsas, T., & Strathdee, R. (2016). The Theory of Planned Behaviour (TPB) in Educational Research using Structural Equation Modelling (SEM). In T. Barkatsas & A. Bertram (Eds.), *Global Learning in the 21st Century. Global Education in the 21st Century Series*. doi:10.1007/978-94-6300-761-0_9

Dean, J. (2014, April 7). *Mindfulness Meditation: 8 Quick Exercises That Fit into Your Day*. https://www.spring.org.uk/2014/04/mindfulness-meditation-8-quick-exercises-that-easily-fit-into-your-day.php

Dewberry, C., & Jackson, D. J. R. (2018). An application of the theory of planned behavior to student retention. *Journal of Vocational Behavior*, *107*, 100–100. doi:10.1016/j.jvb.2018.03.005

Dvořáková, K., Kishida, M., Li, J., Elavsky, S., Broderick, P., Agrusti, M., & Greenberg, M. (2017). Promoting healthy transition to college through mindfulness training with first-year college students: Pilot randomized controlled trial. *Journal of American College Health*, *65*(4), 259–267. doi:10.1080/07448481.2017.1278605 PMID:28076182

EAB. (2020). *Students Spend More Time at Work Than in Class*. https://eab.com/insights/daily-briefing/student-affairs/students-spend-more-time-at-work-than-in-class/

Encyclopedia Britannica. (2015, July 15). *Smṛtyupasthāna Buddhist philosophy*. https://www.britannica.com/topic/smrtyupasthana

Epstein, B. (2015). Providing Psychological Counseling in Community Colleges: Even Greater Challenges and Fewer Resources. *Journal of College Student Psychotherapy*, *29*(4), 289–295. doi:10.1080/87568225.2015.1074020

Esposito, G., van Bavel, R., Baranowski, T., & Duch-Brown, N. (2016). Applying the model of goal-directed behavior, including descriptive norms, to physical activity intentions: A contribution to improving the theory of planned behavior. *Psychological Reports*, *119*(1), 5–26. doi:10.1177/0033294116649576 PMID:27229344

Felver, J., Morton, M., & Clawson, A. (2018). Mindfulness-based stress reduction reduces psychological distress in college students. *College Student Journal*, *52*(3), 291–298.

Fishbein, M., & Ajzen, I. (2010). *Predicting and changing behavior: The reasoned action approach*. Psychology Press.

Flett, J., Conner, T., Riordan, B., Patterson, T., & Hayne, H. (2020). App-based mindfulness meditation for psychological distress and adjustment to college in incoming university students: A pragmatic, randomized, waitlist-controlled trial. *Psychology & Health*, *35*(1), 1–26. doi:10.1080/08870446.2019.1711089 PMID:31204484

Fortin, A. (2018, February 13). *How mindfulness can benefit college students*. https://www.collegian.psu.edu/lifestyles/health/article_58131944-104c-11e8-9e6d-9fd2ac28debc.html

Francis, P. C., & Horn, A. S. (2017). Mental health issues and counseling services in us higher education: An overview of recent research and recommended practices. *Higher Education Policy*, *30*(2), 263–277. doi:10.105741307-016-0036-2

Galante, J., Dufour, G., Vainre, M., Wagner, A. P., Stochl, J., Benton, A., & Jones, P. B. (2018). A mindfulness-based intervention to increase resilience to stress in university students (the mindful student study): A pragmatic randomized controlled trial. *The Lancet. Public Health*, *3*(2), 81. doi:10.1016/S2468-2667(17)30231-1 PMID:29422189

Garriott, P. O., Hudyma, A., Keene, C., & Santiago, D. (2015). Social cognitive predictors of first- and non-first-generation college students' academic and life satisfaction. *Journal of Counseling Psychology*, *62*(2), 253-263. http://dx.doi.org.tcsedsystem.idm.oclc.org/10.1037/cou0000066

Gnilka, P., Ashby, J., Matheny, K., Chung, Y., & Chang, Y. (2015). Comparison of coping, stress, and life satisfaction between Taiwanese and US college students. *Journal of Mental Health Counseling*, *37*(3), 234–249. doi:10.17744/mehc.37.3.04

Goldberg, S. B., & Raymond, P. (2019). Mindfulness-based cognitive therapy for the treatment of current depressive symptoms: A meta-analysis. *Cognitive Behaviour Therapy*, *48*(6), 445–462. doi:10.1080/16506073.2018.1556330 PMID:30732534

Hagger, M. S., Chan, D. K. C., Protogerou, C., & Chatzisarantis, N. L. D. (2016). Using meta-analytic path analysis to test theoretical predictions in health behavior: An illustration based on meta-analyses of the theory of planned behavior. *Preventive Medicine*, *89*, 154–161. doi:10.1016/j.ypmed.2016.05.020 PMID:27238207

Haithcox-Dennis, M., Brinkley, J., Richman, A., DeWeese, A., & Byrd, J. L. (2012). Mhealth on campus: Assessing undergraduates attitudes and utilization of mobile health applications. *Global Journal of Health Education and Promotion*, *15*(1).

Holinka, C. (2015). Stress, emotional intelligence, and life satisfaction in college students. *College Student Journal*, *49*(2), 300–311.

Huberty, J., Green, J., Lee, C., Glissmann, C., Larkey, L., & Puzia, M. (2019). Efficacy of the mindfulness meditation mobile app "calm" to reduce stress among college students: Randomized controlled trial. *Journal of Medical Internet Research*, *21*(6), e14273. Advance online publication. doi:10.2196/14273 PMID:31237569

Ingersoll, J. (2019). *Academic Probation in College Why it Happens & How to Get Back on Track*. https://www.accreditedschoolsonline.org/resources/academic-probation/#:~:text=Students%20on%20academic%20probation%20may,more%20about%20this%20possibility%20below)

Johnson, K. F., & Kalkbrenner, M. T. (2017). The Utilization of Technological Innovations to Support College Student Mental Health: Mobile Health Communication. *Journal of Technology in Human Services*, *35*(4), 314–339. doi:10.1080/15228835.2017.1368428

Juberg, M., Spencer, S. D., Martin, T. J., Vibell, J., da Costa Ferro, A., Kam, B., & Masuda, A. (2019). A mindfulness-based intervention for college students, faculty, and staff: A preliminary investigation. *Clinical Case Studies*, *18*(3), 185–199. doi:10.1177/1534650119836166

Kazemi, D. M., Borsari, B., Levine, M. J., Shehab, M., Nelson, M., Dooley, B., Stinson, B., Fang, F., & Li, S. (2019). Real-time demonstration of a mHealth app designed to reduce college students hazardous drinking. *Psychological Services*, *16*(2), 255–259. doi:10.1037er0000310 PMID:30407059

Key, B. L., Rowa, K., Bieling, P., McCabe, R., & Pawluk, E. J. (2017). Mindfulness-based cognitive therapy as an augmentation treatment for obsessive–compulsive disorder. *Clinical Psychology & Psychotherapy*, *24*(5), 1109–1120. doi:10.1002/cpp.2076 PMID:28194835

Kirmayer, L. J. (2015). Mindfulness in cultural context. *Transcultural Psychiatry*, *52*(4), 447–469. doi:10.1177/1363461515598949 PMID:26264787

Lee, C. J., Geiger-Brown, J., & Beck, K. H. (2016). Intentions and willingness to drive while drowsy among university students: An application of an extended theory of planned behavior model. *Accident; Analysis and Prevention*, *93*, 113–123. doi:10.1016/j.aap.2016.05.002 PMID:27180286

Lin, J. W., & Mai, L. J. (2018). Impact of mindfulness meditation intervention on academic performance. *Innovations in Education and Teaching International*, *55*(3), 366-375. http://dx.doi.org.tcsedsystem.idm.oclc.org/10.1080/14703297.2016.1231617

Linder, A. D., Liu, H.-Y., Woodson-Smith, A., & Jung, J. (2018). Physical activity behaviors among non-traditional and traditional college students: An application of ajzen's theory of planned behavior. *Negro Educational Review*, *69*(1-4), 33–50.

Loiacono, E., Fulwiler, C. E., Cohanim, R., & Davis, M. (2018). An Investigation into College Students' Preferences for Technology Integration into Mindfulness-based Stress Reduction. *AIS Transactions on Human-Computer Interaction*, *10*(4), 188–204. doi:10.17705/1thci.00109

Mahfouz, J., Levitan, J., Schussler, D., Broderick, T., Dvorakova, K., Argusti, M., & Greenberg, M. (2018). Ensuring College Student Success Through Mindfulness-Based Classes: Just Breathe. *The College Student Affairs Journal*, *36*(1), 1–16. doi:10.1353/csj.2018.0000

Mak, W. W., Tong, A. C., Yip, S. Y., Lui, W. W., Chio, F. H., Chan, A. T., & Wong, C. C. (2018). Efficacy and Moderation of Mobile App–Based Programs for Mindfulness-Based Training, Self-Compassion Training, and Cognitive Behavioral Psychoeducation on Mental Health: Randomized Controlled Non-inferiority Trial. *JMIR Mental Health*, *5*(4), e60. doi:10.2196/mental.8597 PMID:30309837

Mani, M., Kavanagh, D. J., Hides, L., & Stoyanov, S. R. (2015). Review and Evaluation of Mindfulness-Based iPhone Apps. *JMIR mHealth and uHealth*, *3*(3), e82. doi:10.2196/mhealth.4328 PMID:26290327

Mann, G., Lambert, L., Chang, Y., & Jurss, M. (2018). P135 - using the theory of planned behavior to understand college student snack choices. *Journal of Nutrition Education and Behavior: Supplement*, *50*(7), 119. doi:10.1016/j.jneb.2018.04.164

Mayo Clinic. (2020, May 15). *Telehealth: Technology meets Healthcare*. https://www.mayoclinic.org/healthy-lifestyle/consumer-health/in-depth/telehealth/art-20044878#:~:text=Telehealth%20is%20the%20use%20of,or%20support%20health%20care%20services

Meilman, P. W., & Weatherford, R. D. (2016). "Wait! Can You Do This Too?!?". College Counseling and Distance Education. *Journal of College Student Psychotherapy*, *30*(1), 1–3. doi:10.1080/87568225.2016.1105636

Ministry of Health – Manatū Hauora. (2017 October). *Guidance on evaluating or developing a health app*. https://www.health.govt.nz/system/files/documents/pages/guidance-evaluating-developing-health-app-oct17-v2.pdf

Mitchell, S. L., Oakley, D. R., & Dunkle, J. H. (2019). White Paper: A Multidimensional Understanding of Effective University and College Counseling Center Organizational Structures. *Journal of College Student Psychotherapy*, *33*(2), 89–106. doi:10.1080/87568225.2019.1578941

Morgan, B. (2017). Stress management for college students: An experiential multi-modal approach. *Journal of Creativity in Mental Health*, *12*(3), 276–288. doi:10.1080/15401383.2016.1245642

Nagatomo, S. (2019). *Japanese Zen Buddhist Philosophy*. https://plato.stanford.edu/entries/japanese-zen/

National Alliance on Mental Health. (2019). *Mental Health by the Numbers*. https://www.nami.org/Learn-More/Mental-Health-By-the-Numbers

Niazi, A. K., & Niazi, S. K. (2011). Mindfulness-based stress reduction: A non-pharmacological approach for chronic illnesses. *North American Journal of Medical Sciences*, *3*(1), 20–23. doi:10.4297/najms.2011.320 PMID:22540058

Nisbet, M. (2020, May 2). *The mindfulness movement: How a Buddhist practice evolved into a scientific approach to life*. https://web.northeastern.edu/matthewnisbet/2017/05/24/the-mindfulness-movement-how-a-buddhist-practice-evolved-into-a-scientific-approach-to-life/

Nobleza, D., Hagenbaugh, J., Blue, S., Vergare, M., Pohl, C., & Stepchin, A. (2018). The use of telehealth by medical and other health professional students at a college counseling center. *Journal of College Student Psychotherapy*. Advance online publication. doi:10.1080/87568225.2018.1491362

Novotney, A. (2017). *A Growing Wave of Online Therapy*. https://www.apa.org/monitor/2017/02/online-therapy

Parmer, R. (2017, May 25). *The Importance of Self-Care for College Students*. https://collegepuzzle.stanford.edu/the-importance-of-self-care-for-college-students/.00000000

Parsons, C. E., Crane, C., Parsons, L. J., Fjorback, L. O., & Kuyken, W. (2017). Home practice in mindfulness-based cognitive therapy and mindfulness-based stress reduction: A systematic review and metaanalysis of participants' mindfulness practice and its association with outcomes. *Behaviour Research and Therapy*, *95*, 29-41. http://dx.doi.org.tcsedsystem.idm.oclc.org/10.1016/j.brat.2017.05.004

Piyadassi, T. (2013, November 13). *The Seven Factors of Enlightenment*. https://www.accesstoinsight.org/lib/authors/piyadassi/wheel001.html

Pooreh, S., & Hosseini Nodeh, Z. (2015). Impact of Education Based on Theory of planned behavior: An Investigation into Hypertension-Preventive Self-care Behaviors in Iranian Girl Adolescent. *Iranian Journal of Public Health*, *44*(6), 839–847. PMID:26258097

Querstret, D., Morison, L., Dickinson, S., Cropley, M., & John, M. (2020). Mindfulness-based stress reduction and mindfulness-based cognitive therapy for psychological health and well-being in nonclinical samples: A systematic review and meta-analysis. International *Journal of Stress Management*. http://dx.doi.org.tcsedsystem.idm.oclc.org/10.1037/str0000165

Rahimdel, T., Morowatisharifabad, M. A., Salehi-Abargouei, A., Mirzaei, M., & Fallahzadeh, H. (2019). Evaluation of an education program based on the theory of planned behavior for salt intake in individuals at risk of hypertension. *Health Education Research, 34*(3), 268–278. doi:10.1093/her/cyz007 PMID:30838389

Ramler, T., Tennison, L., Lynch, J., & Murphy, P. (2016). Mindfulness and the college transition: The efficacy of an adapted mindfulness-based stress reduction intervention in fostering adjustment among first year students. *Mindfulness, 7*(1), 179–188. doi:10.100712671-015-0398-3

Ramos Salazar, L., & Hayward, S. L. (2018). An examination of college students' problem-solving self-efficacy, academic self-efficacy, motivation, test performance, and expected grade in introductory-level economics courses. *Decision Sciences Journal of Innovative Education, 16*(3), 217–240. doi:10.1111/dsji.12161

Roy, N. (2018). *The Rise of Mental Health on College Campuses: Protecting the Emotional Health of Our Nation's College Students*. https://www.higheredtoday.org/2018/12/17/rise-mental-health-college-campuses-protecting-emotional-health-nations-college-students/

Scalco, A., Noventa, S., Sartori, R., & Ceschi, A. (2017). Predicting organic food consumption: A meta-analytic structural equation model based on the theory of planned behavior. *Appetite, 112*, 235–248. doi:10.1016/j.appet.2017.02.007 PMID:28188865

Seyal, A. H., & Abd Rahman, M. N. (Eds.). (2017). *Theory of planned behavior: new research (Ser. Psychology of emotions, motivations and actions)*. Nova Science.

Shatkin, J. P., Diamond, U., Zhao, Y., DiMeglio, J., Chodaczek, M., & Bruzzese, J. (2016). Effects of a risk and resilience course on stress, coping skills, and cognitive strategies in college students. *Teaching of Psychology, 43*(3), 204–210. doi:10.1177/0098628316649457

Shefet, O. (2018). Ultra-Brief, Immediate, and Resurgent: A College Counseling Paradigm Realignment. *Journal of College Student Psychotherapy, 4*(32), 291–311. doi:10.1080/87568225.2017.1401790

Sheil, W. C. (2018, December 11) *Medical Definition of Stress*.https://www.medicinenet.com/script/main/art.asp?articlekey=20104

Shelesky, K., Weatherford, R. D., & Silbert, J. (2016). Responding to the Increased Needs of College Students: A Case Study of Case Management. *Journal of College Student Psychotherapy, 30*(4), 284–299. doi:10.1080/87568225.2016.1219613

Shi, J., & Kim, H. K. (2020). Integrating risk perception attitude framework and the theory of planned behavior to predict mental health promotion behaviors among young adults. *Health Communication, 35*(5), 597–606. doi:10.1080/10410236.2019.1573298 PMID:30810392

Sipe, W. E., & Eisendrath, S. J. (2012, February). Mindfulness-based cognitive therapy: Theory and practice. *Canadian Journal of Psychiatry, 57*(2), 63–69. doi:10.1177/070674371205700202 PMID:22340145

Skyfactor. (2017). *Second-Term Behavior of Students on Academic Probation*. https://www.skyfactor.com/wp-content/uploads/2017/06/2017-Fall-Probation-Students-in-Spring-Research-Note.pdf

Song, Y., & Lindquist, R. (2018). Critical review of mindfulness-based stress reduction for anxiety and depression among patients with heart disease. *Asia Life Sciences. Supplement, 15*(1), 1–8.

Sutter, N., & Paulson, S. (2016). Predicting college students' intention to graduate: A test of the theory of planned behavior. *College Student Journal, 50*(3), 409–421.

The American College Health Association. (2018). *Role of ACHA*. https://www.acha.org/HealthyCampus/About/Role_of_ACHA/HealthyCampus/Role_of_ACHA.aspx?hkey=c825e8f5-670c-4b99-a12f-22b19c4ec863

The RP Group. (n.d.). *Who We Are*. https://rpgroup.org/About-Us

van Emmerik, A. A. P., Berings, F., & Lancee, J. (2018). Efficacy of a Mindfulness-Based Mobile Application: A Randomized Waiting-List Controlled Trial. *Mindfulness, 9*, 187–198. https://doi.org/10.1007/s12671-017-0761-7

Vidic, Z., & Cherup, N. (2019). Mindfulness in classroom: Effect of a mindfulness-based relaxation class on college students' stress, resilience, self-efficacy and perfectionism. *College Student Journal, 53*(1).

Whitehead, P. M., & Wright, R. (2017). Becoming a college student: An empirical phenomenological analysis of first generation college students. *Community College Journal of Research and Practice, 41*(10), 639–651. doi:10.1080/10668926.2016.1216474

Willgress, L. (2020). 12 best mindfulness apps to help you keep calm during a crisis. *The Independent*.

Winerman, L. (2017, November). *By the numbers: Stress on Campus College and University Counseling Centers are Seeing an Uptick in the Number of Students Seeking Help*. https://www.apa.org/monitor/2017/09/numbers

Yagoda, R. (2016, November 9). *College Students and Binge Drinking: When a Rite of Passage Becomes a Path to Destruction*. https://health.usnews.com/health-care/for-better/articles/2016-11-09/college-students-and-binge-drinking-when-a-rite-of-passage-becomes-a-path-to-destruction

ADDITIONAL READING

Kabat-Zinn, J. (2005). *Wherever you go, there you are: mindfulness meditation in everyday life*. Hyperion.

Kabat-Zinn, J. (2013). *Full catastrophe living: Using the Wisdom of Your Body and Mind to Face Stress, Pain, and Illness (Revised and updated)*. Bantam Books Trade Paperback.

Puddicombe, A. (2016). *The Headspace Guide to Meditation and Mindfulness: How Mindfulness can change your life in ten minutes a day*. St. Martins Griffin.

Williams, J. M. G., & Penman, D. (2012). *Mindfulness: An Eight-week plan for Finding Peace in a Frantic World*. Rodale.

KEY TERMS AND DEFINITIONS

Ego Resilience: The ability of people to adapt to changes presented in different environments, self-regulate, and exercise self-control.

Guided Meditation: When engaging in this type of meditation, an instructor leads the meditation exercises and participants responds the guidance provided throughout the practice.

Meditation: A set of mental and physical practices that are used to achieve a state of calmness, improve wellness, and enhance mental clarity.

Mindfulness: Mindfulness is rooted in Buddhism and defined as being present in the moment. Additionally, the field of psychology adopted mindfulness as a coping mechanism practice. Also, there is a focus on being aware of feelings, the environment, physical state, and being open-minded.

Mindfulness-Based Cognitive Behavioral Therapy (MB-CBT): This type of counseling modality merges principles of cognitive behavioral therapy with meditation and is often used to treat a range of mental health concerns such as depression and anxiety issues.

Mindfulness-Based Stress Reduction (MBSR): The American Psychological Association (APA) stated that this intervention occurs over 8 weeks and is usually conducted in a group format with components of mediation and yoga.

Mobile Application: In defining this term, it is a type of software that is formatted for mobile devices such as smart phones and tablets and is commonly referred to as an "app". Users are able to download an app from the either the Apple (iOS) or Android application stores.

Chapter 11
Empowering Adolescents Through Instilling Information Security Culture in Zimbabwean Schools

Trymore Z. Ruvinga
Midlands State University, Zimbabwe

Theo Tsokota
https://orcid.org/0000-0002-7347-515X
Midlands State University, Zimbabwe

Colletor Tendeukai Chipfumbu Kangara
Midlands State University, Zimbabwe

Pamela T. Nyambuya
Zimbabwe Open University, Zimbabwe

ABSTRACT

There is an excellent opportunity to ensure that information security culture (ISU) is embedded in school children before they are employed in the industry. However, for the majority of time spent in primary and secondary school education, pupils are alienated from supervised use of technology, making it difficult to teach pupils proper use of technology. Thus, there is no deliberate effort to empower and impart ISC to school pupils in Zimbabwe. The purpose of this study is to develop a framework for instilling information security culture in secondary school pupils. Based on the literature, the first version of the framework was developed and subjected to a focus group for review. Data from this focus group was analysed, resulting in a second improved version of the framework. Consequently, it was shown that the framework was relevant, useful, and applicable within Zimbabwean settings.

DOI: 10.4018/978-1-7998-8318-0.ch011

1. INTRODUCTION

Information Technology has become the backbone of all business activities across the globe. Over the years, information technology has evolved from being a supporting act in business to become the way of doing business (Mwantimwa 2019). However, the use of Information Technology and the internet poses many security threats to organisational information. Such threats compromise organisational competitiveness and can lead to significant losses. It is, therefore, of paramount importance to ensure that organisational information systems are kept secure at all times. Studies by (Ani, He, and Tiwari 2019; Al-Darwish and Choe 2019) have shown that human behaviour is a key yet often forgotten component in safeguarding information. Researchers have called for Information Security Culture (ISC) to be fostered into organisations to influence employee behaviour towards safeguarding organisational information positively. Previous researches clearly emphasis fostering ISC at the organisational level. Given the competitive pressures that every organisation is subjected to, fostering ISC at this level can be a waste of valuable organisational resources. There is an excellent opportunity to ensure that ISC is embedded in potential employees before they start working.

Most Zimbabwean schools have banned using technologies such as cell phones and note pads hence the difficulties in teaching pupils proper use of technology. For the majority of time spent in primary and secondary school education, pupils are alienated from the supervised use of technology. Information Communication Technology has become the global way of doing business, and such alienation will result in an ill-equipped workforce in the industry. Furthermore, the Zimbabwean secondary school computer studies syllabus does not emphasise information security issues. As pupils graduate from secondary schools, they know about the technology but do not exhibit acceptable behaviour when using it. There is no deliberate effort to empower pupils by imparting ISC in Zimbabwe. Such ignorance is likely to increase the risk of Information Security vulnerability when employed in Zimbabwean organisations. Just as reproductive health is taught at the primary school level to prepare and empower pupils for the future, so should be the case with fostering Information Security Culture (ISC). Thus, there is a lack of effort to foster ISC in Zimbabwean secondary school pupils. Such a scenario leaves private and government organisations at the mercy of computer criminals. This chapter aims to develop a framework for instilling Information security culture in secondary school pupils. The following section briefly outlines the concept of ISC.

2. INFORMATION SECURITY CULTURE (ISC)

Information Security Culture includes all socio-cultural measures that support technical security methods so that information security becomes a natural aspect in the daily activity of every employee (Al-Darwish and Choe 2019). Information security management primarily disregards the human dimension. Technical and procedural mechanisms dominate Information security management. Ani et al. (2019) believe that the way employees act and behave is particularly important in terms of information security because most information security mechanisms heavily rely on human beings. Experts have previously proposed conceptual frameworks for information security management that include information security cultural development based on management initiatives of policy, awareness, training, and education (Hu, Dinev, Hart, and Cooke 2012). Such initiative resembles the realisation that there is a need to address the human side of information security. In past years, several information security culture frameworks have been developed.

3. IMPORTANCE OF ISC

The following sub-section discusses the importance of Information Security Culture. Moving forward, the significant ideas for the importance of ISC these sub-sections will be broken into three as follows.

3.1 When Technology Alone Cannot Provide Total Information Security

Siponen and Oinas-Kukkonen (2007) believe that non-technical aspects are as important as technical aspects in protecting vital organisational information. Competent information security cannot be referred to as a once-off achievement for an organisation. At the present moment, various technologies for barring access to information electronically are in existence. Such technology is not only inadequate in solving information security challenges but also worsening the problem. This can be confirmed by current trends, for instance, the Bring Your Own Device (BOYD) policy, in which employees in an organisation are allowed to use their personal computing devices for work purposes. The same devices are used for leisure and other personal use outside work. The cohabitation of organisational and personal data and the mobility and vulnerability of that data outside corporate confines have given rise to more automated security measures. However, they each share the same fundamental limitation.

3.2 The Human Side Is at the Heart of the Problem

The reason why technology alone cannot provide all the solutions to information security problems is that organisational members will always need access to information to perform their duties. Computer criminals know that the simplest way to access what they want is often via the users of information systems. Information thieves often exploit ignorance and negligence by system users and administrators in using the same technologies that are assumed to safeguard them. In reality, the more organisational information is protected by state of the art technologies, the higher the temptation for its employees to believe that there is no need to be cautious in handling the organisation's information (Hu et al. 2012).

3.1.3 Risk of Significant Losses as a Result of Compromised Data

Organisations whose members are careless in handling information can experience the impact of information security breaches in various ways (Al-Darwish and Choe 2019). Important product information that leaks to rivals, customer data that is disclosed to computer criminals and negative publicity resulting in an image of negligence which in turn can badly influence rapport with significant stakeholders, are all possibilities (Wiley, McCormac, and Calic 2020; Herath and Rao 2009; Colwill 2009). In the most severe of situations, each of these can, in the end, force an organisation to discontinue operations.

4. THEORETICAL UNDERPINNING OF STUDY

Various scholars have come up with frameworks ranging from aiding the understanding of Information Security Culture to fostering Information Security Culture. To develop a well-informed Information Security Culture empowering framework, there is a need to examine existing frameworks. Discussed in the next section is the Model for Information Security Shared Tacit Espoused Values.

4.1 Model for Information Security Shared Tacit Espoused Values (MISSTEV)

Thomson, Von Solms, and Louw (2006) state that it is challenging to achieve assimilation of information into an organisation because of employees' way of behaving or acting. They believe that protecting information must be a part of organisational members' daily routine and second nature. They further stated that negligent and ignorant actions by employees rank among the greatest threats to the success of information security in an organisation. Guarding against this entails members of an organisation learning and integrating acceptable information security practices into their daily behaviour. Their researches address how employees should learn, develop and integrate the proper information security skills into their everyday behaviour.

Thomson and Von Solms (2005) introduced Information security obedience ISO, an amalgamation of corporate culture, information security and corporate governance. Thomson and Von Solms (2005) defined ISO as de facto user behaviour complying with the vision of senior management as defined in the corporate information security policy. The MISSTEV model was proposed to achieve ISO. The model is made up of the Conscious Competence Maturity model and Nonaka's Modes of Knowledge Creation.

The first component of the MISSTEV model is made up of the conscious competence learning model, which describes steps in acquiring new skills and behaviour. It is composed of four stages: unconscious incompetence, Conscious incompetence, Conscious competence, and Unconscious competence. At the unconscious incompetence stage, employees are unaware of specific skills that they are required to have. They are ignorant of the fact that they lack particular skills. Secondly, there is the conscious incompetence stage. Here, organisational members know of the skills that they are required to have. Their shortcomings have been exposed, and they now know what they need to be competent.

Conscious competence comes third. At this level, the members of an organisation have acquired the skills that they need to be competent. However, they are still very cautious in performing the acquired skills as they are not yet embedded in them. Each time they have to think of the skills before they perform them. Finally, there is unconscious competence. Skills acquired at the conscious competence stage have been repeated repeatedly to the extent that they can be performed without thinking about them. The skills are then embedded in people's natural behaviour. This model states that when people need to learn new skills, they naturally go through the discussed stages.

The second component of the MISSTEV model is Modes of Knowledge Creation (Nonaka 1994). Creating knowledge encompasses tacit knowledge and explicit knowledge and knowledge conversion steps that take place for knowledge to be changed from tacit to explicit. Tacit knowledge is based on beliefs and attitudes and is subjective. Explicit knowledge is objective and rational knowledge that can be communicated in words. Nonaka (1994) model aims to single out existing knowledge and change it to new knowledge by categorising four interaction processes between tacit knowledge and explicit knowledge. The categories are Socialisation, Externalisation, Combination and Internalisation.

Socialisation entails a person's tacit knowledge being shifted to tacit knowledge in another person. Here knowledge is captured by directly interacting with people in and out of the organisation. Externalisation transforms tacit knowledge into explicit knowledge. This is achieved by articulating tacit knowledge by expressing beliefs and values in words. When knowledge is explicit, it can be transferred to explicit knowledge by the process of Combination (Nonaka, 1994). Explicit knowledge is communicated in emails and training programs. Through this process, knowledge is transferred among groups in organisations. Internalisation is a stage where explicit knowledge is absorbed into tacit knowledge held by a person.

This model shows the advancement from the creation of the corporate information security policy to the achievement of Information Security Obedience.

This model is primarily based on information security policy. The vision (tacit knowledge) of leaders of the organisation is changed to explicit knowledge through the creation of a security policy. It is transferred to members of the organisation through combination. The explicit knowledge of the organisation's leaders (policy) is transferred to explicit knowledge of employees through awareness programmes. Such programs enable the employees to know the skills they need to protect organisational information. After awareness, employees move to the conscious incompetence stage of the model. At this stage, internalisation will facilitate absorbing the skills learnt. Internalisation paves the way for the unconscious competence stage of the model. By this time, information security skills have been performed repeatedly to the extent of becoming natural behaviour for the employees. For members of an organisation to stay at the unconscious competence stage, information security skills learnt and mastered through the model should be transferred to the tacit knowledge of other members of the organisation through socialisation. In time unconscious competence evolves into information security obedience.

The resultant conceptual model shows the various aspects that should be considered in information security culture. It illustrates that information security culture should be within the organisational culture, and organisational culture, in turn, should be in line with national culture. It also shows aspects that should be considered in developing a secure information culture.

5. METHODOLOGY

In this chapter, a qualitative research paradigm was used firstly by reviewing the literature. This literature enabled the research to gain in-depth knowledge on the subject under study and subsequently develop a draft framework for instilling information security culture. Secondly, the proposed framework was evaluated by a focus group made up of experts in relevant fields.

5.1 Literature Review

Extensive literature studies were used, according to Hofstee (2006), to gain valuable insight into what a cyber-security culture entails and to identify critical components of ISC. Literature played an essential role in this chapter as it provided the basis for creating the proposed framework. Boote and Beile (2005) define a literature review as an evaluative report of studies found in the literature related to your area of study. This review should offer a description, summary, evaluation and clarification of the literature. The review of literature is the source of a theoretical basis for the study to be conducted. A literature review is more than just searching for information. It identifies and articulates relationships between literature and a researcher's field of study. Boote and Beile (2005) further highlight some primary purposes of the literature review relevant to this chapter: Ensuring that the researcher learns from previous theory on the subject matter, showing how the field in question has been studied before and noting faults in preceding studies.

5.2 Focus Groups

Focus groups were the second method used to undertake this research. Tausch and Menold (2016) define a focus group as a moderated discussion among six to twelve people who discuss a topic under the direction of a moderator whose role is to promote interaction and keep the discussion on the topic of interest. He argues that having less than six members in a group may diminish the diversity of views offered. On the other hand, more than twelve participants will make it difficult to air their views and concerns in a comprehensive manner. In a focus group, interaction is between all members of the group and not a case of the researchers asking the participants the questions and just noting down responses. It is more of a discussion in which one participant can respond to a contribution by another participant. Such an approach results in collective and well-informed views on the topic of discussion.

A major advantage of using focus groups was that they were time economical. The research was able to gather many data in a short period. Stewart and Shamdasani (2017) postulate that more time is spent recruiting participants for the focus group. Focus groups also make possible the recording and interpretation of non-verbal data. The first group was composed of experts in various relevant fields. As a result, in-depth and comprehensive data was gathered. The second group was composed of those that deliver the knowledge to the secondary school pupils (i.e. teachers and education officers). However, as with any method of data gathering, focus groups have their drawbacks. Firstly, the success of a focus group is primarily dependent on the skill of the group moderator. The moderator should be able to probe more into opinions raised by the group participants. In this research, however, the moderator was an education expert who has a wealth of experience in conducting focus groups. Focus groups are made up of a small group, and results cannot be generalised to the whole population under study. However, the results of this research are not meant to be generalised to the population from which the respondents were selected but instead to provide a solution for an identified problem.

Tayie (2005) summarised conducting focus groups into seven steps. Following is each of the steps explained concerning this chapter:

5.2.1 Problem Definition

At this stage, the chapter established a clearly defined problem that needed to be addressed. A Framework for instilling information security culture in secondary school pupils was proposed to address the identified problem.

5.2.2 Sampling Technique

For this chapter, a purposive sampling, precisely a judgemental sampling strategy, was employed. This was done to extract in-depth data from the participants who are knowledgeable about the area of interest. The participants were from diverse fields that were thought to be relevant in formulating the framework.

5.2.3 Determining the Number of Focus Groups to Be Conducted

Many scholars recommend that focus groups should be held until no new data is coming from the participants. Some recommend that at least two focus groups should be conducted to at least get the views of two different groups. In this research, two groups were conducted, one that explored the framework and the other that confirmed if the proposed framework is suitable for the application.

5.2.4 Preparation of Study Mechanics

This step involved participant recruitment, making reservations at the facility at which the focus groups were to be conducted, and deciding how the group discussions were to be recorded (video, audio or both). A capable moderator was recruited and enlightened on issues to be discussed in the groups.

5.2.5 Preparation of Focus Group Materials

Preparation of questions to be asked by the moderator was done at this stage. Having such questions prepared for the moderator did not mean that the moderator could not ask his own questions. He could probe to seek clarifications and generate new views from participants.

5.2.6 Conducting the Sessions

Focus group discussions for this research were conducted in a university seminar room. The University seminar room was considered the most suitable location for the discussions because it was easily accessible to all participants. The possibility of distractions was minimal.

5.2.7 Analysing Data and Preparation of Summary Reports

For this research, transcripts of the focus groups only noted issues that are pertinent to the research. Information that was deemed unimportant was ignored. Such information would be contributions that did not affect the structure of the framework.

6. INITIAL FRAMEWORK

Initially, the researchers reviewed the literature on information security and information security culture to develop a draft framework that was to be critically analysed by the exploratory focus group. The following was the draft framework that was tabled b for the focus group to the discussion. All stages in the framework are also explained. This framework is comprised of three levels, namely **Government support, Preparation and implementation.**

Figure 1.

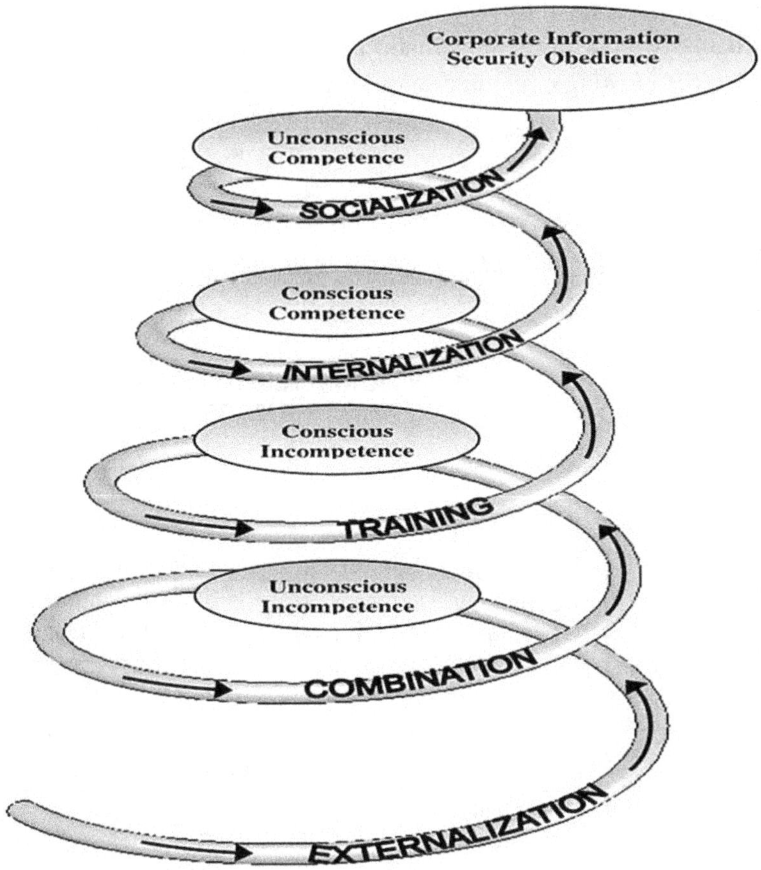

Level 1: Government Support

The MISSTEV model clearly shows that the most important feature in any given framework is the guaranteed support of top management. Without them, any initiative is bound to fail. In the Zimbabwean case, the Ministry of Primary and Secondary Education has to champion fostering an information security culture. As with the fostering of any culture, there should be written down and agreed on guiding principles. Other key ministries are roped in to support this vision, namely the Ministry of Higher and Tertiary Education and Ministry of Information Communication Technology. As shown in **Figure 2**, the Ministry of Primary and Secondary Education, in conjunction with other ministries of interest, draft an information security policy that will be the basis for creating an information security culture.

Level 2: Preparation

After policy creation, the government tasks a team to first come up with amendments to the existing computer studies syllabus to accommodate information security issues. These changes will be applied to syllabi for both secondary schools and teacher training institutions, as shown in **Figure 2**. At this level, it is crucial to ensure that the syllabi will clearly outline the topics that need to be learnt and strongly emphasise the need for information security. These will introduce the pupils to several fundamental technical and administrative aspects of information security. The principal aim of these additions will be to provide pupils with an overview of the information security subject. The following is a proposed list of topics to be added to the existing computer studies syllabi: Information Security overview, Sharing Files, Authenticating People, Connecting Computers, Networks of Networks and Security in Internet Services and Email.

Figure 2. Initial framework for instilling information security culture in Zimbabwean secondary school students.

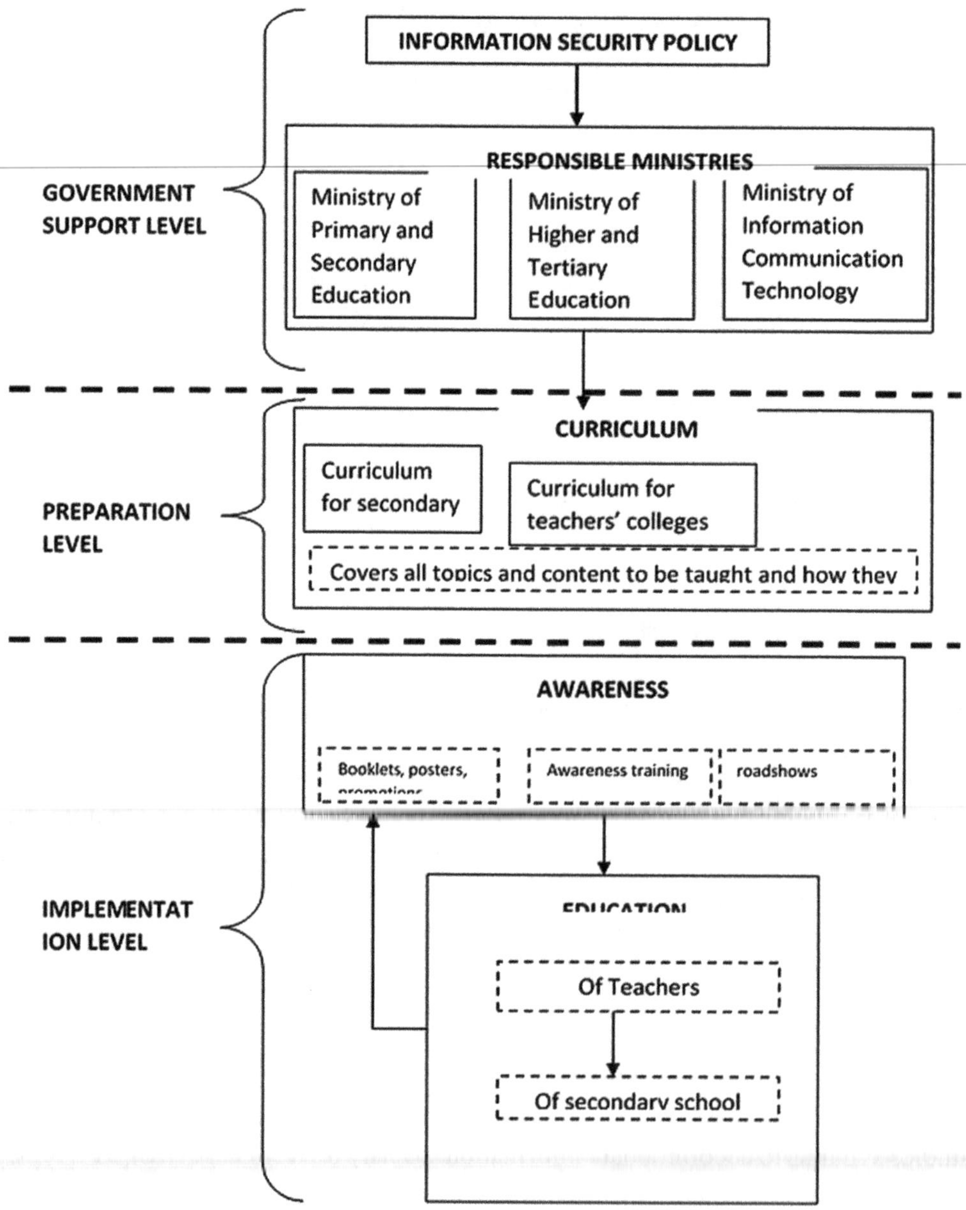

Level 3: Implementation Level

At this layer, the information security message reaches its intended destination; the pupil. It encompasses two concepts **(Figure 2)** that arose from the discussion of other frameworks, namely: Awareness and Education. Firstly, we look at awareness. According to Alnatheer (2014), many studies emphasise that awareness is a critical component in understanding issues to do with information security culture. For the pupils and the general populous to realise the importance of secure information culture, they need to be aware of the principles and skills they ought to have to secure their information systems better. They will only appreciate a given concept when its benefits are made clear. Awareness is the 'what' component of the education strategy of an organisation that tries to change the behaviour and patterns in how the targeted audience (e.g. pupils, the general public) use technology and the internet, and it is a distinct element from training (AlMindeel and Martins 2020). AlMindeel and Martins (2020) believe that awareness of risks associated with ICTs serves as the first line of defence for organisational information security.

Other awareness methods that can be used include community engagement in which a community member concretises the community about information security culture. In our case, we look at the school community in which a teacher will champion securing information. Advocacy meetings that aim to highlight the necessity of information security can further increase awareness of information security. Marketing and communication efforts can also play a pivotal role in awareness. Programs can be spearheaded by role models in given communities (business people, politicians and celebrities). Taking advantage of the high literacy rate in Zimbabwe, information communication materials such as posters, magazines, and pamphlets can also be employed.

The most crucial objective of an awareness program is to enlighten system users on their duty to safeguard the confidentiality, integrity and availability of organisational information systems. McIlwraith (2006) believes that one of the objectives of an awareness program is to convey a simple, clear and presentable message in a format easily understood by the audience. The Global Assurance Certification paper (Insitute 2003) highlights methods of conveying the security awareness message as follows:

- Information security awareness training – an effective way of getting users in a classroom setup. The topic of information security can then be explained interactively.
- Computer-based information security awareness – These are computer applications that are available to users at all times and enable the user to train themselves.
- Awareness services and reminder tools – awareness is an ongoing process, and people need to continually be reminded about information security. Awareness activities that can be carried out include multimedia presentations, security booklets, security posters, and promotional items with security issues.

Before awareness, people are said to be exhibiting unconscious incompetence, as discussed in the MISSTEV model Section **4.1,** where people are unaware of the skills they ought to have. After implementing security awareness programs, they move to a state of conscious incompetence where they have seen the need for security but lack relevant skill and knowledge.

The second component of the action level is Education. Education entails the actual delivery of information security concepts first to the teachers and the teachers, who deliver to the pupils. At this level, a state of conscious competence is reached, as explained by Alnather (2014)'s model. A level when pupils embrace the subject of information into being their second nature is reached by continually running awareness campaigns while, on the other hand delivering in-class education. As illustrated by the conscious competence learning model of acquiring new skills in the MISSTEV model, the education process all four processes of acquiring new skills. The educating process will follow a structured syllabus to cover all essential aspects of information security. Education and awareness are ongoing processes and support each other.

7. ANALYSIS OF THE FIRST FOCUS GROUP

The video recording of the first focus group was transcribed and critically studied. In the first focus group, turnout was low. However, participants were six, which was enough to constitute a recognisable focus group. It consisted of one culture expert, two educators, one Information Technology expert and two parents. In analysing the results, data were grouped into categories. Data affecting a particular level of the framework were grouped into one. This was done for all levels of the framework. Data considered not to fit in the existing levels was also grouped into a separate group.

7.1 Resultant Changes on the Framework

The following are the amendments made to the framework after the first focus group beginning from the first level to the evaluation level. On the governmental support level, other stakeholders outside government like the Computer Society of Zimbabwe and human rights groups were added to ensure that policymaking is all-inclusive. On the preparation level, firstly, the curriculum stage was rephrased to curriculum review to make it clear that changes to existing computer studies syllabi were proposed. Furthermore, a list of proposed topics to be added was included in the framework.

The awareness stage was moved up to the preparatory level of the framework. Each of the campaign strategies was tied to a specific audience to enable customisation of content. Lastly, teacher education was also made a part of the preparatory level. On the implementation level, the awareness phase of the framework was moved to the preparation level. A new concept of IT counselling was included to support the education process.

As reiterated in the focus group, an evaluation level was added to the framework. This would enable the measurement of the success of the framework once implemented. The proposed evaluation technique was an information security awareness survey. The survey was meant to establish whether or not the framework was achieving what it was set out to achieve. This level would provide feedback that would be the basis for continual improvement of the framework.

After all, the adjustments were made, and an improved framework emerged. The next stage was to present the framework to the second focus group to confirm the changes and add any further improvements. The following **Figure 3** is the framework as it looked after the first focus group (exploratory).

7.2 Analysis of the Second Focus Group

This round is the final focus group meant to confirm the changes made to the framework and polish up any outstanding issues. As is in the first focus group, the video recording of this focus group was transcribed. Transcribing was done by watching the video and changing the discussions into a text format. Again data were grouped according to the level of the framework it affected. For the second focus group, more participants were recruited to raise the number of participants. This approach proved beneficial as eleven participants took part in the second focus group. However, the group of participants represented all the fields identified as necessary in this research. The second focus group was primarily made up of two experts in education and the three teachers that deliver the lessons in class, one cultural expert, two Information Technology experts and three parents. However, the results of this focus group were minor because the exploratory focus group had exhausted the framework's most contentious issues. Most participants in this focus group had concerns with the diagrammatic presentation of the framework.

Figure 3.

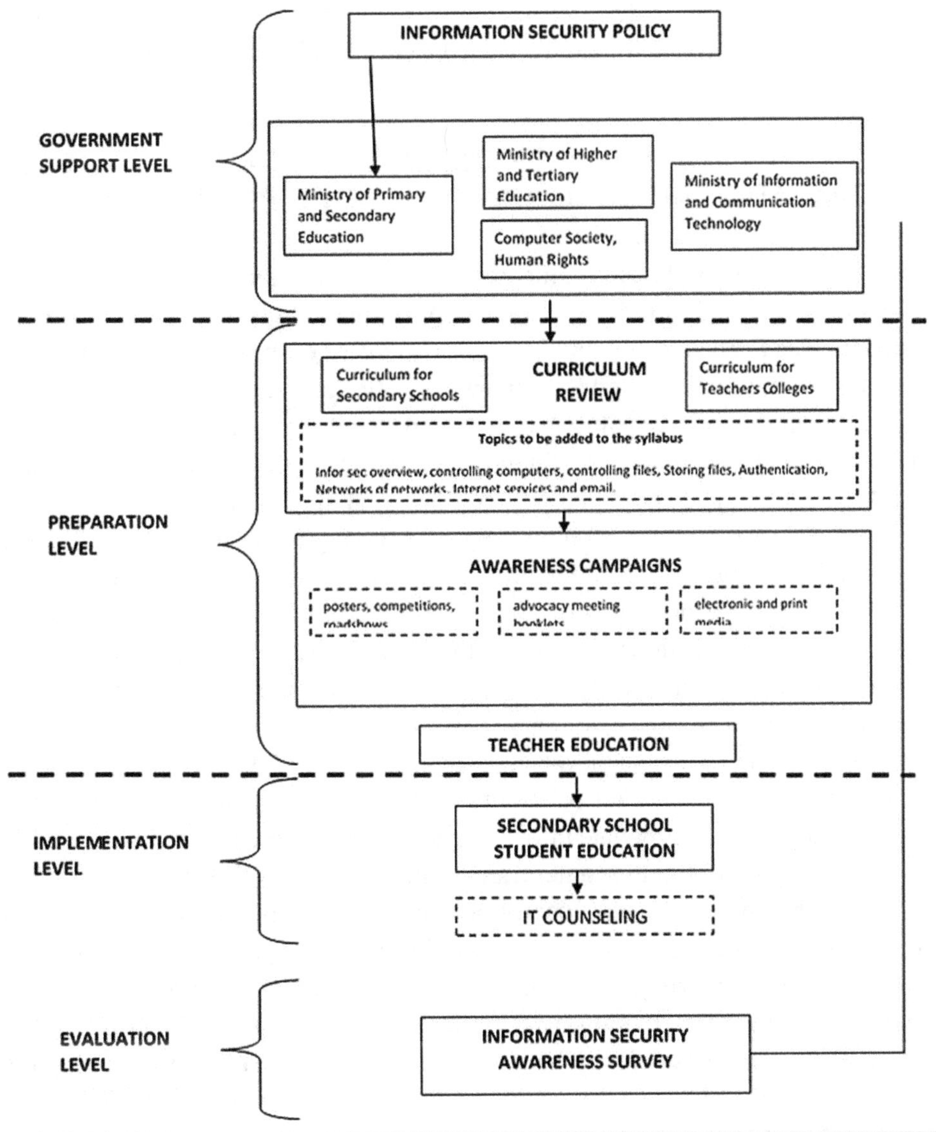

Analysis of government support level, Concerns were raised over the diagrammatically misleading presentation of this level. The stakeholders outside government were shown directly below and connected to the Ministry of Higher and Tertiary Education. Such a setup would mislead in that it will seem as if these stakeholders are within that ministry. It was suggested that all the stakeholders be presented linearly to show their interdependence in coming up with the policy. On the **Analysis of preparation level**

Participants indicated that the Curriculum Development Unit (CDU) should be shown in the framework. It is the unit responsible for formulating syllabi.

In addition, on the Analysis of implementation level, the education and counselling stages were initially presented as separate functions. It was, however, recommended that they are joined together as they complimented each other. Furthermore, one participant noted that an information security awareness survey was insufficient as an evaluation method. It was limiting in that it only identified one way of doing the evaluation. It was suggested that this stage be called Information Security culture evaluation.

8. PROPOSED FRAMEWORK FOR INSTILLING INFORMATION SECURITY CULTURE IN ZIMBABWEAN SECONDARY SCHOOL PUPILS

Having considered the results of the two focus groups, a framework was subsequently developed. The proposed framework consists of four levels, namely: **government support**, **preparation**, **implementation** and **evaluation.** The following are detailed explanations of each of the levels of the framework.

The most important feature in any given framework is the guaranteed support of top management. Without top management support, any initiative is bound to fail. In the Zimbabwean case, the Ministry of Primary and Secondary Education has to champion fostering an information security culture. As with the fostering of any culture, there should be written down and agreed on guiding principles. Other key ministries are roped in to support this vision, namely the Ministry of Higher and Tertiary Education and Ministry of Information Communication Technology. In conjunction with other ministries of interest, the Ministry of Primary and Secondary Education drafts an information security policy that will be the basis for the creation of an information security culture. In drafting this policy, other stakeholders outside government such as the Computer Society of Zimbabwe and human rights organisations are invited to participate. Such an approach ensures a comprehensive policy.

After policy creation, the government, through the Ministry of Primary and Secondary Education, tasks the Curriculum Development Unit (CDU) to firstly come up with amendments to the existing computer studies syllabus to accommodate information security issues. These changes will be applied to syllabi for both secondary schools and teacher training institutions, as shown in **Figure 4**. At this level, it is crucial to ensure that the syllabi will clearly outline the topics that need to be learnt and strongly emphasise the need for information security. These will introduce the pupils to several essential technical and administrative aspects of information security. The primary aim of these additions will be to provide pupils with an overview of the information security subject. The following is a proposed list of topics to be added to the existing computer studies syllabi: Information Security overview, Sharing Files, Controlling Files, Authenticating People, Networks of Networks and Security in Internet Services and Email.

It is also essential to ensure that all stakeholders are aware of the impending changes at this level. Firstly, awareness campaigns can generate feedback that can aid in drafting the amendments to the curriculum discussed above. Secondly, awareness enables people to appreciate issues to do with information security. Awareness is a critical component in aiding the understanding of issues to do with information security culture. For the pupils and the general populous to realise the importance of secure information culture, they need to be aware of the principles and skills they ought to have to secure their information systems better. They will only appreciate a given concept when its benefits are made clear. Awareness of risks associated with ICTs serves as the first line of defence for organisational information security.

Other awareness methods that can be used include community engagement in which a community member conscientise the community about information security culture. In our case, we look at the school community in which a teacher will champion securing information. Advocacy meetings that aim to highlight the necessity of information security can further increase awareness of information security. Marketing and communication efforts can also play a pivotal role in awareness. Programs can be spearheaded by role models in given communities (business people, politicians and celebrities). Taking advantage of the high literacy rate in Zimbabwe, information communication materials such as posters, magazines, and pamphlets can also be employed.

The most crucial objective of an awareness program is to enlighten system users on their duty to safeguard the confidentiality, integrity and availability of organisational information systems. One of the objectives of an awareness program is to convey a simple, clear and presentable message in a format easily understood by the audience. The Global Assurance Certification paper highlights methods of conveying the security awareness message as follows:

- Information security awareness training – an effective way of getting users in a classroom setup. The topic of information security can then be explained interactively.
- Computer-based information security awareness – These are computer applications that are available to users at all times and enable the user to train themselves.
- Awareness services and reminder tools – awareness is an ongoing process, and people need to continually be reminded about information security. Awareness activities that can be carried out include multimedia presentations, security booklets, security posters, and promotional items with security issues.

Each awareness program shown in **Figure 4** will be targeted to a specific audience, and content in each awareness strategy will be tailor-made to suit that specific audience.

Before awareness, people are said to be exhibiting unconscious incompetence, as discussed in the MISSTEV model, where people are unaware of the skills they ought to have. After implementing security awareness programs, they move to a state of conscious incompetence where they have seen the need for security but lack relevant skill and knowledge.

The final stage of the preparation level is the education of the teachers that are going to impart knowledge to the secondary school pupils. This is when the teachers get into a stage of conscious competence, as explained by Alnather (2014)'s model. The teachers must embrace all new aspects of information security they learn to transfer that knowledge to the high school pupils.

At this layer, the information security message reaches its intended destination, the pupil. It encompasses the actual delivery of information security concepts to secondary school pupils. Delivery of information security concepts is done by the trained teachers discussed at the preparation level. At this level, a state of conscious competence is reached, as explained by Alnather (2014)'s model. A level when pupils embrace the subject of information security into being their second nature is reached by continually running awareness campaigns while, on the other hand delivering in-class education. As illustrated by the conscious competence learning model of acquiring new skills in the MISSTEV model, the education process encompasses all four processes of acquiring new skills. The educating process will follow a structured syllabus to cover all essential aspects of information security. Education and awareness are ongoing processes and support each other.

A new concept of IT counselling is introduced at this level to support education. IT counselling is a more intimate way of instilling an information security culture. It accords the teacher a one on one conversation with the pupil in which the former gives advice and clearly explains proper use of information communication technology gadgets. It is mainly targeted to pupils who exhibit deviant behaviour regarding the use of technology gadgets. This has a higher chance of ensuring pupil behaviour change.

Figure 4.

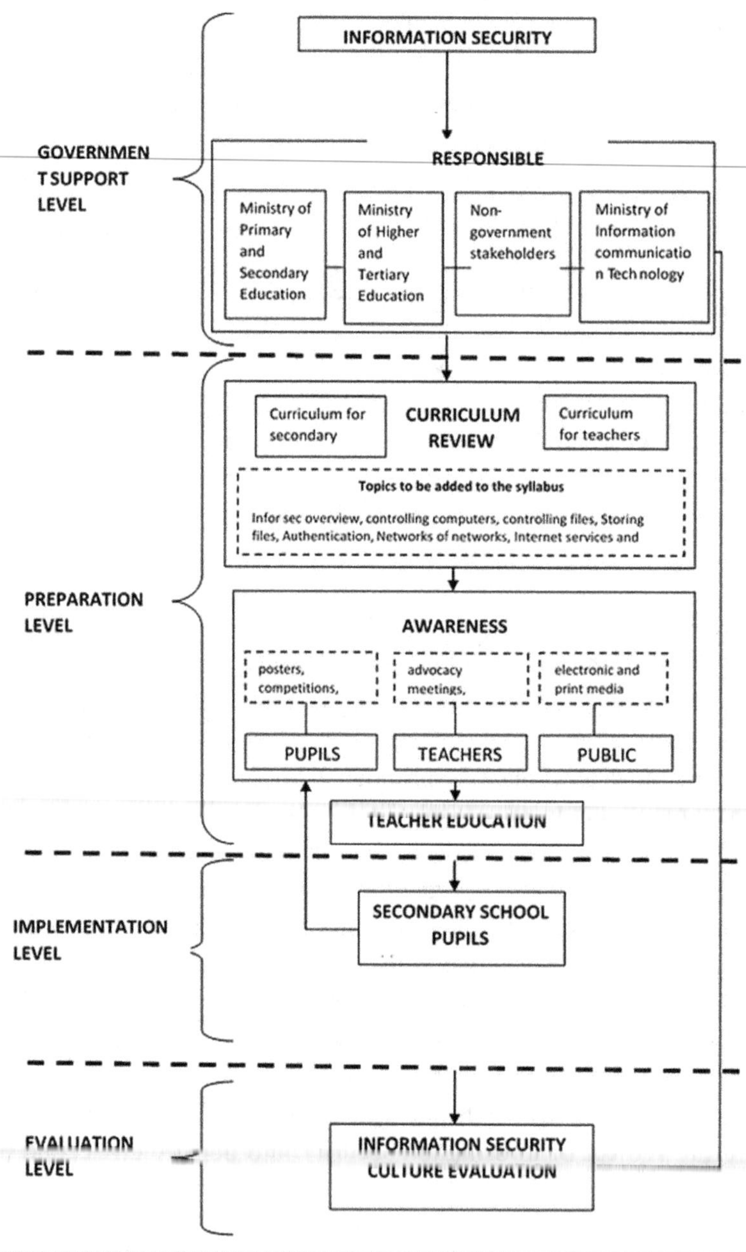

This level aims to establish whether or not the framework is achieving its goal of instilling an information security culture. This level's goal is to evaluate all levels above it and give feedback that can be used for the continual improvement of the framework. An information security awareness survey can be a good starting point. Reviewing and analysing information from the survey will expose shortcomings in the framework and enable the responsible ministry to tackle them. The following is the final proposed framework as a combination of all the levels, as shown in **Figure 4.**

9. CONCLUSION

A culture of information security needs to be cultivated in Zimbabwean secondary school pupils as they are the future workforce in the country's organisations. Such a scenario ensures that by the time they become employees or employers, a culture of securing information will be their second nature. Through the responsible ministries, the government should create policies that govern information security, paving the way for the adoption of the Framework for Instilling Information Security Culture. Noone, to the best knowledge of the researchers, has been carried out on how to inculcate a culture of information security at a stage earlier than secondary school. This entails finding ways to empower pupils to use information communication technologies at the primary school level responsibly.

It is much easier to foster ISC in school pupils as they are still at a learning stage than trying to foster it in adulthood when they have different working experiences. It is therefore essential to ensure that the future Zimbabwean workforce embraces an information security culture to safeguard the country's interests as a whole. At a theoretical level, this study has contributed to the understanding of the ISC by providing a Framework for Instilling Information Security Culture in Zimbabwean Secondary School pupils based on empirical study. The framework provides strategic responses to ISC informed by the centrality of secondary school pupils. Unlike other studies which just looked at an organisational level. Thus, the novelty of this study lies in its development of a context-specific framework informed by the empirical data, unlike previous research, which mainly focused on ISC at the organisational level and ignored the grassroots level.

REFERENCES

Al-Darwish, A. I., & Choe, P. (2019). A framework of information security integrated with human factors. In *International Conference on Human-Computer Interaction* (pp. 217-229). Springer. 10.1007/978-3-030-22351-9_15

AlMindeel, R., & Martins, J. T. (2020). Information security awareness in a developing country context: Insights from the government sector in Saudi Arabia. *Information Technology & People*, *34*(2), 770–788. doi:10.1108/ITP-06-2019-0269

Alnatheer, M. A. (2014). A conceptual model for understanding information security culture. *International Journal of Social Science and Humanity*, *4*(2), 104–107. doi:10.7763/IJSSH.2014.V4.327

Ani, U. D., He, H., & Tiwari, A. (2019). Human factor security: Evaluating the cybersecurity capacity of the industrial workforce. *Journal of Systems and Information Technology*, *21*(1), 2–35. doi:10.1108/JSIT-02-2018-0028

Boote, D. N., & Beile, P. (2005). Scholars before researchers: On the centrality of the dissertation literature review in chapter preparation. *Educational Chapter, 34*(6), 3-15.

Colwill, C. (2009). Human factors in information security: The insider threat–Who can you trust these days? *Information Security Technical Report, 14*(4), 186–196. doi:10.1016/j.istr.2010.04.004

Herath, T., & Rao, H. R. (2009). Protection motivation and deterrence: A framework for security policy compliance in organisations. *European Journal of Information Systems, 18*(2), 106–125. doi:10.1057/ejis.2009.6

Hofstee, E. (2006). *Constructing a good dissertation.* Johannesburg: EPE.

Hu, Q., Dinev, T., Hart, P., & Cooke, D. (2012). Managing employee compliance with information security policies: The critical role of top management and organisational culture. *Decision Sciences, 43*(4), 615–660. doi:10.1111/j.1540-5915.2012.00361.x

Insitute, S. (2003). *Global Information Assurance Certification Paper.* Academic Press.

McIlwraith, A. (2006). *Information security and employee behaviour: how to reduce risk through employee education, training and awareness.* Gower Publishing, Ltd.

Mwantimwa, K. (2019). ICT usage to enhance firms' business processes in Tanzania. *Journal of Global Entrepreneurship Research, 9*(1), 46. doi:10.118640497-019-0170-6

Nonaka, I. (1994). A dynamic theory of organisational knowledge creation. *Organization Science, 5*(1), 14–37. doi:10.1287/orsc.5.1.14

Siponen, M. T., & Oinas-Kukkonen, H. (2007). A review of information security issues and respective researcher contributions. *ACM SIGMIS Database: the DATABASE for Advances in Information Systems, 38*(1), 60–80. doi:10.1145/1216218.1216224

Stewart, D. W., & Shamdasani, P. (2017). Online focus groups. *Journal of Advertising, 46*(1), 48–60. doi:10.1080/00913367.2016.1252288

Tausch, A. P., & Menold, N. (2016). Methodological aspects of focus groups in health research: Results of qualitative interviews with focus group moderators. *Global Qualitative Nursing Research, 3*, 2333393616630466. doi:10.1177/2333393616630466 PMID:28462326

Tayie, S. (2005). *Research methods and writing research proposals.* Pathways to Higher Education.

Thomson, K.-L., & Von Solms, R. (2005). Information security obedience: A definition. *Computers & Security, 24*(1), 69–75. doi:10.1016/j.cose.2004.10.005

Thomson, K.-L., Von Solms, R., & Louw, L. (2006). Cultivating an organisational information security culture. *Computer Fraud & Security, 2006*(10), 7–11. doi:10.1016/S1361-3723(06)70430-4

Wiley, A., McCormac, A., & Calic, D. (2020). More than the individual: Examining the relationship between culture and Information Security Awareness. *Computers & Security, 88*, 101640. doi:10.1016/j.cose.2019.101640

Chapter 12
Perceptual Difference Between Teens and Their Parents About Internet Usage

Suruchi Pandey
https://orcid.org/0000-0003-2752-9468
Symbiosis Institute of Management Studies, Symbiosis International University, India

Swati Tiwari
Independent Researcher, India

ABSTRACT

This chapter is designed to interpret and understand the difference between parents and their children with reference to their usage of internet and its influence in their daily lives. Teenagers between the age of 13 and 16 are the most vulnerable and most prone to cyber bullying and other crimes. Children's effective use of internet relies heavily on the factors surrounding them that not only includes their parents but also their peer and school and neighbourhood. Parents believe that depriving their children of internet access or tools can stop them from accessing the web or restrict their skills and learning about social media. They do not believe that internet use affects family relations negatively. The chapter presents a study undertaken of parents and their children. This presents differences in perception and understanding of internet usage.

1. INTRODUCTION

India's journey on the path of economic reforms has transformed it to one of the world's fastest growing economies. This development and growth have come at its own cost. Internet has become the part and parcel of our daily lives. Especially for the people living in cities from travel to food to education to office to banking – everything is now linked with a smartphone. Children are also not excluded from the grip of this tech environment. Even from before their birth to everyday monitoring of a new born, from school admission to school assignments – everything is now online. And with time we are seeing more

DOI: 10.4018/978-1-7998-8318-0.ch012

children even of very tender age being exposed to digital world in the name of online studies and learning. Post Covid 19, internet usage has spread to every stratum of society including the children of all ages.

Interesting study by Bianchi and Robinson 1997 discussed time children spend on reading or recreational activities. The study also found difference in the time spend on watching television among various groups. Very obvious that time television was more popular electronic device for information and entertainment access. As time changed this source also changed. Chapman, 2002 commented on Internet as a channel with vast pool of resources, scholarly articles and information where it varies a depending upon the contents, aim, target group, reliability etc. User must be very carefully accessed the information and use the information available with due care. The technology has enabled the reach of information, webpage, internet, sites, blogs to every home, every institute and street.

The Internet has a major impact on us and with time its impact are becoming clearer. Specially children where it's providing immense material for learning and knowledge while exposing children to potentially negative content. Internet has penetrated into lives of people and offers great utility. It is important for inclusion of access to information and technology. However this comes with harmful effects to name few exposure to violence, sexually coloured content, hate messages. The websites causing more harm than good include those dedicated to negative content such as pornography, violent online games, online gambling, and so forth. These sites are easily accessible without restrictions. Kids are also exposed to numerous obscene pop-up banner ads and extensive pornographic content on the touch of seemingly innocent key words into a search engine, for example, fly, cock, pussy, Spears etc. For teens, their school research projects demand lots of internet surfing and that's where the risk is maximum. Many sites that play free videos mostly have obscene popup ads that take kids to porn sites. Unaware of the foreseen risks, children tend to explore the dark web.

Despite potentially severe negative impact of internet, parents fail to communicate with their children about the risks involved. Ignorance is to the extent that parents are not even aware of what their children are doing online. There is a huge gap between what parents think their children are doing online and what their children are doing. There are major concerns that inappropriate and explicit Internet contents may jeopardize the health and safety of our children. The study of risks of exposure can only be understood by knowing the level of exposure, what, when and how much are the children exposed. It helps identify the extent, direction, angle and mental impact internet is having on our future generation. Therefore in the landmark study done by Malta Communication authority commissioned by University of Malta, recommended supervised or trained use of internet.

2. REVIEW OF LITERATURE

Review of related literature is an important characteristic of research. It comprises of a synthetic and a synoptic understanding of research work undertaken previously in the same area over a period. It serves as a guidepost not only about the quantum of work, but also enables one to perceive the gap in the concerned field of research and thus empowers the researcher to improve his own investigation. According to the ITU (The World Telecommunication/ICT Indicators database), the Internet passed the 3 billion user milestone in 2015, with just over 3.2 billion users worldwide by the end of the year. Official statistics for India show that 28% of Indian population is not connected. The total number of users has increased to 350.48 million at the end of June 2016. % of children having internet access either at home/ school or through friends has also grown drastically over the past few years. Number of Facebook users alone in

India is over 195 million (2017). Current data suggests that almost 545 million users have internet access with a penetration rate of 45%, age group 12 to 15 made almost 14% internet user. The report also mentions digital gender gap by having 70:30 ratio of male to female ratio among users (Keelery 2021, statista.com). And now with Service providers offering very cheap/ free internet, its access is seen on every smartphone, tablet, gaming device and PC. For the modern society internet has become the most powerful socializing medium as well. Post covid 19 internet usage increased as life shifted to digital mode including education system. It is suspected that in the past few years the number of hours children spend online has increased multifold with respect to the number of hours spent with family and friends or for any other social activity.

Internet is very much part of life now for everyone like electricity, food and cloth. It is a source of information and entertainment. It is enabler of facilities and knowledge. Still majority users are madly addicted to this technology, in which young generation, men and women are also involved. About millions and millions of user are growing to use internet on daily basis. Access of Internet has becoming widespread; affordable by everyone, easy and comfortable to use.. Not only in advance country but internet is also in boom in developing countries. Internet does not discriminate user. It provides the same information and content to user across globe. The same is the strength and weakness of using internet.

The study conducted by Kubey, Lavin and Borrows (2001) mention that about "68% of parents and 70% of teachers said that they have shelter higher grades of students because of internet usage". Internet is appeared to useful in areas user can use it for entertainment or education, business or travel, socialising, or blogging. Kumar and Kaur (2006) found that about 69% of the Users are using internet only for entertainment purpose. D'Esposito and Gardner 1999 among other following authors discussed internet use in connecting people, culture, and education. Junni 2007 established relationship of Internet usage of masters students with the quality and quantity of information required for thesis writing.

Internet as resources are also being utilized to access information and gain knowledge. One can know about planets and universe even without stepping in class. It has changed the landscapes of classroom learning (Pandey and Tiwari, 2014). Many other authors in their research and blogs emphasised that internet can also be used for scholarly purpose, contact information, purchase of travel tickets, communication, and entertainment. This usage may vary from country to country for various reasons of access.

Weitzner (2007) questioned reliability of information available on internet. Pirated literature, fake news, misrepresented data, biased blogs, privacy breach are cause problems for educators, students, and Researchers. Numerous fake profiles and ids exists that provides misleading information. Also search engines may show up the result based on algorithm or optimisation. Users in education, students, teachers and researchers are getting depending on the information available online while the same is found non reliable. This creates confusion and at times found misleading knowledge being passed on. In study conducted by Luban (2000) it was found that student's performance could be linked to the access of information they had from internet. Quality of written work done by undergraduate students showed the positive influence of internet users. However, it was further questioned that students are not able to discriminate right or wrong content and correctness of information.

Most of the time online is spent on exploring (surfing and searching), followed by games, and e-commerce, learning and homework- education related, then comes music, video, etc., that's multimedia, e-mail, chat, and instant messages – mostly for communication. And most of this is in the comfort of their home with full privacy and comfort using connectivity through laptops/ PC/ tablet/ smartphone etc. that have complete access to the world wide web. There have been several claims of a possible negative impact of the Internet on children's daily lives. Negative impact on the children, Mid-teens and adolescents have been concern of parents and schools. These concerns include violence, disrupted social relationship and exposure to sexual content. Reduction in physical activity, leisure time and spending quality time with friends, elder members and family are also observed. Hyper active, attention deficiency, lack of concentration, depression and feeling of loneliness are few other consequences of internet exposure and time suggested Kraut, Patterson, Lundmark, Kiesler, Mukophadhyay, & Scherlis (1998). LaRose, Eastin, Gregg (2001) established social support linkages to decreased depression. Inline Wästlund, Norlander, & Archer (2001) also argued psychological wellbeing will reduce with internet usage not necessary in proportion of usage is directly to be linked with psychological wellbeing. The fact remained physical, social and psychological wellbeing of children is questionable with excess and irrelevant exposure to internet. The discussion of such issues was conducted in the study of Gross, Juvonen, & Gable (2002) and Robinson, Banda, Hale, Lu, Fleming-Milici, Calvert, & Wartella (2017).

This resource has become a double-edged sword for children – a necessary evil. The concern for parents, teachers, counsellors On one side is School books, assignments, notes, projects, research access, socialization, entertainment, and a communication tool with families, and on the other end is exposure to violence, pornography, hate sites, isolation, predators, and commercialism. Pornography, violent online games, online gambling have been found to be the most harmful medium online affecting the entire generation of internet enabled young community. Unwanted and unexpected exposure to explicit content online even through simplest of search words is a huge problem for the parents and control on the same has become trickier by the day (Çankaya & Odabas, 2009)

Lancaster and Sandore (1997) outlined the roles of the Internet. Articles, books, blogs, social websites achieve are useful as library content. Electronic library as repository of e books and e journals and getting popular among teachers, students and researchers. Kamja 2008, discussed that internet acts as bridge to knowledge gap. Making information available to vast users and person needing that information. This was also supported by Adegboji and Toyo (2006) in their study about lack of information, inadequate resources and outdated material is leading to knowledge gap.

Hatch (2011) found out students believed that all actions were appropriate at younger ages. The first question asked what age it becomes appropriate for children to play portable video games while out to dinner with their families. "A resounding ninety-five percent of adults stated that it was never appropriate, some of them even laughing or scoffing at the question. The remaining five percent was distributed among the lower age categories ranging in ages from 1 year old to 14 years old. When the students were asked the same question, only 76% said it was never appropriate and 15% said it becomes appropriate when the child is between the age of 1 and 6."

"The student's leniency was displayed even further when asked at what age it becomes appropriate for children to text while out to dinner with their families, or during family time. Again, the majority of both adults and students said it was never appropriate at 89% and 68% respectively. However, 25% of students felt it became appropriate when the child was fifteen years old, and 7% thought it was appropriate when the child is between the ages of 11 and 14, garnering a total of 32%."

Awais, Bilal, M Usman, M Waqas, Sehrish (2013) in their dissertation found children use internet for data collection and information retrieval most of the time. There is an increase of 1-3 kg weight in them if they use internet for over 2 hours per day. Majority of the students have been using internet for more than 3 years and also many rate their internet skills to be high.

Daraha (2013) found that "there is a significant relationship between public high school and private high school by using the internet on three broad aspects: physical well –being, cognitive and academic skill development and social development and relationship. There is a significant between high school students 'level of using the Internet." Its almost uninterrupted exposure of internet for almost 5-6 years. Most of the high school students are well exposed to technology and internet for well over 3 years. "It was also discovered that over fifty percent of the high school students had eye strains, back pain and headache (tension or migraine) due to spending long hours in front of the computer screen. Clarke (2003) and Wettlaufer (2003) discovered that aching, burning, or stabbing pain in back arms, wrist or fingers were symptoms relating to Repetitive Strain Injuries (RSI) caused by repetitive activities of computer users." Chang (2002) also stated that the time sitting at a computer can aggravate existing back problems or prompt new ones.

Subrahmanyam, Greenfield, Kraut, Gross (2001) discussed "The impact of computer use on children's and adolescents" development state Understanding the impact of computer use requires good estimates of both the time children spend on computers, and the time taken away from other activities. Time use data on children's use of computers has been gathered mostly through self-reports and reports by parents. "Despite their overall usefulness, particularly for sampling a large number of people, self-report data are beset by problems of accuracy and reliability stemming from memory limitations and inaccurate estimations on the part of respondents; these problems are further accentuated when studying children."

Serious concerns from parents were raised in survey report on Pennsylvania (Woodard & Gridina, 2000). "Teens in the study were much heavier users of the Internet and all its services than were their parents. Media time at home was concern and parents were worried about guidance." Small children also spend long time in form of televisions, video games, computer and use internet. Media has penetrated the bedrooms spoiling the normal life. Similiarly Yusuf,, Osman, Hassan, Teimoury (2014) highlighted need for parents trust on children with respect to internet usage. The study suggested that Malaysian children are aware of the risk associated with the internet usage and parental guidance will be helpful to them.

The teens used the Internet for multiple purposes. They use net to do their school projects, for chatting and communication with their friends from neighbourhood and also the ones they meet online. They use it to have fun, especially by finding information related to their interests and hobbies. They were also more likely to use the Internet to listen to music, play games, and download software. Teens also used the Internet to explore and find answers to most of their unanswered questions – that they are either shy of asking others or to just quench their curiosity. Teenagers are more likely than adults to report using the Internet for social purposes. To extend take example teenager may like to use internet and devise to play online, be on social media to interact with friends, get help or join interest group . Adults may have constructive purposes like purchasing, paying, information or job search. Pandey, Chawla and Mathur (2021) in the study reported feeling of isolation and deprivation from human interaction while studying online for PG students.

Rideout, Foehr, Roberts, & Brodie (1999) found that "the time a particular child spends on a computer and their activities on the computer may depend on age, gender, ethnicity, and social class. In a national survey of children and teenagers from 2 to 18, the percentage of children who reported (or were reported by their parents) to have used a computer out of school the day before rose with age: from 26% in the 2 to 7 age range, to 44% among the 14 to 18 year old. Interestingly, while more boys than girls reported using (or were reported to use) computers in school the day before, there were no gender differences in percentages using a computer out of school." It was also found that difference in the time of accessing internet exists . Difference in race was also one of the factors discovered.

Difference in the access of internet games was associated with the gender and boys were found having more access as compared to girls in teen age. (Subrahmanyam, Greenfield, Kraut, Robert & Gross 2001). Difference in the Usage of internet existed in purposes of home study, emails and any other aspects

. Other evidence suggests a more even gender distribution in nongame uses of the computer. Roberts et al 1999 supported evidence of difference among teen aged boys and girls. The differences is found usage of computer and interest varying in its purpose and hours. Some further research also looked at gaming as computer beginning to computer literacy, the prominent question remained the same of having gender difference in computer gaming. To encourage gaming among girl the core of the game was changed by software designer on inputs of barbie fashion designer. This picked up gaming among girls as was discussed by Subrahmanyam and Greenfield (1998). All these done to lure and temp children to stay online longer and learn more is a positive stand. Designer along with the growing popularity of the Internet among girls suggests that they are not turned off by computer technology. Instead, they just need applications that appeal to their interests.

2.1 Internet and Its Effect on Children

There is an increasing concern about the negative influences of using the Internet on the physical, mental and social development from educators, psychologists, and parents. Fears of its impact on children's health, mental development, skills and level of patience has become an everyday problem. One of the most serious concerns including the availability and easy accessibility of all kind of content. Exposure to harmful content online of all sorts like pornography, violence, hate speech, gambling, sexual solicitation etc. These types of negative content harm children and destroy their development. They may not have anyone to discuss or guide with this kind of content. At times they may feel shy to ask for help in such a situation. This leads to various sorts of ill effects ranging from poor academic performance, to aggression, feeling of being deprived, fear, insensitive behaviour towards people and other living being. Taking the list forward it can even cause psychological trauma, lack of self confidence antisocial behaviour, negative self-attitude, identity crises, non-acceptance of reality. Furthermore, the anonymity of the Internet makes it easier for paedophiles to approach children through online chatting. These are formative ages of transforming into a responsible citizens of world. They are neither matured nor child. At such a time Sexually explicit materials on the Internet can desensitize children to deviant sexual stimuli and encourage them to enact antisocial aggressive or sexual behaviours. Occasions of social stalking can make children uncomfortable in their social interactions with world outside. Children who spend hours in chat rooms looking for friends, blind dates, time engagement are easy target of sexual offenders and get prey to them. This leads to several other gender issues in the society. Online gambling has also been cited as a serious Internet problem affecting children, disrupting a child's emotional and social development. They may ask parents to pay for the games or get into theft mode to meet their financial needs. These negative effects of violent games on children are even more serious as their access is much easier these days. Violent online games are another major concern. It is known that violent computer games increase children's physical, verbal, relational, and antisocial aggressions without realising boundaries.

William Powers (2010) author of Hamlet's Blackberry, was cited by many authors for his book . His description of technology and its impact on world around presents interesting facts. He mentions "our new technology driven world as a giant room, in which everyone is standing within reach of their neighbours. In this room, people are constantly poking one another and then asking them questions or telling them about their days. In this world, it becomes impossible to escape the constant connection with others, aside from completely disconnecting from it, and leaping out into the unknown, or what's outside the room." He further stated that all of us live in this room i.e our world we are connected all the time with each other. His words made many think about critical points he is trying to address.

2.1.1 Negative Effects of the Internet

1. Addiction to Games

Children, young adults and adults get addicted to the online gaming. The games are designed in a way that player gets fully engrossed in it in a flow without even realising the time . This takes away the opportunity time and cost of doing something which is more productive and healthy option.

2. Access to Pornography Material

This is really threatening to know how internet has rampant pornography material easily available. The delicate minds of adolescents and Mid-teens get negatively exposed to adult content. These content can come to them in form of advertisement or send for fun. Personal encounters are made public to spoil the image of the public personalities with malicious motives.

3. Away From Human Touch

The biggest threat online socialisation and internet access is taking away people from human. By spending more time on media real people in life are miss. Virtual meetings, celebrations are considered more important than knowing own neighbours and people in the surroundings.

4. Being Trapped in Phishing or Victimised

There has been increasing in online crimes through phishing or getting fooled by illegitimate websites. Confidential and private information leakages leads to financial frauds online. Identity theft along with financial crimes are other alarming issues of being online too much.

5. Neglecting of Family

People loose sensitivity while being absorbed on internet. They are guided to wrong doings and forget the reality of life. In one instance, a couple in Korea was preoccupied with their virtual baby forgetting to feed their real baby who died of hunger (CNN news 2016).

Technology and Internet has both positive and negative consequences on users. To great extent it has become reality of life however security, controls and restrictions have to be enabled to make Internet a secured place to be and do the constructive work.

2.2 Parents Exposure and Need for Internet

For most of the parents who are resourceful and enlightened, Internet comes as a handy tool taking care of most of their work processes. Benefits of technology has surpassed the negative and therefore there is increasing use and demand of tools and resources using internet.

1. Information Access and Research

Before electronic means of gathering information, students or researchers need to pay a visit to libraries to access materials relevant to their research topics. Nowadays, it will take just a subscription to a journal online or just browsing online to search for literature required to develop one's conceptual framework.

Preparing the review of literature is easier and faster. Thus, more research gets published and knowledge generation is greatly facilitated.

2. Communication With Family, Friends and Relatives Is Faster

Gone are the homesick days because a son or daughter can now easily contact their parents online. Services such as that of meet, Skype and whatsapp enable a loved one to communicate with another in almost any part of the globe where internet services are available.

3. There Is a Great Possibility to Earn While Working From Home

It is now possible to earn online through writing, filling up forms, or even just clicking away on advertisements in the internet. Information is a valuable thing that people are willing to pay. Through advertisements posted online, manufacturers and sellers are able to increase the possibility of having their products purchased.

4. Faster Business Transactions and Cheaper Products

Sellers and buyers can now transact through online paying facilities like banking apps, Paypal or credit cards. Buyers can surf online and select the products they want to buy. There is no need to have a physical store to maintain as the products can be shipped directly from the factory.

5. Savings on Travel Cost

Using the internet, it is possible for executives or business managers to do teleconferencing. They do not need to meet physically just to discuss matters. This could save a lot on travel cost.

2.3 Common Utilities and Knowledge Sharing Through Technology

There are multiple utilities where parents and children are on a common platform and use internet together. From social media to mail accounts to game zones both parent and children have been known to be working together- sharing and tagging each other on events and occasion. All this is very interesting and fun so long till it does not affect the privacy of the other person. When a person is tagged, information like his presence at a certain location, at a certain point of time doing certain activity is relayed and this can be used by others to gather close and personal details about him.

Parents share their passwords of various account with the children which is yet another threat as children are known to have accessed their parents accounts which may have some very essential information and access details. Essential mail login details, bank details, transaction passwords are all stored in mobiles for convenience and at times are misused by children Cases where the child has used the online bank access to make in app purchases for their games are not uncommon. There are also instances where children for fear of being punished also end their lives.

2.3.1 Variation Between Parents and Children Usage

With its presence around us, we are known to be online almost all the time. Even during sleep our mobile is working for us recovering messages and delivering notifications stored by us for later.

There is and should be a variation between the usage of an adult and a child. However, these days children surpass the usage of internet. They are also seen online most of the time, if not for chatting or playing games then it is for some academic use as schools also promote use of technology and from questions to guides to notes – all are shared on school groups or pages.

2.4 How Is This Usage Going to Affect Our Future?

Internet is still a new tool that's evolving with time. It's very dynamic and is spreading around us with much faster pace than imaginable. Now even electronic gadgets around us are controlled online. Internet usage has suddenly multiplied in the last 10 years. The world is changing and has fast pace movement towards global connectivity that will make changes in how and where people associate, gather and share information, and consume media.

The Internet usage is effortless and most people tap into it so easily. It flow through their lives "like electricity." Mobile, wearable, apps, Televisions, Alexas' and embedded computing is tied together in the Internet of Things, allowing people and their surroundings to tap into artificial intelligence-enhanced cloud-based information storage and sharing.

Some of the trends that experts predicted and have grabbed world around Internet quickly:

- A global, immersive, invisible, ambient networked computing environment built through the ongoing supply of smart sensors, cameras, software, databases, and massive data centres in a world-spanning information fabric known as the Internet of Things.
- "Augmented reality" enhancements to the real-world input that people perceive using portable/wearable/implantable technologies.
- Disruption of existing business models established impacting every segment education, finance, entertainment, media, retails, hospitality etc.
- Tagging, databasing, and intelligent analytical mapping of the physical and social realms.

Disruption in use of technology is also due to post pandemic digital new normal. Technology and digitisation impacting negatively and positively. This will transform almost all sorts of human interaction in every aspect like healthcare, education, work environment, political campaign, banking, entertainment and shopping experience . Technology is considered to be biggest boon for mankind and helping people connect. It has simplied life for everyone and make learning click away for. Positive effect of technology can be seen and been predicted however, when it comes to describe the good and bad aspects many of the experts can also clearly identify areas of concern, some of them extremely threatening and intimidating. Heightened concerns over interpersonal ethics, surveillance, terror, and crime, may lead societies to question how best to establish security and trust while retaining civil liberties.

2.4.1 Internet Governance

Internet governance is an area where developing counties like India can, and should, significantly improve its role, presence and influence given the fact it has a large community of internet users. An even larger community here would start using it in the next several years. Governing internet is a complex and challenging task given its globally distributed nature, lack of centralised authority, jurisdictional issues, diversity of stakeholders and services and plurality of content.

Although the multilateral model - loosely structured based on UN system - had been proposed, it is the multi-stakeholder model that is accepted today as the most appropriate. Here, all stakeholders - businesses, governments, research institutions, non-government organisations and civil society - are given formal roles in the governance process. It is thus possible for any user of the Internet located anywhere on earth to provide his or her inputs to the Internet policy and governance processes.

Every country needs its own laws and range to make sure the guidelines are implementable and usable. The norms should be in accordance to the local usage guidelines.

3. SIGNIFICANCE OF STUDY

Society is sincerely unaware of the actual amount of exposure our children have to vulnerable and inappropriate content and activities on the Internet. Parents are not even remotely aware of children's activities online. They tend to underestimate the time spent online and the type of content to which they are exposed. The difference in awareness of parents leads them to underrate the impact of internet on their children. This cause limited control and more risk for the children. We have little on records about the possible discrepancy between parental estimates and children's actual Internet usage. With this ground we are trying to detect the level to which children are exposed to these sources of negative content and how much do parents know about this exposure. This study shall provide a concrete platform to understand how children's exposure to such negative Internet content affects and impacts the children's physical, psychological, emotional and social wellbeing.

This study becomes a prerequisite to the understanding of parental impact and control on their child's overall development. Children cannot be devoid of internet access and limiting or controlling access is yet to be established in a concrete manner. Children live within a family boundary; therefore, parental influence on children's media usage and effect is very important.

The outcomes of this research will add to the body of knowledge about the difference between how parents and children look to the modern tools and how they use/ misuse or even abuse access to online resources. Additionally, the findings of this research study can be used by parents/ educationists/ counsellors/ psychologists by understanding how much children are already exposed to and to some extent exposed themselves to vulnerable elements online. Privacy factors of not just the individual but the whole family is compromised by one individual who can share confidential information about oneself or family. Shy and meek ones will have an opportunity to open up and have a conversation without worrying. The awkward silences and lack of conversational topics can well be resolved.

Recent instances of suicides by youngsters while playing online game is a proof that they have access and resources that can harm them badly without the knowledge of their parents. There is much needed understanding and communication needed between parents and children that is seriously lacking and for their answers children are looking up to a source that can lead them into a world that is more scary than imaginable. Children are in a developmental stage and are ending up spending more time in front of a computer using the Internet. Their claim that using it for educational or informational purposes is arguable, all the more with post pandemic education system moving online across the world. The fact that there is no specific regulation about the Internet, nor proper knowledge and guidance of children's Internet use by government, school, or parents, induces more concern.

With modern requirements, convenience and access, Internet has penetrated deeper into our lives than we ever thought. Unfortunately, being a relatively new and a very smart tool, both parent and children are learning to use it at the same time in most of the houses. However unfortunately the negative effect on children is much worse and the main challenge observed is parents being unaware of what children are doing online or what are they being exposed to.

Hence, it is extremely essential that we understand the level of exposure and information delivered to our children. Right level at the right time is the key. Wrongful exposure and over exposure can have a severe impact on their overall development.

This study is undertaken to identify the possible causes and consequences of perceived parental control over children's Internet usage. Recognizing the possible underestimation of, or ignorance about, their children's Internet usage and its effects, parents are unaware of the ever-serious negative aspects of children using the Internet. This study explores the degree of exposure that children may have to negative content and detects the possible discrepancy between what parents think their children are doing online and their children's actual activities.

Present study aims:

- To examine children's access, use and perceptions of the Internet.
- To examine the parents' perceptions of their children's Internet use
- To compare parents' perceptions with patterns reported by children.

3.1 Scope for Study

As seen earlier it is visible that it is essential to understand the variation between what we believe about our children usage and what they actually do online. Checking through all the resources available it is clear and obvious that not much study has been conducted to understand this difference in perception between parent and children.

This is first conscious effort made to understand and compare Indian parents with their school going children in the age range of 13 – 16 who have computer access in some form or the other. There are various parameters that can be studied in detail and the variation can be analysed.

This needs to be converted into a country wide analysis where sample schools from each zone / region considered and parents along with children studied for their difference in their belief, knowledge and understanding.

4. OBJECTIVES AND METHODOLOGY

The descriptive research method are used used for this research specially to understand the difference.

4.1 Problem Statement

With modern convenience and access, Internet has penetrated deeper into our lives than we ever thought. Unfortunately, being a relatively new and a very smart tool, both parent and children are learning to use it at the same time in most of the houses. However unfortunately the negative effect on children is much worse and the main challenge observed is parents being unaware of what children are doing online or what are they being exposed to.

4.2 Objectives

The primary objective of this study is to access the exposure of children (mid-teens) to internet content by understanding their perception.

1. Perception of parents about use of internet by their children (mid-teens).
2. Exploring the discrepancy between parents and children about internet usage.

The questions this research intents to answer are -

RQ. 1: How exposed are our children to Internet?
RQ. 2: What do parents know about their children's internet use?
RQ. 3: How much is the discrepancy between internet use by children and parents' perception of the same?

4.3 Hypothesis

To fulfil the above-mentioned objectives and to deal with the research problem, hypothesis was framed for this research study.

H0: There is no significant difference between Parents perception and Children's Internet usage.
H1: There is a significant difference between Parents perception and children's Internet usage.
Variables – There are 2 variables that are involved –
Independent Variable – The parents and the children.
Dependent Variable – Parental perception about internet usage with respect to their children.

4.4 Sampling and Participants

All the items under consideration in any field of inquiry constitute a "Universe" or "population". The researcher must decide the way of selecting a sample – also known as sample design. Before any data is collected from a given population – a plan is designed and accordingly the sample is extracted.

Clustered Random sampling was used by selecting the participants from their school. In children records, as per the request of the school name was omitted, however age and gender were recorded in the questionnaire. For parents too, the personal information collected was limited to parents age group and gender. To avoid any complexity names were totally omitted. For parents there was an option to tick the working vs the home maker however, most parents did not tick on either therefore making it anon usable.

Though recorded gender has not been taken as a parameter in study. This can be used for further analysis at a later date. Sample size was limited to a short number of 30 sample population all seated in the same class for answering the question.

Personal visit to a private school after scheduled appointment and planning was made. This visit was linked with a parents and teachers meeting (PTM) at the end of year where both parents and students were present. Pre-coded Questionnaire was distributed to children (from 10-15 years of age, referred as mid-teens) and their parent (father or mother). This questionnaire was the tool used for further study and analysis. Appropriate instructions were given to both parents and children about filling the questionnaire. The message included "This questionnaire is essentially to capture the internet usage and preferences. Please feel free to choose the answers that you believe is true from your perspective. Parents have their own separate set of questionnaires that is independent of their children. There is no time limit to answer the questions, but generally one can complete the same in 15-20 minutes."

4.5 Analysis

A well designed and developed questionnaire is important to achieve the objectives and purposes of this survey. To collect the information, a questionnaire was constructed by the researchers. Items of questions chosen were from the review of literature. Questionaire comprised of 30 variables / Items to seek the perception of the parents as well as children. Each question had Yes/ No response . Information related to participant`s Age, gender, preferences were collected for the deeper analysis.

The primary data collected was sorted, classified, edited, tabulated in a proper format and analysed by utilizing appropriate statistical tools. The researchers use Windows Excel Spreadsheet for recording and classification of samples. The following statistical tools will be used for analysing the data procured from the respondents from different locations selected for the study

Simple percentage analysis is one of the basic statistical tools which is widely used in the analysis and interpretation of the main data. It deals with the number of respondents' response to a question in percentage arrived at the total. Simple percentages was used in the study to analyse the factors like demographic and other details of the respondents. Data collected are edited and coded by using the tally bars. This helps in converting the gathered data into a tabulated grouped data. Percentage Analysis is applied to create a contingency table from the frequency distribution and represent the collected data for better understanding. Each question is referred as set (capturing response of both parent and child).

5. RESULTS AND DISCUSSION

Results shows that there is significant variation between parent's perception and children's actual internet usage. There is a significant difference between Parents perception and children's Internet usage.

Two of the most important results of the study are that the internet in general, and social networks, are very much present in the lives of our school going teenagers. Between the ages of 13 and 16 (referred as mid-teens). Parents' perceptions of their children's internet use is, in several cases, is very different from what internet use is.

1. Most children participating in the survey between 13 and 16 have internet access in some form or the other. 83.33% parents said that their children have availability of internet access through some

owned/ shared devices. The devices children used mostly for internet access are computers, laptops, tablets, and most of all mobile phones. 50% of children use the internet more than 4 days a week. 63.33% of children report that they use the internet more than 2 hours every day.

2. Parents (53.33%) think that their children use the internet for schoolwork much more than their children (66.67%) do. On the other hand, children engage in several activities online more than their parents believe they do. Such activities include chatting, social networking, emailing, video calling, downloading films or music, watching videos online, blogging and online shopping.
3. 13.33% Parents believe Children put fake information online to create account or to use social media. Parents and children also differ in their assessment of risk. Parents (16.67) think that their children interact with strangers online. While only 13.33% have accepted doing the same.
4. The public posting of things on the Social Networking Sites provides another divide: 33.33% parents believe children share location and other details online while only 23.33% of children agreed to be doing the same.
5. 33.33% children believe Internet is a safe place, while only 23.33% parents believe the same. 20% parents believe it is safe to share personal details online while only 13.33% children think it is safe to share such details. 26.67% parents agree that their children know their account passwords. 24% children agreed to knowing their parents' passwords. These parents also claim their children have accessed their accounts. 93.33% children have declined doing the same.
6. Both parents and children have been of the same consensus on their children accessing any other person's account. 86.67 of both parents and children have declined doing the same. 33.33% Parents believe their children stay online for too long, while only 26.67% children agree that they do stay online for long.
7. There is a very interesting observation when it comes to talking about internet safety with their children. Where only 30% parents agreed they talk to their children about internet safety 53.33% children said their parents do talk about it. Looking at the effect of internet on academic performance 76.67% parents agree that internet does affect the academic performance of the child. 66.67% children also agreed that internet affects their academic performance.
8. 46.67% Parents believe internet has affected the social life of their children. They are more on internet than playing outside in the open. 33.33% children also agreed that they do enjoy being on internet more than going out and playing with friends. 83.33% parents believe children come to them for their online issues, while only 70% children agreed to going to their parents for support concerning online issues. 30% children claimed they prefer going to friends for their online issues.
9. 63.33% parents believe their children do not know how to block spam. 60% children did agree to the fact that they do not know how to block spam. 36.67% parents believe their children do not know how to delete history while 26.67% children confessed, they do not know how to delete history.
10. 73.33% Parents welcome the app notification settings of their children phones while 60% were actually happy with their app notification setting. 26.67% parents have changed their app settings whereas 46.67% children claim their app settings are changed by their parents.
11. 50% parents believe their children know how to change privacy settings so that only their friends can contact them, 56.67%children agreed to knowing how to change privacy settings to secure so that only friends can contact them online.
12. Perception of parent and children also varies with respect to the usage time are presented in the table -1. The aggregate perception score about time match among children and parent.

Table 1. Perception Variation On time usage.

How Many Hours Do You Think Your Child Is Online?	Less Than 2 Hours		More Than Two Hours		Total	
		Percent		Percent		Percent
Children:	19	63.33%	11	36.67%	30	100.0%
Parents:	19	63.33%	11	36.67%	30	100.0%

Variation between Parent and Child is 0.

13. 46.67% parents believe their children are able to surf the net and are able to find what they are looking for. Same 46.67% children also believe they can find required data online and decide what information is true and relevant to their requirement. Only 13.33% parents fear their child may have shared his password with his friends, but almost 20% children agreed to have done the same. They did the same to edit, delete or change a post. 43.33% parents and children claim parents know the passwords of the child's account. 56.67% parents do not know the password of their child's accounts.

14. The table-2 indicate response of parent and child to access of parent account. The variation is almost 17%.

Table 2. Perception Variation On Parent account access.

Has He/She Tried to Access Your Account?	Yes		No		Total	Percent
Children:	2	6.67%	28	93.33%	30	100.0%
Parents:	7	23.33%	23	76.67%	30	100.0%

Variation between Parent and Child is almost 17%.

15. 30% parents agreed to the fact that their child has visited a restricted site at some point of time. 26.67% of the children accepted that they have visited restricted sites. 43.33% parents accept that their children know about dangerous games like Blue Whale, while 56.67% claim their children do not know of such games. 46.67% of children agreed to know such games. Variation here is a huge risk to the young generation who are unaware of the threat of such games.

16. The table-3 presents perception variation on discussion about internet safety and restricted sites among parent and children.

Table 3. Perception Variation On Discussion about Internet Safety and Restricted Sites.

Do You Talk To Your Child About Internet Safety and Restricted Sites?	Yes		No		Total	Percent
Children:	16	53.33%	14	46.67%	30	100.0%
Parents:	9	30.00%	21	70.00%	30	100.0%

Variation between Parent and Child is almost 20%.

17. 10% Parents agreed their children have seen obscene and violent images inline while 13.33% children agreed to have seen such images. Only 6.67% parents believe their child may have been treated in a nasty way online while 13.33% children agreed to have been treated in a nasty way online.
18. 70% of parent population believes online games can be threatening and dangerous for children while 56.67% children believe the same. 43.33% population children believe online games are same and not as dangerous as predicted.
19. Table -4 below presents variation on online games being threat and danger. It is evident that less number of children consider it to be threat as compared to the parents.

Table 4. Perception Variation On Online Games being threat and Danger.

Do You Believe Online Games Can Be Life Threatening or Dangerous for You?	Yes		No		Total	Percent
Children:	17	56.67%	13	43.33%	30	100.0%
Parents:	21	70.00%	9	30.00%	30	100.0%

Variation between Parent and Child is almost 13%.

20. The variation with in the parents and children perception kept varying for each questions. The blow figure -1 is the pictorial presentation of the same.

Figure 1. Number Of Answers variation between parent and child.
X Axis – Number of Participants
Y Axis- Variation in the number of Answers.

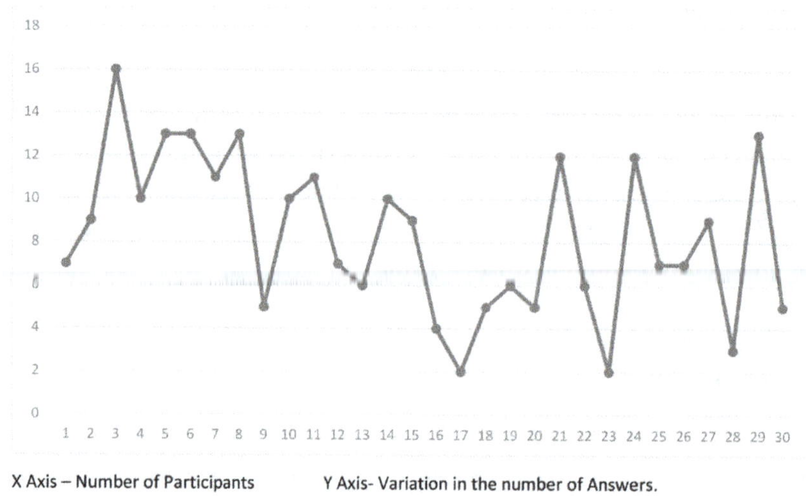

21. Analysis of Variance Between Parent and Child Perception - This segment discusses the variance among the response of child and parent for each question set. Variance of response among the parents and children and also presented for discussion purpose.

Set 1 - Parent and Child had a Perception variation of 7 out of 30 questions. Their difference was 23.33%. Parent knowledge of the number of hours the child is online was different. Also the child believed that internet is a safe place through parent disagree. Parent believe the child stays online for too long though the child disagrees. They also believe the child's academic performance is affected by it staying online for too long, against what the child believes. Where parents believe internet can be a life-threatening place the child disagrees.

Set 2 - Had a perception variation of 30%. Where the parents believe the child is not talking to strangers online – the child agreed to be doing so. They believe their child has never accessed anyone's account, but the child agreed to have accessed another person's account. While parents agree staying online affects their child's academic performance – the child disagrees to the statement. Parents believe the child discusses internet safety issues with them, but child claims he is comfortable discussing them with friends. Parents believe their child has not seen violent/ sexual images online – the child accepts to have done that. He also agrees to having been bullied online of which his parents are not aware. Again, the child believes online game are not threatening but the parents think otherwise.

Set 3 - had a huge perception variation of over 53.7% between parent and child. Here the parent believes the child spends more time, but the child claims he is online less than 2 hours a day. Parent claim the child has a fake account with incorrect DOB or other information, but the child has declined. Similarly, parents believe the child has made online friends they do not know of but the child has declined. Parents believe internet is not a safe place for children, but the children think otherwise. They even believe child knows their account passwords and has accessed their account. They have not spoken to their child about internet safety, but he claims they have. They also believe the internet has affected their child's performance, but the student of not of the same opinion. The student even agrees to have shared his password with his friend which theparents are not aware of.

Set 4 - had a perception variation of 33.33% with 10 out of 30 questions answered differently. Wherein the parents acknowledge the child has access to internet device the child claims to not owning or sharing one. Parents also claim he shares his location when online which he declines doing. They believe staying online affects the child's academic performance and they talk to him about internet safety, but the child declined either. Parents believe he does not know how to block spam or change his settings – the child claims he is aware of these.

Set 5 - had 13 answer variation that was with 43.33% variation. Set 6 - also has 13 answer variation. Set 7, 8 and 9 had 11, 13 and 9 answer variations. Set 10 had 33.33% and Set 11 had 36.67% variation. Set 12 had 23.33% and Set 13 had 20% variation with 7 answers different between parent and child. Set 14 had 10 answers different and Set 15 had 9 answers variation.

Set 16 has 4 answer variation.

The smallest variation has been observed with 2 answers difference between parent and student. This is in set 17 where the parents believe child accessed their account using their password and the child declined the same. Also that child knows how to secure privacy settings which they declined.

Set 18 and 20 had 5 answer variation while Set 19 had 6 answer variation. Set 21 had 12 answers difference. While Set 22 had 6 answers that were different, Set 23 had only 2 answers that were different. Here the parents claim the child stays online for more than 2 hours and he claims he stays for less than 2 hours. Also parents claim to have changed his privacy settings which he probably is not aware of.

Set 24 had 12 answer variation, Set 25 and 26 had 7 answers variation. Here they differ in opinion if Internet is a safe place for children or not? Parents agree while children disagree. Also for the question if the children stay online for too long – parents answer is yes and child answer is No. Set 28 had 3 answers variation. Set 29 and 30 had 13 and 5 answers variation respectively.

The question with most variation is - Whom does he discuss with any of his internet safety/ online issues with? Here most parents claimed children talk to them, but most children claimed they talk to friends about it. Similarly, for the question -Do you believe online games can be life threatening or dangerous for you? 70% parents believed it is but only 56.67% of children accepted it. Interestingly for the question - Has your child seen obscene, violent and sexual images/ videos online? There is not much variation between parents and children. Major agree to not having seen such content online.

22. Difference in the response of parent for each question is presented in Image 2 . Difference in the response of the children for each question is presented in Image 3.

Figure 2. Number Of Answers variation between parents.
X Axis – Number of Parents
Y Axis – Choice option A or B

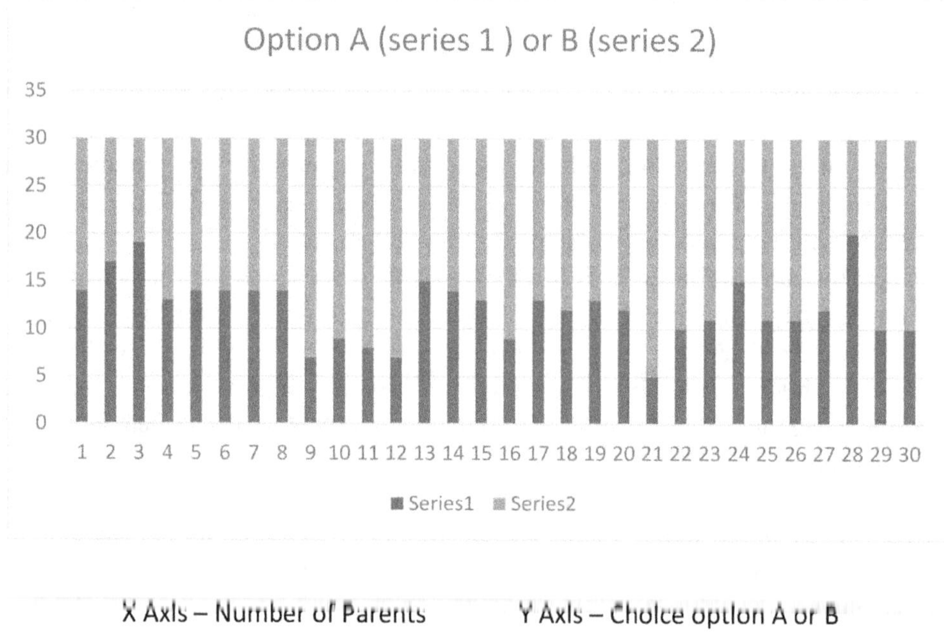

The table 5 present parent and child perception for each question set.

Figure 3. Number Of Answers variation between Children (mid-teens)
X Axis – Number of Children
Y Axis – Choice option A or B

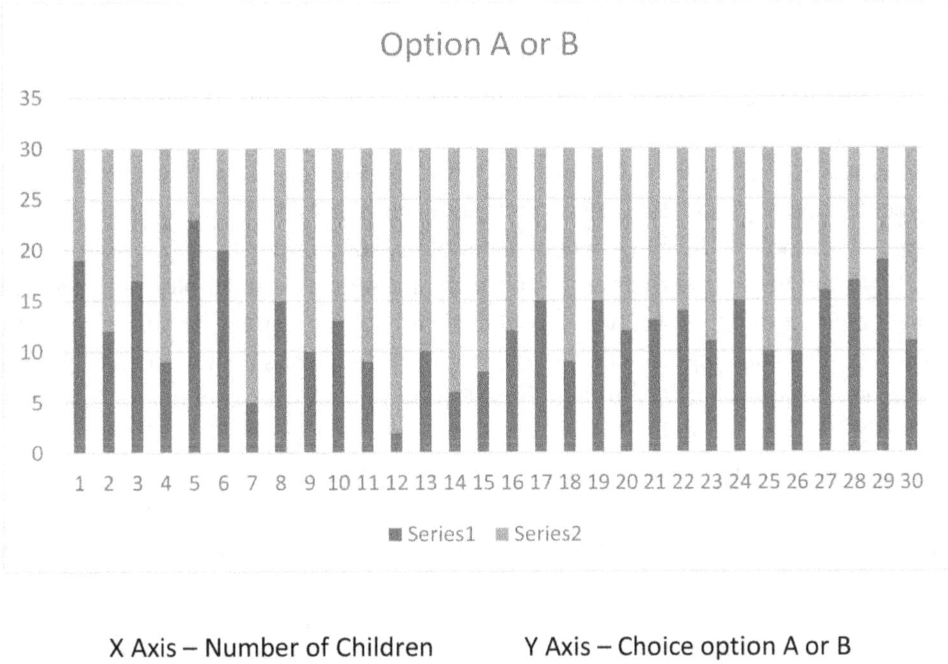

6. SUMMARY

It is clear and visible that Internet has severely affected our lives and research has shown that technology is affecting our day to day life in both positive and negative manner. It is easy to see that it gives great benefits and helps a lot with our daily activities, while also posing great risks. It is equally as simple to argue why we have to award the use of internet to our children, as it is to point out why they shouldn't. Pointing the finger in either direction will do no good as it shall never come to a conclusive result. and the reason for this it simple. When it comes down to finding the results, technology has been around and assisted us through time, it is here to stay and grow and assist with our daily lives. It is an integral part of our society and be an inseparable part for the rest of our foreseeable lives. Technology is going to keep advancing with time, and we will most certainly tap more into its potential, find more ways to use and interact with it. To keep kids away from it entirely would require them to be shielded from the cell phone, the television, the computer, and iPods. Also, it is because of the fact that we are still growing and learning about its impact on us. We cannot conclude at the moment *how* technology is impacting us or our children. This generation is the first to have capabilities and resources that would assist with social networking and provide instant and intimate connectivity with friends. We has so far never tried to understand the depth and implications of this exposure on our children. These teenage kids are learning the same technology and visiting same sites as we are during our search for knowledge. We are yet to see the overall impact of this, because these children have a different level of absorption and understanding than adults. To safely allow our children to have our modern technologies and resources like fast internet so that they can tap into its tremendous potential, and also prevent them gaining the risks, there are two things that can be done. Parental controls and moderation. Parental control is all about training the children as to how they interact with technology. This can to a great level protect them from the risks of modern technology and internet. It is easier said than done. It is slightly unrealistic to expect that controls over technology or children is going to alter or reduce the adverse impact on our future generation. The vast amount of technology that is now available would require a huge shift in our life. We need to shift from adults, who are neck deep in use of technology and have already exposed our children to its risks to protecting them from the repercussions of its gruesome impact on them. We are already feeling the impact of internet on our children and how much it is controlling them. Now we are also seeing the impact on the mental and psychological development of our teens as a result.

Modern parents who know the prospects of technology also understand the limitations that come along with it. They want to prepare their children for the future that is technology-driven and must also understand that there is a delicate balance between what can and cannot be shared with them. There should be some sort of monitoring and control to ensure that they are safe both mentally and physically. There is a difference between letting child surf the Internet and letting a child surf the Internet under some sort of filter and control system. There is a difference between allowing a child to have a computer and allowing a child to have a computer that the adult monitors and has some sort of limitation over.

Allowing a child to play video games on computers or tablet, or mp3 player when they are young is not a crime. However, it does become a concern if the access is uncontrolled and when there is no time restriction on how much they can sit in from of the blue screen. When a teenager who is already at the peak of puberty, adolescence hormonal force, physical and mental changes is given uncontrolled access to the web which is becoming darker by the day is when problems begin to arise. Their confusion added to the raw knowledge out there is what will cause blunders.

Children are taught from a very tender age that they should not talk to strangers. We make sure to remind them over and over again that it is for their own safety and that they should abstain from having any sort of interaction with strangers. However, that same parents forget these rules and allow their children to activate social media accounts with which they can do whatever they want. They have allowed their child access and resources that helps them to talk to anyone and everyone and share their private information to whoever has gained their confidence and asked for it. The ability and freedom to share their location and photos and videos and tell everyone exactly where they are and doing what is like providing a free gateway to perverts who can exploit and harm these vulnerable young immature kids. Their curious minds accepts the information shared without logic and reasoning and ends up causing more harm than imaginable.

When a child is expected to access internet and has some activity related to computers, parents can allow them to do so, as long as they consciously monitor what their children are doing and clearly explain to them what the risks and benefits of Internet are. Use of toolbars or blocking sites that can be harmful or using filters that restricts use can help to some extent. Keeping a child safe while online is not easy as there are many ways the obscene are accessible to them by just use of our everyday names. It is also important that we all respect the age restrictions that have been imposed by various sites. For sites and apps used for social networking, parents should respect the age limit and allow their child to be mature enough to have such accounts. Yet another step towards keeping an eye on our children is by staying as friend on these apps and sites. Parents can put a limit on how long the online video games can be played per day by their children and under what conditions.

Table 5. Summary Question wise parent and child Response scores

Q.N.	Mention of Broad Issue	Parent Perception				Child Perception			
		A	%	B	%	A	%	B	%
1	Access to hardware	25	83.33	5	16.67	18	60.00	12	40.00
2	Internet access per week	16	53.33	14	46.67	15	50.00	15	50.00
3	Internet usage a hours per day	19	63.33	11	36.67	19	63.33	11	36.67
4	Type of online activity	16	53.33	14	46.67	20	66.67	10	33.33
5	Fake account / information creation	4	13.33	26	86.67	2	6.67	28	93.33
6	Interact with unknown	5	16.67	25	83.33	4	13.33	26	86.67
7	Provides location access	10	33.33	20	66.67	7	23.33	23	76.67
8	Views on Internet safety	8	26.67	22	73.33	10	33.33	20	66.67
9	Sharing of personal details	6	20.00	24	80.00	4	13.33	26	86.67
10	Knows passwords etc	8	26.67	22	73.33	7	23.33	23	76.67
11	Access to parent account	7	23.33	23	76.67	2	6.67	28	93.33
12	Access to others account	4	13.33	26	86.67	4	13.33	26	86.67
13	Duration of internet access	10	33.33	20	66.67	8	26.67	22	73.33
14	Discussion about internet safety and restricted sites	9	30.00	21	70.00	16	53.33	14	46.67
15	Academic performance and Internet	23	76.67	7	23.33	20	66.67	10	33.33
16	Outdoor games and time	14	46.67	16	53.33	10	33.33	20	66.67
17	Internet issues	25	83.33	5	16.67	21	70.00	9	30.00
18	Block spam	11	36.67	19	63.33	12	40.00	18	60.00
19	App notification information	22	73.33	8	26.67	18	60.00	12	40.00
20	Editing App notification	8	26.67	22	73.33	14	46.67	16	53.33
21	Deleting history	19	63.33	11	36.67	22	73.33	8	26.67
22	Privacy setting	15	50.00	15	50.00	17	56.67	13	43.33
23	Information seeing and comparing website	14	46.67	16	53.33	14	46.67	16	53.33
24	Sharing of password with friends	4	13.33	26	86.67	6	20.00	24	80.00
25	Parents have info of child's account password	13	43.33	17	56.67	13	43.33	17	56.67
26	Visiting restricted websites	9	30.00	21	70.00	8	26.67	22	73.33
27	Games like blue whale	13	43.33	17	56.67	14	46.67	16	53.33
28	Seen obscene, violent, sexual content	3	10.00	27	90.00	4	13.33	26	86.67
29	Treated others in nasty way	2	6.67	28	93.33	4	13.33	26	86.67
30	Views on online games as life threatening	21	70.00	9	30.00	17	56.67	13	43.33

The physical risks of technology also cannot be ignored. Developmental problems and obesity are the first in line. Children who overuse and abuse the resources staying for extended hours in front of the blue screen are bound to miss on their physical exercises, games and activities not to mention the impact on their eyes. Children who play games online for extreme amounts of time per day and do not indulge in outdoor activities are the ones who are dealing with obesity, overweight, lethargies, vision impairment and stress related to playing video games. Putting a limit on how much children use technology and how much they share online will help tremendously to prevent the health hazards linked to their online exposure. The similar concern are raised by researchers Robinson et al 2016.

Second in line is lack of privacy, and loss of social skills. This is especially important as there is still little research about the effects of technology on children. The electronic media market is currently heavily targeting our vulnerable and overexposed children. The modern-day sales mostly impacted with the amount of marketing done and with new emerging products everyday the competition is getting tougher. It is the same be it food or games or clothing – product types doesn't matter.

It is extremely important that we keep our children safe from the threats of society that loom around, and their private information is kept private be it online or otherwise. Children's' privacy online is all the more important as they are most vulnerable and have turned into a commodity themselves. Parents can take several measures to protect their kids from threats of technological advancements by applying quality filters on their systems, closely monitoring their children's use of internet/ computers and openly communicating with their children about the dangers of ever-expanding technology and internet.

It is also important that care and caution be laid in schools as much as home. Children spent almost half a day most weekdays in schools where their monitoring and control is under teachers who are solely responsible for their routines. In most modern schools' internet access is well restricted and monitored. In the classroom environment, technology has proven to be a boon in student learning, however as with most things, too much of a good thing can be detrimental. In most modern schools teachers are encouraged to frequently engage students in group activities or work with partners to complete assignments, to help build group identities and facilitate social interaction. There are many students who do not engage in such group activities/ discussions, they do not prefer having to debate and discuss issues, such children prefer hiding behind the blue screen and engage more in solo activities.

Everything in life should be in balance. And the same applies to how much should our children be exposed to technology. Keeping control over duration of exposure and its type is the best way to keeping our children safe from the risks of modern advancements of technology that can have negative impact.

This study presented discussion to understand how exposed are our children to Internet. "What do parents know about their children's internet use? Is there any discrepancy between children's actual exposure to Internet content and parents' perception of children's exposure?"

The conclusion derived from the study here shows that there is a significant difference between Parents perception and children's Internet usage.

6.1 Implication of the Study

If put to practical use, there is the possibility of a major long-term implication of such studies. Understanding the difference and then resolving the differences is essential. What is expected of the children and to what extent is the freedom needs to be understood. The boundaries are very thin and very vulnerable. Where on one end parents need to understand that internet is the need of the hour and is important for the children to gain access to it to be sure they are able to complete their academic requirement and learn and know more, it is important for the children to know their limitations and set their own boundaries. Exposing themselves to the evil black of the online world can be a severe risk to life. The administrators of school education and governing bodies will have plan for sensitisation sessions for parents as well as children. This is no different than what is done for sex education at school levels. All India council of technical Eduction has started 'e-saksham' programme for students and facilitators in higher education . the programme focusses on digital literacy. Similar programmes should be encouraged for school going students.

There is tremendous scope of further study on the topic. Not much has been done in this line and this research has just managed to scratch the surface of the problem. Understanding in debth the root cause of variation and then finding ways to eliminate the same is a must. Enlightenment of the new generation and knowledge sharing is the need of the hour.

7. LIMITATIONS AND SUGGESTIONS

It is acknowledged that the work here is exploratory study because previously not much has been done in Indian environment. Also, with changing time the trend is changing fast. Under circumstances this study was done with various limitations. The survey took some time due to time conflict. Bringing children and the parents on the same platform at the same time in and atmosphere that was amicable and welcoming was tough. Waiting for a school event where this could be done was the only option. Time constraints and cost are also the limitation of our study. This limited the number of response to 30. More the number of response and further better statistical analysis could enhance the study.

This Study is to subject to several limitations because of small sample size and self-reported data collection technique. Children hesitated to mention their true response for fear of their inner truth being leaked. May be the response was over stated by children to some extend but have nominal variation. There has been little research concerning the understanding difference between the two generations. How our new upcoming generation is using the internet and how we believe they are. This study investigated children's (mid teens)Internet usage preferences and their knowledge base about it and what their parents think about their children about these points.

Results revealed that there is immense variation between the two. Children are using internet more than what their parents believe. They are more knowledgeable about the handsets/ laptops and its features. They know much more about deleting history/ spam settings and notification settings than what their parents believe. It was also interesting to know that many children not only knew about their parents account passwords but also accessed them. They have used fake name/ date of birth and other details to open accounts that parents are unaware of. Children were very clear and confident about their response during the survey and did not hesitate to give their inputs about their preferences and response.

Concerning the relationship between the Internet use and physical activity, it was interesting to see how children believe internet is not affecting their academic skill and it not hampering their physical and social lives. Parents have shown concern about the same where they believe it is not only affecting their child's physical, mental, social but also psychological development. It was interesting to see that the amount of Internet use was not associated with time spent with organized physical activities that involving competitiveness, while positively associated with doing own physical activity or just hanging out. Internet has become a strong influence on our daily lives.

REFERENCES

Adegboji, O. B., & Toyo, O. D. (2006). The impact of the Internet on research: The experience of Delta State University, Abraka, Nigeria. *Library Philosophy and Practice*, *8*(2), 1–8.

Bianchi, S. M., & Robinson, J. (1997). What did you do today? Children's use of time, family composition, and acquisition of social capital. *Journal of Marriage and Family*, *59*(2), 332–344. doi:10.2307/353474

Çankayaa, S., & Odabas, H. F. (2009). Parental controls on children's computer and Internet use, World Conference on Educational Sciences 2009. *Procedia: Social and Behavioral Sciences*, *1*, 1105–1109.

Chang, R. (2002). *Healthy computing*. Retrieved September 20, 2006, From http://www.slais.ubc.ca/courses/libr500/01-02- wt2/www/R_Chang/index.html

Chapman, L. (2002). Russian roulette or Pandora's box: use of Internet as a research tool. In *VALA 2002. 11th Biennial Conference and Exhibition*. Victorian Association for Library Automation Inc. Retrieved 24 December, 2006, from http://www.vala.org.au/vala2002/2002pdf/18Chpmn.pdf

Clarke, M. (2003). *Healthy computing. It's in your hands*. Retrieved September 20, 2006, from http://www.slais.ubc.ca/courses/libr500/02-03-wt2/www/M_Clarke/Webpages/index.htm

CNN. (n.d.). http://edition.cnn.com/2010/WORLD/asiapcf/04/01/korea.parents.starved.baby/index.html

D'Esposito, J. E., & Gardner, R. M. (1999). University Students' Perceptions of the Internet: An Exploratory Study. *Journal of Academic Librarianship*, *25*(6), 456–461. doi:10.1016/S0099-1333(99)00078-6

Daraha, K. (2013). The Effect of the Internet Use on High School Students: A Case Study of Pattani Province of Thailand. *Procedia: Social and Behavioral Sciences*, *91*. Advance online publication. doi:10.1016/j.sbspro.2013.08.422

Delen, E., & Bulut, O. (2011). 'The relationship between students' exposure to technology and their achievement in science and math'. *The Turkish Online Journal of Educational Technology*, *10*(3), 311–317.

Gross, E. F., Juvonen, J., & Gable, S. L. (2002). Internet use and well-being in adolescence. *The Journal of Social Issues*, *58*(1), 75–90. doi:10.1111/1540-4560.00249

Hatch, K. E. (2011). *Determining the Effects of Technology on Children Senior Honors Projects*. University of Rhode Island. https://digitalcommons.uri.edu/cgi/viewcontent.cgi?article=1212&context=srhonorsprog

ITU-D. (n.d.). *ICT statistics*. Available online www.itu.int/en/ITU-D/Statistics/Pages/stat/default.aspx

Junni, P. (2007). Students seeking information for their Masters' theses: the effect of the Internet. *Information Research, 12*(2) paper 305. Available at http://InformationR.net/ir/12-2/paper305.html

Kamba, M. A. (2008). *The changing roles of researchers in Nigeria: The internet as an alternative future to modernity*. Library of Philosophy and Practice.

KeeleryS. (n.d.). www.statista.com/topics/2157/internet-usage-in-india/

Kraut, R., Patterson, M., Lundmark, V., Kiesler, S., Mukophadhyay, T., & Scherlis, W. (1998). Internet paradox: 'A social technology that reduces social involvement and psychological well-being?'. *The American Psychologist, 53*(9), 1017–1031. doi:10.1037/0003-066X.53.9.1017 PMID:9841579

Kubey, R., Lavin, M., & Barrows, J. (2001). Article. *Journal of Communication, Internet Use and Collegiate Academic Performance Decrements: Early Findings, 51*(2), 366-382.

Kumar, R., & Kaur, A. (2006). Internet Use by Teachers and Students in Engineering Colleges of Punjab, Haryana, and Himachal Pradesh States of India: An Analysis. *Elect J Acad Special Libra, 7*(1). Available from URL: https://southernlibrarianship.icaap.org/con-tent/v07n01/kumar_r01.htm

Lancaster, F. W., & Sandore, B. (1997). Technology and Management in Library and Information Services. Univ. of Illinois Graduate School of Library and Information Science.

LaRose, R., Eastin, M. S., & Gregg, J. (2001). Reformulating the Internet paradox: Social cognitive explanations of Internet use and depression. *Journal of Online Behavior, 1*. Retrieved January 21, 2003, from http://www.behavior.net/job/v1n2/paradox.html

Lauri, M. A., Borg, J., & Farrugia, L. (2015). *Children's Internet Use and Parents' Perceptions Of their Children's Online Experience.* Report of Malta Communications authority commissioned by University of Malta. Accessed online from www.mca.org.mt/

Pandey, S., Chavla, M., & Mathur, M. (2021). Good and Not so good of learning using Digital platforms: Experience sharing from PG students. *Turkish Journal of Computer and Mathematics Education, 12*(7), 749–757.

Pandey, S., & Tiwari, S. (2014). Education and Prosperity through Technology – Case of Virtual Education Limited Nigeria. *Procedia: Social and Behavioral Sciences, 157*, 55–62.

Powers, W. (2010). *Hamlet's Blackberry: A practical philosophy for building a good life in the digital age*. Harper.

Rideout, V. G., Foehr, U. G., Roberts, D. F., & Brodie, M. (1999). *Kids & media @ the new millennium.* Kaiser Family Foundation.

Robinson, T. N., Banda, J. A., Hale, L., Lu, A. S., Fleming-Milici, F., Calvert, S. L., & Wartella, E. (2017). Screen Media Exposure and Obesity in Children and Adolescents. *Pediatrics, 140*(Suppl 2), S97–S101. doi:10.1542/peds.2016-1758K PMID:29093041

Subrahmanyam, K., & Greenfield, P. (1998). Computer games for girls: What makes them play? In J. Cassell & H. Jenkins (Eds.), *From Barbie to Mortal Kombat: Gender and computer games* (pp. 46–71). The MIT Press.

Subrahmanyam, K., Greenfield, P., Kraut, R., & Gross, E. (2001). The Impact of Computer Use on Children's and Adolescents' Development. *Journal of Applied Developmental Psychology, 22*, 7–30. 10.1016/S0193-3973(00)00063-0

Norlander, T., & Archer, T. (2001). Internet blues revisited: Replication and extension of an ...udy. *Cyberpsychology & Behavior, 4*(3), 385–391. doi:10.1089/109493101300210295

Wettlaufer, V. (2006). *Computer & your health*. Retrieved September 20, 2006, From www.slais.ubc.ca/courses/libr500/03-04-wt1/assigments/www/V_Wettlaufer/index.htm

Woodard, E. H., & Gridina, H. (2000). *Media in the Home The Fifth Annual Survey of Parents and Children*. The Annenberg Public Policy Center of The Univeristy of Pennsylvania.

Yeora, K. (2003). *The Impact Of The Internet On Children's Daily Lives: Physical, Social And Psychological Well-Being*. University of Georgia.

Yusuf, S., Osman, M. S., Hassan, S. J., & Teimoury, M. (2014). Parents' Influence on Children's Online Usage. *Procedia - Social and Behavioral Sciences, 155*, 81-86. doi:10.1016/j.sbspro.2014.10.260

Chapter 13
Reform of Education Services for Adolescents Through Digital Technology:
HIV/AIDS Prevention

Nana Noviana Nadarsyah
Regional Research and Development Agency, South Kalimantan Province, Indonesia

ABSTRACT

The problem of HIV/AIDS cases affects the number of human resources in education such as teachers, administrators, principals who are reduced due to illness or death. Increasing cases of HIV/AIDS that spread among adolescents make the government pay more attention to the increase in these cases. Empirical facts found during the study show that one of the stakeholders in the world of education stated that learning about HIV/AIDS prevention has not been considered an urgent matter. The research is qualitative research using a descriptive approach based on the perspective of public policy. In this case, researchers have developed a basic health promotion system about HIV/AIDS and have been introduced to sufferers newly infected with HIV/AIDS. The results of studies in the city of Banjarmasin explored policy issues in education for the prevention of HIV/AIDS. Study findings assist policymakers in making decisions regarding HIV/AIDS prevention policies.

INTRODUCTION

The spread of HIV/AIDS cases is not only a problem in Indonesia but has become a worldwide problem. In some countries, it is reported that HIV/AIDS affects the number of human resources in education such as teachers, administrators, principals who are reduced due to illness or death. The problem of HIV/AIDS is a health problem that cannot be separated from social problems so that the handling needs to be integrated with other sectors. Increasing cases of HIV/AIDS that spread among adolescents makes the government should pay more attention to the increase in these cases. To explore the current social conditions in the development of HIV a policy has been developed by Vass in A social perspective, 1986.

DOI: 10.4018/978-1-7998-8318-0.ch013

The high quality of community welfare in an area is an indicator of the success of government performance in health development. To develop health to maintain and improve the highest health status of the community, it must be carried out jointly by the government and the community optimally by the mandate of Health Law No. 36 of 2009. Efforts to prevent HIV/AIDS can be done together with the public and the private sector, this is by what Collin (2016) stated that the role of social responsibility of public companies in making invaluable contributions to reduce the negative effects of HIV/AIDS.

Likewise, the importance of advocacy to local government and education authorities, and school management is the key to the success of HIV prevention programs in schools, Pohan, et al. 2011. For HIV/AIDS prevention programs to run smoothly, reforms in education services are needed, especially regarding HIV/AIDS prevention. As in Japan, health care reforms have been discussed since 1997. Japan has a picture of a good quality of life compared to countries such as Canada, United States, United Kingdom, etc. According to the World Health Organization (WHO) (2000), good health care systems in Japan are highly rated in the world.

Empirical facts found during the study that one of the stakeholders in the world of education stated that learning about HIV/AIDS prevention has not been considered an urgent matter. This understanding of stakeholders is very influential in efforts to prevent HIV/AIDS in adolescents. This education for HIV/AIDS prevention must start as early as possible and continue into adolescence. Based on the results of research in Guatemala, shows the results that more than a third of orphans have dropped out of school due to HIV/AIDS (UNAIDS,). The impact of HIV/AIDS on education systems in severely affected countries is particularly acute (UNAIDS). With this situation, the education management system is threatened by illness and death caused by HIV/AIDS. Therefore a competent education system is needed to survive the effective effects of HIV/AIDS.

To achieve effective prevention efforts can be through the world of education because the world of education is a strategic approach to the policy of applying knowledge about HIV/AIDS. The problem of HIV/AIDS is not only faced by the world of education but also related to health, socio-cultural, economic, and religion as well. Therefore, efforts to prevent HIV/AIDS require the participation of all parties and the community to achieve the desired goals. The results of studies in the city of Banjarmasin explored policy issues in education for the prevention of HIV/AIDS in the city of Banjarmasin. Study findings to assist policymakers in making decisions regarding HIV/AIDS prevention policies.

LITERARTURE REVIEWS

UNICEF conveyed about how to work by adapting to the workplace when we are facing the Covid-19 pandemic. In this case, the information and communication technology division of UNICEF (ICTD) is thinking about how to get back to work. Therefore ICTD dedicates digital services as a transformation design in building exponential growth in digital activities and utilizing digital and utilizing technology to meet children's needs.

Since its establishment in 2017, the Technology for Development (T4D) function within ICTD, has been dedicated to strengthening health, education, child protection and other systems to utilize digital technologies more effectively. That means building national enabling environments for digital transformation; scaling up digital platforms; strengthening coordination across platforms and among partners; and enhancing the viability and maturity of digital public goods across the world. Over the past year, ICTD intensified its collaborations across UNICEF and with partners to rapidly develop and expand programme-centred technologies to meet the circumstances and needs of the day. We were instrumental to the organization's risk communication and community engagement effort, which leveraged digital technologies to reach communities with the information they needed to protect themselves against COVID-19. We adapted UNICEF's health and education programmes using real-time information technology and digital education solutions. And we were quick to scale our existing technologies for child protection case management, cash transfers and other systems to ensure that our government partners could continue to reach children in need, despite the global shutdown. (UNICEF, 2021)

Recognizing the integral role of technology (eHealth, mHealth, Cloud, Internet of things, wearables, etc.) in healthcare delivery, a National Digital Health Authority (NDHA) will be established to regulate, develop and implement digital health. The policy advocates the widespread deployment of digital tools to improve the efficiency and outcomes of the healthcare system to serve the needs of all stakeholders and enhance efficiency, transparency, and citizen experience. Providing better health outcomes in terms of access, quality, affordability, reduced disease burden, and efficient monitoring of health rights for citizens, is the goal. This digital form of linking systems across public and private healthcare providers at the State and national levels consistent with Metadata and Data Standards (MDDS) & Electronic Health Records (EHR), will be supported by this policy.

Based on the results of research by Shelley Jones, 2013 entitled Digital literacy, HIV/AIDS information, and English language learners in Uganda, access to information about the causes, symptoms, and treatment of this disease remains a challenge for many people, especially young people. In contexts where discussion of sexuality is often taboo, and information about HIV/AIDS is often in English. This paper reports on an action research study conducted in a rural Ugandan village in 2006, in which 12 young women participated. The focus of this research is a digital literacy course, which was developed with the main aim of helping participants gain access to information about HIV/AIDS through a global health website, available in English, Uganda's official language. The findings of this study confirm that digital languages increase access to HIV/AIDS information and provide many opportunities for learning English. Furthermore, the conditions under which digital language occurs are best understood regarding theories of investment, imaginary identity, and language learning (Norton, 2000; in press). We address the challenges and possibilities associated with digital languages in African communities and conclude with a discussion of the implications of our research for poorly resourced communities in Africa and elsewhere.

LIMITATIONS OF ADOLESCENT ACCESS TO REPRODUCTIVE HEALTH SERVICES AND HIV/AIDS

Adolescents are more vulnerable because of their role in a family. The proportion of adolescents (15-19 years) for HIV infection in Indonesia has increased from 827 in 2010 to 1,119 in 2015. In the case of HIV in 2015, there were children aged 15-24 years of 5,989. The Ministry of Health has predicted an increase in HIV infection in children due to an increase in HIV infection in women that is increasing. UNICEF and the National AIDS Commission in the initial findings of the last study showed the difficulties faced by children affected and infected with HIV/AIDS. They experience limitations and discrimination in accessing education and health services. Other difficulties are family financial difficulties due to illness, worsening children's health, and the need to care for sick parents. The projection for the number of infected children is increasing from 1,070 in 2008 to 1,590 in 2014.

In our society, most parents, teachers still consider it taboo to discuss sex issues openly even with health care providers. So that unmarried young people find it difficult to access sexual and reproductive services because of the limiting legal provisions. Dissemination of information about sex education by law can be said as a criminal act. In certain areas, the promotion of condom use is a problem and this promotion is opposed on religious and moral grounds.

Availability and health services for young people and unmarried adults are very difficult to access. This is stated in the Population and Family Development Act (Law No. 52/2009) and the Health Act (Law No. 36/2009) which stipulates that only legally married couples can access sexual services and reproductive health. However, condoms can be obtained easily in stores and markets except in remote areas.

HEALTH PROMOTION SERVICES THROUGH DIGITAL MEDIA

The development of the technological era is expected to be able to bring changes in a more positive direction, such as changing the ability to receive access to information. Alvin Toffler (1980) predicts and analyzes that the era of communication society is an era of humanity. The development of information technology is an innovation that can be useful for community needs, such as a health promotion system on HIV / AIDS. This digital media health promotion system is one of the proposed approaches for disseminating information on HIV / AIDS control.

The health promotion system using applications on mobile phones makes it possible to disseminate information about the early treatment of HIV infection and minimize stigma from those around you. Besides, this digital media is also able to have a positive effect on reducing transportation costs to visit health care centers to obtain basic information on HIV / AIDS.

This health e-promotion application uses mobile devices because increasingly advanced technology and internet networks can be accessed in various areas, not only in cities but also in rural areas, this provides easy access to information for communities with HIV / AIDS who are far from cities.

The following is a digital media model offered by the author in his research to provide health services in HIV / AIDS prevention efforts for adolescents:

INCREASING ADOLESCENT KNOWLEDGE ABOUT HIV/AIDS BUT STILL LIMITED AND DO NOT GUARANTEE SAFE BEHAVIOR.

The ministry of health conducted a study in five provinces and the results showed that there was a comprehensive increase in HIV and AIDS. The increase in HIV/AIDS among young people (aged 15-24 years) in the general population, from 11.4 percent in 2010 to 20.6 percent in 2011, with the same proportion for men and women. More than half of young people know that AIDS cannot be transmitted through food sharing, and two-thirds answer correctly that people who look healthy can be infected with HIV. In another 2011 study, only 22 percent of junior high school students in grade 2 of senior high school had comprehensive knowledge about HIV transmission, and 64 percent still had misconceptions about HIV.

Although adolescents in Indonesia have experienced a comprehensive increase in knowledge of HIV/AIDS, it does not guarantee adolescent behavior. This can be seen from the results of a survey that showed a consistently low number of condom use, which was less than 20% in the period 2007-2009. And the results of a 2011 survey of high school adolescents stated that 49% of them had had sex without using a condom. Besides, the knowledge of adolescents 15 years and older is still low about HIV/AIDS. Based on the results of the 2010 Riskesdas survey, showed that approximately 42% of the population above 15 years had never heard of HIV/AIDS. Only 10% of women and 13% of men have comprehensive knowledge about HIV prevention.

Figure 1.
Sources: processed by researchers, 2019

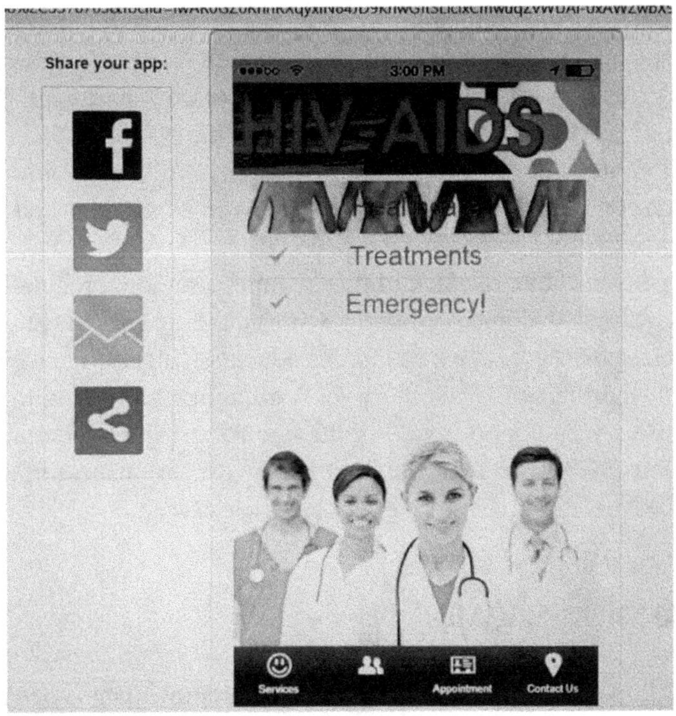

HIV/AIDS EDUCATION FOR ADOLESCENTS

The attitude of ignorance and rejection of HIV sufferers arises from the perception that HIV only occurs in groups that have questionable morals. The public should be aware that the current spread of HIV/AIDS is not only for high-risk groups. This happens because many members of the community at a young age have been active in sexual activities but lack knowledge about HIV/AIDS prevention and do not have access to reach sexuality services for their protection.

Therefore, special strategies are needed for young people affected by HIV/AIDS or at high risk where the strategies and programs provided must be tailored to the special needs of young people.

STAKEHOLDER PERCEPTIONS ABOUT HIV/AIDS CONTROL POLICY

Increasing HIV/AIDS is not only a problem in the city of Banjarmasin but also a world problem. Therefore, efforts are needed to prevent the increasing spread of HIV/AIDS, especially among young people and adolescents. One way for HIV/AIDS prevention efforts to succeed is to collaborate with the world of education. In the field of education, prevention efforts can easily be implemented because an understanding of HIV/AIDS can be given in learning. But in fact, HIV/AIDS material is not yet included in the subjects in schools. Some schools only insert knowledge about reproductive health, while there is no specific explanation for HIV/AIDS.

As a result of research in the city of Banjarmasin that the increase in cases of HIV/AIDS in the city of Banjarmasin in the response of the city government of Banjarmasin by issuing Regional Regulation No. 11 of 2012 concerning controlling HIV/AIDS in the city of Banjarmasin. From the interviews, it was found that the Education Office did not know about the local regulations governing prevention efforts involving the education office as the implementor. So in the absence of these regional regulations, they do not carry out HIV/AIDS prevention efforts in the education department. Besides, they explained that the activities they carried out were in collaboration with the National Narcotics Agency regarding drug explanation. Besides, the opinion of the Banjarmasin city AIDS prevention commission said that the regional regulation on controlling HIV/AIDS was unclear in its contents and was still confused with the regional regulation.

Besides, it was also conveyed that the lack of clarity on this regional regulation was also due to the absence of technical guidelines (technical guidelines) or guardianship (Mayor regulation) so that the implementation was unclear. So there is no basis for the education office to make efforts to prevent HIV/AIDS in the educational environment. The objectives written in a regulation that has been formulated carefully and clearly will be very important as a guideline for the implementing agency of the policy. With the clarity of the objectives to be achieved it will show the seriousness of the role of stakeholders in achieving policy objectives.

ADVOCACY REGARDING HIV/AIDS

In advocating for HIV/AIDS in adolescents / young people, credible data is needed, not only data from high-risk groups of young people but also data on young people in general. During this time, the HIV information system emphasizes high-risk group data. However, trends show that information about HIV prevalence and epidemiology, risk behavior, and knowledge among young people in the general population is increasingly needed. For example, the National AIDS Commission (2012) in the UNGASS report mentions the lack of data on some indicators that the government considers relevant to Indonesia, but are not available. Likewise, there is an important need to obtain disaggregated data among young people (ages 15-24) to tailor programs to their specific needs. Therefore, support is needed for the district level to build a strong information system, sorted by age, gender, location, and ethnic group

STAKEHOLDER SUPPORT FOR EDUCATION ABOUT HIV/AIDS

The rate of spread of HIV/AIDS has increased from year to year and the increasing number of people with HIV/AIDS and many people with HIV/AIDS that have not been identified is because the characteristics of HIV/AIDS is an iceberg phenomenon that is identified only on the surface. With an increase in HIV/AIDS cases, efforts should be made to control HIV/AIDS by making policies for controlling HIV/AIDS. Research results in Banjarmasin City show that the implementation of regional regulations regarding HIV/AIDS control has not been maximized, because several agencies involved in the regional regulation do not know the regional regulations. Therefore we need support from all parties involved, including the public and the private sector to carry out HIV/AIDS prevention efforts. However, due to the lack of socialization regarding HIV/AIDS prevention policies, the Banjarmasin city education office has stated that they do not specifically provide knowledge about HIV/AIDS because they are considered non-specific. Seeing from the perspective of the education department stakeholders, efforts to make HIV/AIDS material subjects in schools are very unlikely. This is because the views of stakeholders are weak and do not support the existence of HIV/AIDS subjects.

In policy implementation, many factors will influence implementation in the field, both factors that will deliver better policies and which will make policy implementation more difficult to implement. Besides, policy implementers need clear objectives of a policy as a source and understanding for field implementation. The more clear the purpose of a policy in providing guidance, the more likely it is to achieve the goals set by the policy. As stated by Goggin et al (1990), policy messages that contain direct statements about their standards or targets - especially those without qualifications, contradictions, or ambiguous language - are clear and related to goals.

Achieve a successful policy implementation not only requires accuracy of objectives but also requires supporting resources. In this case, the active role of the bureaucrats is needed to socialize the regional regulations to achieve the objectives precisely. The active role of bureaucrats can be achieved by the involvement of many government resources involved in the implementation of policies that have formed patterns of collaboration in efforts to control HIV/AIDS.

The process of implementing public policy follows a cycle of support and attention from relevant stakeholders and the community. The stakeholder perspective that is related to minimal information and health problems is a deviation of the objectives of policy implementation. The stakeholder perspective that is related to minimal information and health problems is a deviation of the objectives of policy implementation. When the stakeholder's point of view in education does not support local regulations on HIV/AIDS control, then the implementation of the policy will not succeed. With such conditions, HIV/AIDS prevention efforts for school students will not be implemented. By deviating the purpose of policy implementation, will cause unresolved problems and can hamper the successful implementation of HIV/AIDS control policies, especially HIV/AIDS prevention efforts for school students.

In efforts to prevent HIV/AIDS in adolescents, it is necessary to have better coordination between sectors related to policies and programs for young people. In efforts to prevent HIV/AIDS in adolescents, it is necessary to have better coordination between sectors related to policies and programs for young people. Most of the problems faced by young people are interconnected. National policies promote cross-sectoral approaches to combating HIV/AIDS, but coordination and collaboration must reach the district and provincial levels. To develop policies and programs intended for young people, a greater level of participation is also needed from young people and other stakeholders.

Education should be one of the media efforts to prevent HIV/AIDS. Because efforts to spread knowledge about HIV/AIDS can be through formal and nonformal education is one alternative that can be chosen. Efforts to increase knowledge have been known as IEC (Communication, Information, and Education). However, if stakeholders in education do not provide support for HIV/AIDS prevention efforts, the policy objectives to prevent the spread of HIV/AIDS, especially among young people will not be achieved. The commitment of the policy implementor is very influential in implementing the policy. A policy implementor that has a high commitment will contribute to the success of a policy. From the results of research at Banjarmasin City, several implementors that are related and have been determined in the Regional Regulation as the ones responsible for HIV/AIDS control efforts, both as coordinators and supervisors in HIV/AIDS control activities, do not yet have a high commitment in controlling HIV/AIDS. This is based on the absence of technical guidance or guardians in the implementation of the policy so that it is unclear the role and duties of agencies related to HIV/AIDS control so that HIV/AIDS control efforts cannot be carried out. And this affects the perception of the relevant agencies that efforts to control HIV/AIDS are not a priority for the institution because there are still other activities that are more priority.

It was stated that the spread of HIV can be through the community, the indirect causes are poverty, low education, low policy, and political support, lack of women's role in the decision-making process, low community involvement, and accessibility to the healing process is still low (State Ministry of Women's Empowerment, 2008: 32). Low education is one of the factors causing the spread of HIV. Therefore it is necessary to prevent HIV/AIDS in the field of education by making HIV/AIDS material as one of the school subjects at least as an additional subject at school.

REFERENCES

Fischer, F. (2003). Public Analysis Handbook. Academic Press.

Goggin, M. (1990). *Implementation, Theory, and Practice. Toward a Third Generation*. Forasmann and Company.

Jones, S. (2013). *Digital literacy, HIV/AIDS information and English language learners in Uganda*. https://www.researchgate.net/publication/292424756

Ministry of Health of the Republic of Indonesia. Health Law No. 36 of 2009 concerning health.

National Health Policy. (2017). *Ministry of Health and Family Welfare Government of India*. Author.

Noviana, N. (2018). *Kesehatan Reproduksi untuk Mahasiswa Kebidanan (Revisi)*. Trans Info Media. https://scholar.google.co.id/citations?user=ac_7qq0AAAAJ&hl=en

Noviana, N. (2020). Development of Health Promotion System in Industrial Era 4.0 to Support E-Service. *International Joint Conference on Science And Technology*.

Noviana, N., Suwitri, S., Supriyono, B., & Sutopo, P. J. (2019). Implementasi Kebijakan Kesehatan Dalam Pengendalian HIV/AIDS. Pena Persada. https://scholar.google.co.id/citations?user=ac_7qq0AAAAJ&hl=

Regional Regulation No.11 of 2012 concerning HIV / AIDS Control in Banjarmasin City.

Schneider, S. (2001). *Implementing AIDS Policy in Post-Apartheid South Africa*. Academic Press.

Sri, Sagala, & Santoso. (2009). *Implementation of HIV and AIDS Prevention Policies in Central Java (Review of Central Java Provincial Regulations Number 5 of 2009)*. Academic Press.

UNAIDS. (2015). *HIV/AIDS & education*. UNAIDS.

UNESCO. (2006). *Good Policy and Practice in HIV/AIDS and Education. Overview Booklet 1*. UNESCO.

UNICEF. (2021). *Harnessing the Power of Technology and Digital Innovation Fpor Children. Technology for Development Report, Information Communication and Technology Division.* www.unicef.org

Vass, A.A. (1986). *AIDS: A Plague in Us: A social perspective. The Condition and its Social Consequences.* St.Ives: Venus Academica.

Chapter 14
Effects of Digital Technologies on Zimbabwean Adolescent Lives During the COVID-19 Era:
Pros and Cons

Wilfreda I. Chawarura
Midlands State University, Zimbabwe

Munyaradzi Zhou
https://orcid.org/0000-0002-2184-0290
Midlands State University, Zimbabwe

Cyncia Matsika
Midlands State University, Zimbabwe

Tinashe Gwendolyn Zhou
Midlands State University, Zimbabwe

ABSTRACT

The chapter focuses on the effects of using digital technology amongst Zimbabwean adolescents. The research was necessitated by the fact that adolescents are a vulnerable group still in the self-discovery age. COVID-19 pandemic resulted in the widespread use of digital learning to facilitate learning, communication, and social interaction among adolescents. Focus groups and interviews were used to gather data in primary schools. There is a disparity in technological device use and access to the internet between urban and rural areas dwellers. Ninety percent of students in rural areas did not have access to technological devices, and during the pandemic, they hardly used them for learning purposes but mostly used them for communicating with relatives. In Zimbabwe, digital technology devices have not yet replaced social relationships as most adolescents are restricted to 2-3 hours on their devices. A model for effective use of digital technologies in adolescents can be developed and implemented.

DOI: 10.4018/978-1-7998-8318-0.ch014

INTRODUCTION

Digital technologies use has become an important feature of the 21st century and has shaped how societies function and operate and there is more prevalent use of Internet among adolescents. Digital technologies use play a significant part in adolescents' behaviour (Cenameri & Taçi (Luga), 2015). This study seeks to analyse how adolescents have been affected socially, mentally, psychologically and in terms of literacy through the use of digital technology. New knowledge will be unearthed with the context of Zimbabwe through primary data collection. A systematic literature review informed the research through accessing journal articles via the databases ISI Web of Science platform, and other Scopus-Indexed databases within 5 years from the year 2021. This research is unique as it looks at the effects of digital technologies among Zimbabwean adolescents. Muchemwa (2017) conducted a similar study but concentrated on teachers' attitudes towards digital technologies use in classrooms. The book chapter explicitly focuses on the advantages and disadvantages of using digital technologies amongst adolescents in Zimbabwe. Specifically, it looks at how adolescents are affected behaviourally and socially, in skill development, learning, mental well-being and psychological developments. This was attained through literature review and primary data collection.

BACKGROUND

The development of the first computers in the 1950s gave birth to digital technologies as computers became household gadgets in the 1980s. The advent of the Internet completely revolutionized the way people communicate. The birth of social media had a profound effect especially on generation z, as digital technologies are shaping and influencing their lives. COVID-19 stimulated the use of digital technologies as a means of communication, entertainment, and learning. However, the effects and impact of digital technologies on this generation during the COVID-19 haven't been ascertained.

Digital technologies are both a cause and a symptom of change in society. Changes in society are reflected during technology use and at the same time, technology influences new social cultures and norms (Martin, 2008). People are now dependent on digital technologies for their everyday lives activities such as banking, education, and communication (Dufva & Dufva, 2019). Little study has been done on adolescents and digital technologies in low-income countries and the researches done in Sub-Saharan Africa have mostly been done in South Africa thus it can't be regarded as a true representation of all Sub-Saharan countries like Zimbabwe (Livingstone et al., 2017).

Adolescents are growing up in an era where information, learning, and media are influenced by digital technologies. However, adolescents are at a stage where they are developing their emotional and social identities whilst using digital technologies (Magis-Weinberg et al., 2021). Whether adolescents are going to have healthy emotional and social boundaries is yet to be determined.

Inequality in access, use, and adoption of digital technologies in low income countries can magnify prevailing economic cracks since adolescents are early receptors of digital technologies and social media platforms (Magis-Weinberg et al., 2021). However, there has been little or no research on how adolescents are affected mentally, psychologically, and emotionally through social media usage in low-income countries such as Zimbabwe. In the developed countries, the constructs have been explored. Most researches conducted on digital technologies use by adolescents in African countries have mainly concentrated on accessibility, skills and practices, and opportunities (Livingstone et al., 2017). This research is unique as it seeks to analyse the effects of digital technologies on adolescent lives in Zimbabwe, which is one of the least or 'never' researched countries on this notion.

MAIN FOCUS OF THE CHAPTER

The book chapter focuses on the constructs, access to digital technology; the impact of digital technologies on adolescent learning; the uses of digital technology and social skills development on adolescents; digital technologies' effect on adolescents' mental well-being; digital technologies' effect on adolescents' psychological development; and pros and cons of digital technologies on adolescents.

Access to Digital Technology

Access to digital technologies is still a challenge in African countries. Only 28% of Africans could access the Internet in 2019, Gunnlaugsson et al. (2020) yet 95% of American adolescents had access to smartphones (Paat & Markham, 2021). Sixty percent of African adolescents do not have access to digital technologies whilst only 4% of European adolescents do not have access to digital technologies (The Lancet Child & Adolescent Health, 2018). In a survey conducted in Guinea Bissau, access to digital technologies is a challenge amongst adolescents in the country. Adolescents could only improve their capabilities and proficiency through improved access to digital technologies (Gunnlaugsson et al., 2020). Adolescents living in remote rural areas and those from minority groups are the worst affected in accessing digital technologies (The Lancet Child & Adolescent Health, 2018). The findings are in line with Livingstone et al. (2017) who observed that access to digital technologies amongst adolescents was a function of affluence in most low to middle-income countries. Adolescents from marginalized families either did not have cellphones or they shared technological devices with other members of the family. Many pieces of research have been done on the effects of using digital technologies in day to day lives of adolescents and it is critical to study the notion in Zimbabwe. The rapid use of digital technologies such as networked technologies has foundationally changed homes, schools, and societies (Gerben, 2018).

The effect of technology on adolescents depends on the type of technology being used and the purpose of the technology thus it is important to understand how the technology was used and why they were using that form of technology when analysing the impact of digital technology amongst adolescents(Gottschalk, 2019). This book chapter focuses on understanding how the use of digital technology affects adolescents, and this will enable parents to set limits and controls on the use of digital technology on adolescents.

Impact of Digital Technology on Adolescent Learning

Digital technologies enable browsing through information to be faster, learning new information to be faster and it influences learning significantly amongst adolescents (Griffiths, 2020). Tripathi & Shahri (2019) concluded there was the need to understand child psychology and discuss regularly with adolescents what they were learning and watching. Traditional care and face-to-face communication were still relevant in the face of digital technology (Tripathi & Shahri, 2019). In Zimbabwe use of smartphones in classrooms is prohibited as teachers believe that it leads to poor performance amongst primary and secondary school students (Muchemwa, 2017). Completely discarding the use of digital technologies amongst adolescents is impossible given that the society is becoming more and more virtually connected.

The outbreak of COVID-19 has indirectly contributed to the urgency of participation in the digital age (Tripathi & Shahri, 2019).

Social media has the potential of increasing learning among adolescents. Zu et al. (2017) observed that social media needs to be combined with face-to-face class interactions to effectively improve the exchange of information amongst adolescent students. In contrast, it was concluded that the use of social media in education could increase absenteeism, destroys students' ability to communicate effectively in real-life situations. Digital technology is useful as a 'temporary' tool of learning. Nevertheless, the advent of the COVID -19 pandemic made the use of digital technology a predominant method of learning amongst adolescent students. Digital technologies have made phenomenal changes in educating students thus its influence is visible in all walks of life (Abdulellah Alaboudi, 2020).

Digital Technologies Uses and Social Skills Development on Adolescents

Most adolescents use social media to communicate with their social circles and this has led to improved social relationships (The Lancet Child & Adolescent Health, 2018). Adolescence is a phase characterised by the increased need for social interaction with peers as humans transits from childhood to adulthood. Social distance restrictions measures imposed globally to manage the COVID-19 pandemic removed the opportunity to engage in face-to-face social contact which is an essential need for adolescents' brain development. Orben et al. (2019) concluded that the use of digital platforms for interaction during the COVID-19 pandemic could have significantly reduced the negative effects of social deprivation on brain development amongst adolescents. The use of digital technology such as social media was mostly used to articulate previously established relationships amongst adolescents (Zu et al., 2017).

Digital Technologies' Effects on Adolescents' Mental Well-Being

Hormonal, developmental, and emotional changes that take place naturally in adolescents result in their need to interact outside the confines of families, and digital technologies have a significant role in the interaction process, and especially through the use of social media (Gunnlaugsson et al., 2020). Cenameri & Taçi (Luga) (2015) observed that there was a relationship between the time spent on the Internet and the adolescents' behaviour. On the other hand, Orben et al. (2019) found that social media use had no direct relationship with adolescents' ability to enjoy their lives.

Earlier research conducted in the 1990s and early 2000 concluded excessive use of digital technology lead to mental health issues (Odgers, 2018). However, recent research had found technology to be beneficial to adolescents' development. The moderate use of digital technology is beneficial to adolescents' mental well-being. However, excessive use can have detrimental effects on adolescents (The Lancet Child & Adolescent Health, 2018).

There is a convergence between physical and online playing fields among adolescents (Gottschalk, 2019). In America, adolescents spend on average seven to eight hours a day on technology gaming, social networking, or chatting (Gottschalk, 2019). This is worrying as children are susceptible to 'plasticity' or experience dependent change which can have serious negative effects on the brain and behavioural development. According to Crone (2018), there was massive structural changes in adolescents' brain and many neural networks responsible for functions such as decision making, cognitive, flexibility, and maturity traits during the adolescence period.

The use of digital technologies at night is associated with sleeping disorders amongst adolescents and is significantly related to depression and anxiety (Gottschalk, 2019). Furthermore, it has been found that there is an increase in grey matter in amygdala area of the brain due to an increase in social networking amongst adults (Gottschalk, 2019). George et al.(2018) found a positive relationship between the number of hours spent using digital technologies and mental conditions such as attention deficit hyperactivity disorder (ADHD) and conduct disorder amongst adolescents. However, George et al. (2018), observed digital technologies to be good to children already suffering from mental conditions, where he observed that the greater the time spend online in a day the lower the chance of internalizing problems, thus digital technology gave adolescents a platform to socialise. Gottschalk (2019) concluded that there is no real relationship between the time spent on screen and the health of children. There is still need for more research on the effects of digital technologies on their mental health.

Digital Technologies' Effects on Adolescents' Psychological Development

Modern technologies such as computers have many positive or significant results in children's development even though they pose health risks to adolescents. Mustafaoğlu et al. (2018) found that inappropriate use of digital technological devices in terms of duration, content, posture, and so on posed a variety of clinical challenges on adolescents' development hence there is a need to monitor adolescents' use of the technological devices. A similar study conducted by Dienlin & Johannes (2020) found insignificant negative effects on adolescent wellness due to the use of digital technologies. Procrastination and passive use had pronounced negative effects on mental health and active use had more positive effects on adolescents' development.

On the other hand, the negative effects of digital technologies were found to be similar in both adults and adolescents but more severe in adolescents (Dienlin & Johannes, 2020). Parents need to be less judgmental when adolescents fall victim to cybercrime to enable adolescents to recover from the shame and trauma associated with the incident. Schools and higher authorities should devise strategies to curb cyberbullying such as implementing programs that concertize adolescents on how to form healthy relationships online (Paat & Markham, 2021). However, a healthy parent-adolescent relationship is equally critical for the health and developmental well-being of adolescents (Gehan et al., 2017).

Adolescents are doing most of their emotional and social developments on social media. Since it is an experimenting and self-discovery stage of their lives, adolescents are susceptible to peer pressure and could end up being abused when they use the Internet and social media wrongly (O'Keeffe et al., 2011). Adolescents are increasingly becoming more independent as they spend more time on digital technological devices. Parents have to evolve to understand digital technologies as the mode of communication and to deter from relying on traditional forms of communication. This will help them to bond with their children (Gehan et al., 2017). There is no causal relationship between social media use and psychological outcomes in teenagers. This outcome is different from the widely speculated negative relationship between the use of digital technologies and psychological threats (Ma, 2018). Increased use of digital technologies among teenagers has equally exposed and increased the number of minors who are being sexually abused (The Lancet Child & Adolescent Health, 2018).

Pros and Cons of Digital Technology on Adolescents

This section focuses on the pros and cons of digital technology on adolescents, namely, improved communication, identity building, educational purposes, access to wide range of information, and entertainment.

Improved Communication

Digital technologies have improved communication tremendously amongst adolescents especially during the COVID-19 pandemic. Social media has been used to disseminate information and enables adolescents to interact with families and friends (Paat & Markham, 2021). However, there was an increase in cyberbullying and physical social detachment as some adolescents became glued to their digital technological devices "bedroom syndrome" (Gunnlaugsson et al., 2020). Excessive use of social media has been observed to lead to psychological and mental disorders amongst adolescents such as an increase in suicides. Excessive use of digital technology was destroying social circles hence parents need to educate themselves in the use of technology for effective communication with their adolescents. Conversely, parent-to-child discussions are still relevant and will assist in reduced exposure to online crimes (Gehan et al., 2017).

Identity Building

Digital technologies assist adolescents in developing independent characters, which is critical in developing their individuality (Erstad, 2013). Access to technology also shapes adolescents in the way they think, function, and make decisions (Tripathi & Shahri, 2019). On the other hand, they observed there is a need for regular discussion with adolescents on what they used their technological gadgets for. Lack of self-control and supervision of adolescents is detrimental. It could lead to an increase in Internet addiction, cyberbullying, and abuse amongst adolescents.

Educational Purpose

Generally, there is an increase in the use of digital technologies for educational purposes amongst adolescents (Odgers, 2018; Mishra et al., 2015). The COVID-19 pandemic has further intensified the use of digital technologies due to the banning of gathering as it became the readily available, and socially safe mode of teaching students (Witt et al., 2020). Google classroom became a quite popular teaching tool too. This has made it easier for the teachers to disseminate information quickly, and easily to the students (Gerben, 2018). The use of digital technologies has expanded the knowledge base for adolescents and during the COVID-19 pandemic. Online learning became an integral part of adolescents' school life (Paat & Markham, 2021).

Contrary to general beliefs that digital technologies are bad for adolescents, it resulted in the attainment of 84% pass rate mark in graduation in high school in 2016 and a decline in substance abuse, pregnancy, and alcohol abuse amongst American adolescents (Odgers, 2018). It can be coined that digital technologies are not bringing new vulnerabilities amongst adolescents but it only exacerbates existing challenges that affect adolescents (Odgers, 2018). Online educational programs can increase science skills, mathematics, and problem-solving skills (Gottschalk, 2019). However, some educators are worried that the learning applications and gadgets process information faster than the learning speed of adolescents.

Access to Wide Range of Information

Digital technologies have made information freely and readily accessible to adolescents, irrespective that not all information is good. There is a need for guidance and adult restrictions on what adolescents read online. If unsupervised, adolescents can end up copying bad habits such as smoking. Conversely, access to smartphones does not necessarily lead to unwanted behaviours (Odgers, 2018).

Entertainment

Digital technologies provide adolescents with a lot of entertainment. Ultimately, it can lead to social disconnection, and at the same time, it enables them to develop their individuality. However, using digital technologies for entertainment means adolescents will spend many hours on their devices. Research has shown that adolescents who spend many hours on their gadgets are generally unhappy, and some have challenges during personal and social development. Given these inconclusive outcomes, the research set to find out the practical examples of advantages and disadvantages in the use of digital technologies amongst Zimbabwean adolescents. The elements are discussed in the section on solutions and recommendations.

Methodological Gap

Systematic literature was used and it was complemented using interviews and focus group discussions to analyse the effects of digital technology on adolescents in Zimbabwe. The study sample consists of grade seven primary school students who were engaged using focus group discussions to analyse how these adolescents had been affected through the use of digital technology. School teachers facilitated the focus group discussion sessions. The adolescent age group of 12-15 years was purposively selected and the respective teachers. Since the age group consists of minors, the research was facilitated via school teachers as they conducted the focus group discussion. In addition, teachers were interviewed on what they had observed in their students' development during the pandemic. Ten primary schools were targeted, 8 of the schools were in urban cities and 2 in remote rural areas, and at each school, two classes were purposively selected. Rural and urban school children and teachers were all targeted to have a balanced view of the effects of digital technology in Zimbabwe. The authors made use of a descriptive research design to collect data. Twenty teachers were interviewed to understand their views on the impact of technology on their adolescent students during the COVID-19 pandemic and they facilitated the focus group discussions which had an average of 8 learners per session for the 3 sessions giving an average of 480 students.

SOLUTIONS AND RECOMMENDATIONS

Three-quarters of adolescents in cities had access to either a smartphone or a tablet whereas only one-tenth of children in rural areas had access to any form of digital technology device. The finding is in line with Gunnlaugsson's et al. (2020) findings that remote areas had little or no access to digital technology. Ninety percent of city adolescents confirmed that their parents bought them smartphones at the beginning of the COVID -19 pandemic to facilitate their online learning. Adolescents who stay in town said they used their devices for online learning mostly. On the other hand, adolescents from rural areas never used digital platforms for learning as only a few students had access to technological gadgets. At one of the rural schools, only three grade seven students had access to cellphones out of 70 students. Most of the rural students used their cellphones to play games and communicating with friends and relatives.

Seventy-five percent of the students have access to social media accounts and most of them use cellphones for a duration of 2 to 3 hours a day usually during the study. Twenty percent use their devices for chatting with friends and relatives for over an hour every day. The majority of students who are using social media accounts have their parents working outside the country. The platforms are mainly used for socialization. All students coined that their parents had access to their technological devices. They use their devices under their parents' supervision. Five percent of the adolescents had access to their devices during the evening. Mostly, they use them to do homework. Adolescents who slept late at night using cellphones said the light affected them and they were not able to sleep very well at night. This reflects to the experiences they had when they slept late watching television (Gottschalk, 2019). Most of the adolescents said that they felt bored if a day passes by without looking at their cellphones or tablets. Ironically, others felt indifferent about using their cellphones or tablets.

Students maintained their existing physical friendships on social media. One point zero four percent of the adolescents had experienced bullying on social media (Odgers, 2018). One of the students faced social media bullying. It was done by animating his photo by a fellow student. The incidents coin back to the students who once physically bullied him. Eighty percent (80%) of the adolescents who had technological devices used them to play games. Gaming resulted in addiction, so they didn't want to be disturbed. A few adolescents admitted to missing their cellphone games than playing with their physical friends during the COVID-19 pandemic.

Ninety percent (90%) of the teachers acknowledged that digital technology was important and had assisted them tremendously in disseminating information to their students during the COVID-19 pandemic. However, they were worried about the ability of the students to comprehend the notes and or materials disseminated (Zu et al., 2017). Teachers mainly in urban areas were using social media such as WhatsApp to teach students. Zimbabwe recorded a very poor grade seven pass rate in 2020 of only 37.11% (ZIMSEC, 2020). Rural areas were the most affected as the teachers were not able to conduct online teaching due to the lack of technological devices and the Internet for both students and teachers in most remote areas (Gunnlaugsson et al., 2020). The teachers also observed inappropriate use of technology and a lack of concentration during the online lessons. All teachers believed that digital technology was not a good way to teach students as a stand-alone method but needed to be used in conjunction with face to face learning as these adolescents were still young and it was critical to stimulate their minds in the right way and not overburden them with information, this finding was in line with finding from (Zu et al., 2017). In general, they concurred that blended learning is the way to go for adolescents.

The teachers observed that the lack of free Internet in Zimbabwe also assisted in ensuring that those adolescents who had access to technological devices could not visit social media accounts. However, they insisted that these children need to be monitored on how they use technology, since they may end up being abused or initiated into the wrong way of living (Tripathi & Shahri, 2019). Digital technologies play a significant role in maintaining social relationships and teaching students during the COVID-19 pandemic and in the future. Internet access remained a challenge amongst adolescents and teachers that dwell in rural areas in Zimbabwe hence the need to invest in internet infrastructure.

There is a need for supervision and monitoring of how adolescents use technological devices in the social media streets and avenues. Cyberbullying was not yet prevalent in this age group in Zimbabwe.

The authors observed that adolescents were maintaining their existing social circles when using social media. Social media enabled most adolescents to keep in touch with their friends and peers during the total lockdown. Digital technologies supported them to communicate with their loved ones who stayed outside the country at an affordable price.

Using digital technologies as a sole method of teaching is not appropriate for adolescents. Most teachers observed poor performance amongst students during the pandemic. The poor performance resulted from the lack of adequate supervision during the use of mobile devices and the poor or unavailability of the Internet. There is an unequal distribution of technological devices and the Internet between urban and rural areas adolescents. Hence, this made it difficult to effectively judge or generalize the effects and impacts of digital technology countrywide. There is a need to improve Internet access for adolescents living in rural areas. The use of mobile Internet Cafés is essential. Availability of digital technologies greatly enables continued learning during and after the post-COVID-19 pandemic. Digital technologies are now equally an important playing ground amongst adolescents as most of them use their devices for playing games and keeping themselves occupied. Adolescents extremely addicted did not see the need to have friends anymore since they viewed their phones as their 'friends'. The adolescents will become anti-social; thus it has detrimental effects on adolescents' ability to develop social skills. It is essential to come up with a governance framework for the use of digital technologies in schools and at home.

FUTURE RESEARCH DIRECTIONS

Future research may need to make follow-ups on the effects of excessive use of digital technologies by adolescents in their adult lives. It is critical to come up with a model to curb cyberbullying and a governance framework for the use of digital technologies in schools and at home.

CONCLUSION

The researchers concluded that the effects of digital technologies on adolescents varied from country to country depending on accessibility to digital technology in the adolescent lives. In Zimbabwe, the negative impact of digital technologies is still minimal as there is a challenge of inaccessibility in terms of mobile devices and the Internet. There are a lot of positive outcomes in terms of access to education during the COVID-19 pandemic. The long-term effects of digital technologies use on adolescents' lives can only be ascertained in the long run.

REFERENCES

Abdulellah Alaboudi, A. (2020). *Impact of digital technology on Saudi students*. Academic Press.

An Assessment of Zimbabwe Secondary School Teachers' Attitudes towards the Use of Smart Phones in the Classroom: A Case of Midlands Province, Zimbabwe. (2017). *International Journal of Social Sciences & Educational Studies, 4*(1). doi.10.23918/ijsses.v4i1p12

Cenameri, M., & Taçi (Luga), P. (2015). The Impact of Technology on Adolescent Behavior. *Academic Journal of Interdisciplinary Studies*. doi:10.5901/ajis.2015.v4n2s2p288

Crone, E. A., & Konijn, E. A. (2018). Media use and brain development during adolescence. *Nature Communications*, *9*(1), 10. doi:10.103841467-018-03126-x PMID:29467362

Dienlin, T. (2020). The impact of digital technology use on adolescent well-being. *Dialogues in Clinical Neuroscience*, *22*(2), 135–142. doi:10.31887/DCNS.2020.22.2/tdienlin PMID:32699513

Dufva, T., & Dufva, M. (2019). Grasping the future of the digital society. *Futures*, *107*, 17–28. doi:10.1016/j.futures.2018.11.001

Erstad, O. (n.d.). *Identity, Community, and Learning Lives in the Digital Age*. Academic Press.

George, M. J., Russell, M. A., Piontak, J. R., & Odgers, C. L. (2018). Concurrent and Subsequent Associations Between Daily Digital Technology Use and High-Risk Adolescents' Mental Health Symptoms. *Child Development*, *89*(1), 78–88. doi:10.1111/cdev.12819 PMID:28466466

Gerben, C. (2018). Factoring in Student Technology Use in Literacy Education: A Review of *The Class: Living and Learning in the Digital Age*. *Journal of Adolescent & Adult Literacy*, *61*(5), 593–595. doi:10.1002/jaal.725

Gottschalk, F. (2019). *Impacts of technology use on children: Exploring literature on the brain, cognition and well-being*. Academic Press.

Gunnlaugsson, G., Whitehead, T. A., Baboudóttir, F. N., Baldé, A., Jandi, Z., Boiro, H., & Einarsdóttir, J. (2020). Use of Digital Technology among Adolescents Attending Schools in Bissau, Guinea-Bissau. *International Journal of Environmental Research and Public Health*, *17*(23), 8937. doi:10.3390/ijerph17238937 PMID:33271847

Livingstone, S., Nandi, A., & Stoilova, M. (n.d.). *Young adolescents and digital media*. Academic Press.

Ma, C. M. S. (2018). A Latent Profile Analysis of Internet use and Its Association with Psychological Well-Being Outcomes among Hong Kong Chinese Early Adolescents. *Applied Research in Quality of Life*, *13*(3), 727–743. doi:10.100711482-017-9555-2

Magis-Weinberg, L., Ballonoff Suleiman, A., & Dahl, R. E. (2021). Context, Development, and Digital Media: Implications for Very Young Adolescents in LMICs. *Frontiers in Psychology*, *12*, 632713. doi:10.3389/fpsyg.2021.632713 PMID:33967899

Mustafaoğlu, R., Zirek, E., Yasacı, Z., & Razak Özdinçler, A. (2018). The Negative Effects of Digital Technology Usage on Children's Development and Health. *Addicta: The Turkish Journal on Addictions*, *5*(2). Advance online publication. doi:10.15805/addicta.2018.5.2.0051

O'Keeffe, G. S., & Clarke-Pearson, K. (2011). The Impact of Social Media on Children, Adolescents, and Families. *Pediatrics*, *127*(4), 800–804. doi:10.1542/peds.2011-0054 PMID:21444588

Odgers, C. (2018). Smartphones are bad for some teens, not all. *Nature*, *554*(7693), 432–434. doi:10.1038/d41586-018-02109-8

Orben, A., Dienlin, T., & Przybylski, A. K. (2019). Social media's enduring effect on adolescent life satisfaction. *Proceedings of the National Academy of Sciences of the United States of America*, *116*(21), 10226–10228. doi:10.1073/pnas.1902058116 PMID:31061122

Paat, Y.-F., & Markham, C. (2021). Digital crime, trauma, and abuse: Internet safety and cyber risks for adolescents and emerging adults in the 21 st century. *Social Work in Mental Health*, *19*(1), 18–40. doi: 10.1080/15332985.2020.1845281

The Lancet Child & Adolescent Health. (2018). Growing up in a digital world: Benefits and risks. *The Lancet Child & Adolescent Health*, *2*(2), 79. doi:10.1016/S2352-4642(18)30002-6

Tripathi, D., & Shahri, M. A. (2019). Digital Communication Controlling Youngsters in Delhi, India, and Salalah, Oman: A Case Study. *International Journal of Communication and Media Science*, *6*(3), 7–14. doi:10.14445/2349641X/IJCMS-V6I3P102

Witt, A., Ordóñez, A., Martin, A., Vitiello, B., & Fegert, J. M. (2020). Child and adolescent mental health service provision and research during the Covid-19 pandemic: Challenges, opportunities, and a call for submissions. *Child and Adolescent Psychiatry and Mental Health*, *14*(1), 19. doi:10.1186/s13034-020-00324-8

Zu, N., Chen, Z., Duan, Z., Li, L., & Meng, Z. (2017). *A review of social media in education: Effects and attitudes*. Academic Press.

KEY TERMS AND DEFINITIONS

Adolescence: Is a phase characterised by the increased need for social interaction with peers as humans transit from childhood to adulthood.

Cyberbullying: The use of digital technologies in bullying a person.

Digital Technologies: Is the use of technologies in different disciplines like education, healthcare, insurance and so on through the incorporation of software, hardware and services that generates, stores, and processes data.

Education: Is the process of acquisition of knowledge.

Mobile Device: Refers to portable computing devices like cellphones, tablets, smartphones, laptops and so on.

Social Media: Is a digital networking platform which consists of websites and systems (mobile, desktop, embedded) to facilitate communication, engaged interactions, content (text, video, audio, pictures, files and so on) sharing and collaboration. These include Facebook, YouTube, Instagram, Twitter and so on.

Socialisation: Is the connection or mixing among family members, community, friends, and entities for development or belonging.

Chapter 15

Impact of COVID-19 on Adolescent Online Learning in Bangladesh:

Insights From Government School Teachers

Fahmedur Rahman Himel
Bangladesh University of Professionals, Bangladesh

Fariha Jahan Prima
Jahangirnagar University, Bangladesh

ABSTRACT

In this pandemic circumstance, the Bangladesh government started distance learning using broadcasting and online classes. This is a comparative study exploring the impact of online classes on adolescent learning during COVID-19 lockdown in Bangladesh. The study conducts a telephonic interview with 10 headteachers of urban and rural government schools through a semi-structured questionnaire. The study uses descriptive and documentary methods for analysis. The study finds the limitations and challenges facing by the teachers and the students to attend online classes. However, online examinations and assignments are more impactful than online classes to engage students and practice lessons but not enough to compensate for the learning gaps. Even parents are not satisfied with these. Many parents admitted their children to madrasah, and many parents want to continue their children's education after reopening the schools.

DOI: 10.4018/978-1-7998-8318-0.ch015

INTRODUCTION

A school is a place of learning and a commonplace of social relationships, communication, and physical activities. Besides, school is the only place where adolescents never asked for permission to go (Hamad et al., 2020). But, the Lockdown has a drastic impact on every aspect of human life around the world. Due to prolonged Lockdown, everything is happening online. Similarly, academic institutions started online education through different mediums like Zoom, Google Classroom, and Google meets (Tabassum et al., 2021).

In developed countries, the practice of the online education system is familiar and adaptive to the students. Furthermore, (Fernandes et al., 2020) scholars find that social media and streaming services significantly increase during the pandemic. The study's findings also disclosed that the students who are getting addicted to gaming, using social media, and compulsive internet use are high depression, loneliness, escapism, poor sleep quality, and anxiety related to the pandemic. In contrast, (Alabdulkareem, 2015) discusses that the current use of social networks positively impacts teaching and learning. The study finds that in Saudi middle schools, 95% of teachers and 85% of students have the internet at their homes (Alabdulkareem, 2015). As a way of education, they use WhatsApp as a central tool. The writer mentioned that both the teachers and the students are comfortable using social media for education from this study.

In developing countries like Bangladesh, the online education system is quite a new concept, and a sluggish stream of internet service is a challenge. Consequently, many rural students lag, struggling to get primary education, and many students dropping out unaccountably. For these reasons, relevant ministries of the Government of Bangladesh decided to implement distance learning for the students. These classes are broadcast on the *Shangsad Bangladesh* tv channel. More recent evidence from (Biswas et al., 2020) study reveals that 86% of the surveyed students who respond are aware of Shangsad tv classes. Among these students, 92% responded to having access to it.

In contrast, the analysis by (Baird et al., 2020) finds less than 10% of the adolescent students watch the government-supported Shangsad tv classes. Less than 10% of students use media to continue learning. Moreover, (Biswas et al., 2020) find that 50% of the surveyed students are aware of providing online educational materials, only 21% of the students respond using the internet for educational purposes.

A recent literature review on the related subject ((Baird et al., 2020; Farhana et al., 2020; Shama, 2020; Bishwas, 2020) mentions adolescents' anxiety and mental health issues. School closures have curtailed adolescents' learning and, contrariwise, increased involvement in household chores. Students are not concerned with study gaps, even those who have better internet access. In addition, lack of internet access and awareness of using apps among the teachers are also observed. In the case of the female adolescents whose families are not economically affluent, their future learning is uncertain.

Analogously (Hamad et al., 2020) study finds that the girls' adolescents felt like a loser as the school was closed, and they also lost contact with their teachers and friends. The examination of (Bishwas, 2020) testifies a model where most university students are not satisfied with their online classes. The study also finds that the majority of the students are depressed and concerned about their future. A statistical analysis by (Fernandes et al., 2020) shows that internet use increased in lockdown time. Use of social media sites increased 5.37 to 7. 81; gaming hours increased .73 to 1.54, Streaming services use increased [illegible] to [illegible] hours per day.

Considering the pandemic circumstance, finite literature has been found that analyses or measures adolescents' learnings. It is estimated that the study's findings will help academicians, researchers, policy practitioners, and others working in the educational sector conduct further investigation.

THE OBJECTIVE OF THE STUDY

The study intended to explore the impact of online classes on adolescent learning during the Covid-19 lockdown in Bangladesh. Based on the broad objective, the study compares the benefits and limitations of online classes for adolescent learning between the students of urban and rural government schools in Bangladesh through the lens of the Government headteachers.

RESEARCH METHODS

The study collects primary data through the *Telephonic Interview Method* with the headmasters of the Government schools. A total number of ten headteachers from different urban (5 headteachers) and rural (5 headmasters) Government schools were interviewed for the study.

A semi-structured questionnaire was developed and designed in three parts for the interview. The first part is designed to collect information about the teachers' experiences in online classes. In the second section, the questions contain information about the teachers' perception of students' participation in online classes and academic activities. The final part of the questionnaire collects information about family contributions to online learning participation and support during the Covid-19 lockdown.

The study investigates adolescents between 10 to 16 years old. The study deploys a descriptive method to analyze the primary data using MS Excel to evaluate adolescents' online learning factors.

FINDINGS FROM THE SURVEY

Teaching Experience on Online Classes

The general background of the teacher shows that 50% of the teachers have 26-30 years of teaching experience, whereas 30% have 21-25 years and 20% of teachers have 36-40 years of teaching experience (Figure-1). The teachers respond that the majority of the teachers used the smartphone for taking online classes. In rational, the teacher said that using a phone for online classes is much friendly and handy.

The teachers mentioned that they have heard about digital learning, but they have no prior experience of taking online classes in their entire teaching career. For these reasons, they and their assistant teachers face different challenges such as Connection problems, sluggish internet streaming, functioning apps, and lack of smart devices (Figure-2). In addition, the teachers express sorrowfully that no teacher gets any training on taking online sessions from the government or NGOs—consequently, many of their assistant teachers guiding the students over phone calls.

Figure 1. Teaching experience of headteachers
Source: Telephonic Interview, 2021

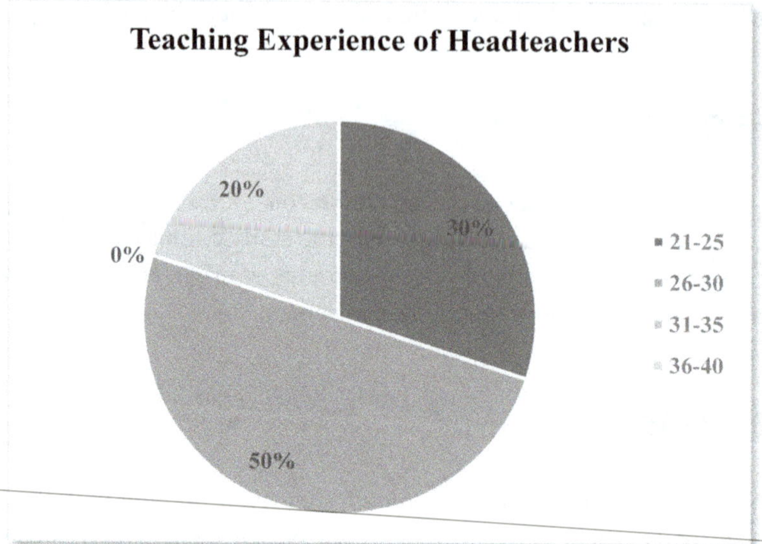

Figure 2. Challenges for teacher for attending online classes
Source: Telephonic Interview, 2021
Note: Multiple answer were possible

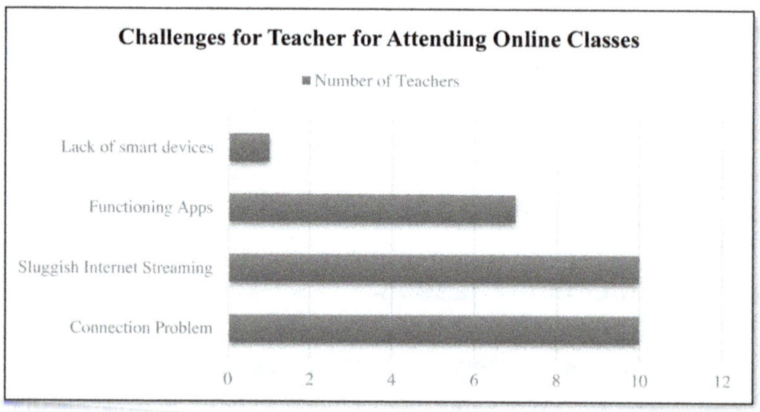

Besides, the headteacher mentions some subjects that teachers often face difficulties making students understand studies over online classes (Figure-3). All the headteachers said that they and their assistant teachers enjoy teaching in classrooms rather than online classes. Because online classes are unidirectional where a teacher cannot meet and communicate in person with the students and colleagues.

Students Are Lagging Despite Online Classes

This section discusses the Teachers' Perception of Students Participation in Online Classes amid CO-VID-19 lockdown.

Figure 3. Challenges for teacher for teaching subjects
Source: Telephonic Interview, 2021
Note: Multiple answer were possible

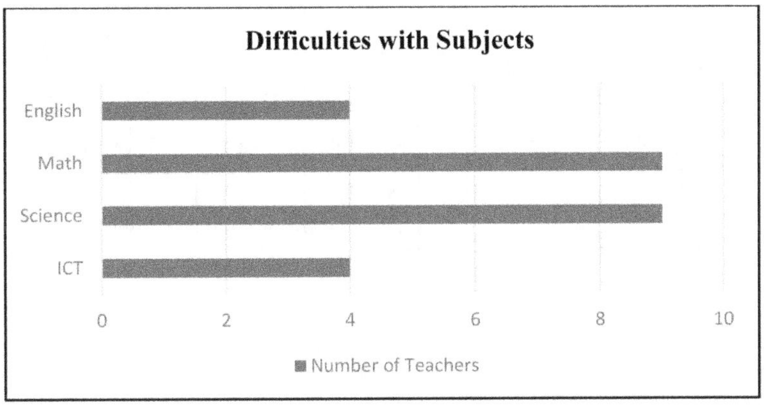

The teachers mentioned that no students regularly attend online classes by excusing different reasons (Figure-4). Teachers also note that as the students participate in online classes by their parent's devices, many students cannot join the classes if parents take their phones at work. Besides, 90% of teachers respond that students don't have internet access and lack interest in online classes, whereas 80% of teachers respond that students face connection problems. 20% of teachers respond to not having smart devices. One of the great responses from the teachers was that the students felt bored and don't enjoy the online classes as there are no other activities like a classroom.

In urban areas, headteachers respond that the percentage of attending online classes is between 10-39%, whereas, in rural areas, it is 10-29% (Table-1). In the case of attending online examinations and assignments, the percentage of attending is between 60-89% in urban areas and 50-69% in rural areas (Table- 2).

Figure 4. Students challenges for attending online classes
Source: Telephonic Interview, 2021
Note: Multiple answers were possible

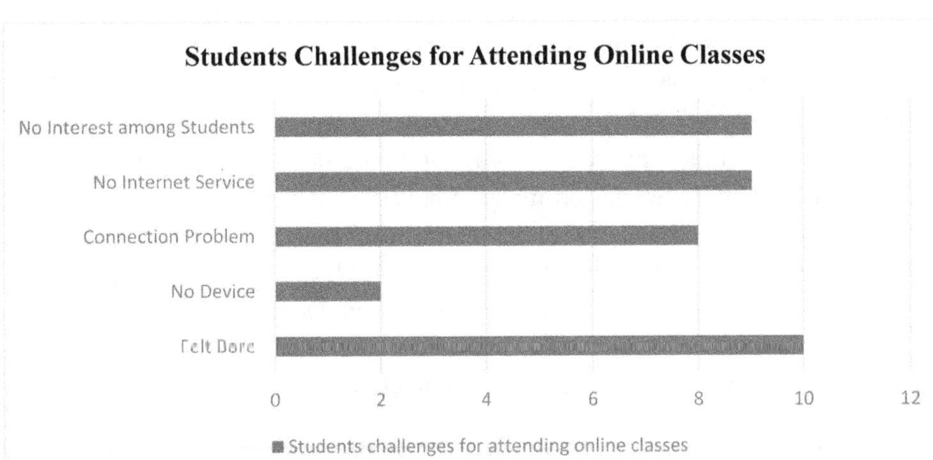

Table 1. Percentage of attending online classes

		10-19%		20-29%		30-39%		40-49%		50-59%		60-69%		70-79%		80-89%		90-99%	
		Count	Column N %	Count	Column N %	Count	Column N %	Count	Column N %	Count	Column N %	Count	Column N %	Count	Column N %	Count	Column N %	Count	Column N %
Headteachers	Urban School	3	57.1%	1	50.0%	1	100.0%	0	0.0%	0	0.0%	0	0.0%	0	0.0%	0	0.0%	0	0.0%
	Rural School	4	42.9%	1	50.0%	0	0.0%	0	0.0%	0	0.0%	0	0.0%	0	0.0%	0	0.0%	0	0.0%

Rationally all teachers respond that the students are not spending the same time on the study. The teachers also mentioned that the parents also get frustrated with the children. Because, Most of the time, they are spending their time on rooftops, garages (mostly urban areas), or going to the fields. On the other side, the escalation use of social media was also observed.

Table 2. Asking about online exam and assignments

		Asking About Online Exam and Assignments																	
		10-19		20-29		30-39		40-49		50-59		60-69		70-79		80-89		90-99	
		Count	Column N %	Count	Column N %	Count	Column N %	Count	Column N %	Count	Column N %	Count	Column N %	Count	Column N %	Count	Column N %	Count	Column N %
Headteachers	Urban School	0	0.0%	0	0.0%	0	0.0%	0	0.0%	0	0.0%	1	25.0%	3	60.0%	1	100.0%	0	0.0%
	Rural School	0	0.0%	0	0.0%	0	0.0%	0	0.0%	3	75.0%	2	40.0%	0	0.0%	0	0.0%	0	0.0%

The teachers also state that most of the girls get busy with household chores. Many female adolescents from rural areas aged between 15-16 years or above are getting married, or their parents are thinking of marrying them off. In contrast, Male students are working with their fathers to support the family financially. In both cases, the teachers observe less time spent in learning and a lack of learning motivation.

In measuring the impact of online learnings on students (Table-3), 50% of the headteachers respond that online learnings are helpful for the students. 40% of the headteachers respond to believing slightly beneficial, and 10% of teachers answer as online learnings are not helpful.

In the case of measuring the impacts of online exams (Table-4), 40% of the headteachers respond that somewhat online examination and assignments are helpful for the adolescents between 10-16 years old, whereas 30% of the rural headteachers respond that online examination and assignments have no impact on the students. But 20% of the headteachers believe that these are very helpful and extremely helpful headteachers for the students for engaging and practicing the lessons.

Table 3. Effect of online learnings on students

		Effect of Online Learnings on Students									
		Not at all Helpful		Slightly Helpful		Somewhat Helpful		Very Helpful		Extremely Helpful	
		Count	Column N %	Count	Column N %	Count	Column N %	Count	Column N %	Count	Column N %
Headteachers	Urban School	0	0.0%	2	50.0%	3	60.0%	0	0.0%	0	0.0%
	Rural School	1	100.0%	2	50.0%	2	40.0%	0	0.0%	0	0.0%

Table 4. Effect of online examination and assignments on students

		Effect of Online Examination and Assignments on Students									
		Not at All Helpful		Slightly Helpful		Somewhat Helpful		Very Helpful		Extremely Helpful	
		Count	Column N %	Count	Column N %	Count	Column N %	Count	Column N %	Count	Column N %
Headteachers	Urban School	0	0.0%	1	100.0%	2	50.0%	1	100.0%	1	100.0%
	Rural School	3	100.0%	0	0.0%	2	50.0%	0	0.0%	0	0.0%

Consequently, 50% of the headteachers respond that distance learning or online classes can fulfill the learning gaps created by the COVID-19 lockdown. In addition, 30% of the teachers who are teaching at rural schools respond that online education cannot fulfill the learning gaps. But 20% of the teachers teaching at urban schools respond that online education can fulfill the learning gaps (Table-5).

Table 5. Teachers perception on learning gaps

		Teachers Perception on Learning Gaps					
		Yes		No		Maybe	
		Count	Column N %	Count	Column N %	Count	Column N %
Headteachers	Urban School	2	100.0%	0	0.0%	3	60.0%
	Rural School	0	0.0%	3	100.0%	2	40.0%

Parental Support for Online Learning Amid COVID-19 Lockdown

The section describes the headteachers' experiences with the parents for supporting adolescents' online learning and observed satisfaction level amid COVID-19 Lockdown.

In the case of parental financial support for online classes amid COVID-19 Lockdown (Table -6), 60% of headteachers from urban schools respond that the parents are a little more responsible; they financially supporting their children by giving private tuitions, school fees, providing required reading materials. At the same time, the response is 40% in the case of the rural scenario.

The headteachers from the urban schools mentioned that the parents are a little more responsible, financially supporting their children by giving private tuitions, school fees, providing required reading materials. But the headteachers from the rural schools mentioned that parents are not interested in investing in their children's online education. The parents think that the teachers are taking money from them by excusing online classes, assignments, and exams.

Table 6. Parental financial support for online classes amid COVID-19 lockdown

		Parental Financial Support For Online Classes Amid COVID-19 Lockdown									
		Highly Disagree		Disagree		Neither Agree nor Disagree		Agree		Highly Agree	
		Count	Column N %	Count	Column N %	Count	Column N %	Count	Column N %	Count	Column N %
Headteachers	Urban School	1	33.3%	1	50.0%	3	60.0%	0	0.0%	0	0.0%
	Rural School	2	66.7%	1	50.0%	2	40.0%	0	0.0%	0	0.0%

In the case of non-financial supports (Table -7) like motivating the students, guiding them, helping to attend online classes, 66.7% of headteachers from urban schools respond positively, whereas 33.% of headteachers from rural schools. In addition, 66.7% of headteachers from rural schools (33.3% of urban headteachers) respond that perhaps the parents support their children. Because they found parents are not interested in investing in their children's online education. The parents think that the teachers are taking money from them by excusing online classes, assignments, and exams.

Table 7. Parental non-financial support for online classes amid COVID-19 lockdown

		Parental Non-Financial Support For Online Classes Amid COVID-19 Lockdown					
		Yes		No		Maybe	
		Count	Column N %	Count	Column N %	Count	Column N %
Headteachers	Urban School	4	66.7%	0	0.0%	1	33.3%
	Rural School	2	33.3%	1	100.0%	2	66.7%

Consequently, most of the teachers respond that the parents expressed high dissatisfaction and dissatisfaction with the online education system (Table -8). The teachers also said that parents think that online classes are not bringing any changes to students. For this reason, most parents from rural areas admitted their children to madrasah for continuity of education. Besides, many of the parents called headteachers that they are not economically feasible amid lockdown. They will continue their child's education when schools reopen. The headteachers and the assistant teachers have to understand the importance of online classes amid COVID-19 lockdown to continue children's learning.

Table 8. Parental satisfaction of online classes amid COVID-19 lockdown

		Parental Satisfaction Of Online Classes Amid COVID-19 Lockdown									
		Highly Dissatisfied		Dissatisfied		Neutral		Satisfied		Highly Satisfied	
		Count	Column N %	Count	Column N %	Count	Column N %	Count	Column N %	Count	Column N %
Headteachers	Urban School	2	40.0%	1	33.3%	2	100.0%	0	0.0%	0	0.0%
	Rural School	3	60.0%	2	66.7%	0	0.0%	0	0.0%	0	0.0%

CONCLUSION

Due to a prolonged lockdown for continuing the learnings, the relevant ministries of the Bangladesh government started online classes. The study's objective was to explore the impact of online classes on adolescent learning during the Covid-19 lockdown.

This study reflects an image of how urban and rural area students are facing problems with online classes. In Bangladesh, in rural areas, students cannot participate in online classes due to internet service, smart devices, and other technical problems. In addition, many students get involved in working with their fathers, and females become busy with household chores. On the other side, the students from the urban areas are more concerned about tuition and other private study activities along with online classes. But still, the percentage of attendance is mediocre.

The teachers mentioned that teachers' training is essential for more effective online classes. In addition, the teachers suggest that the government should count scholarships for financially challenged students and offer student internet packages on easy terms. Besides, launching apps for recorded lectures according to classes and strictly monitoring online examinations could be helped to cover the learning gaps.

REFERENCES

Alabdulkareem, S. A. (2015). Exploring the use and the impacts of social media on teaching and learning science in Saudi. *Procedia: Social and Behavioral Sciences, 182*, 213–224. doi:10.1016/j.sbspro.2015.04.758

Baird, S., Seager, J., Sabarwal, S., Guglielmi, S., & Sultan, M. (2020). *Adolescence in the Time of COVID-19*. Academic Press.

Bishwas, P.C., Hakim, M.U., Alam, M.S. (2020). Online Class and Its Psychological Impact on Satisfaction of University Students in Bangladesh during COVID – 19 Pandemic. *Ioer International Multidisciplinary Research Journal, 2*(4), 101 – 109.

Farhana, Z., Tanni, S. A., Shabnam, S., & Chowdhury, S. A. (2020). *Secondary Education during Lockdown Situation Due to Covid-19 Pandemic in Bangladesh: Teachers' Response on Online Classes.* Academic Press.

Fernandes, B., Biswas, U. N., Mansukhani, R. T., Casarín, A. V., & Essau, C. A. (2020). The Impact of COVID-19 Lockdown on Internet Use and Escapism in Adolescents. *Revista de Psicología Clínica con Niños y Adolescentes, 7*(3), 59-65.

Hamad, S., Abu Hamra, E., Diab, R., Abu Hamad, B., Jones, N., & Małachowska, A. (2020). *Exploring the Impacts of Covid-19 on Adolescents in the Gaza Strip*. ODI.

Shama, S., & Ikbal, J. (2020). *Education during A Pandemic: A feasibility study of online classes in Bangladesh to counteract potential study gaps caused by COVID-19 related lockdowns*. Academic Press.

Tabassum, M., Mannan, S. E., Parvej, M. I., & Ahmed, F. (2021). Online Education during COVID-19 in Bangladesh: University Teachers' Perspective. *Aquademia, 5*(1).

Chapter 16
COVID-19:
The Effects and Implications on Students Mental Health and Wellbeing

Gillala Rekha
https://orcid.org/0000-0003-2688-2323
Koneru Lakshmaiah Education Foundation, India

ABSTRACT

Due to the COVID-19 pandemic, governments around the world closed all the educational institutions to control the spread of disease, which is creating a direct impact on students, educators, and institutions. The purpose of this study was to analyze the perception of academic stress experienced by students during current online education and coping strategies using emotional intelligence adopted by them. The study aims to conduct a timely assessment of the effects of stress due to COVID-19 pandemic on the mental health of college students. The authors conducted interview surveys with 227 students at a private university in India to understand the effects of online education during pandemic on their mental health and well-being. The data were analyzed through quantitative and qualitative methods. Of the 227 students, 71% indicated anxiety and stress due to ongoing pandemic.

1. INTRODUCTION

A lot of awareness has been created by the government of India on the spread of the COVID-19 pandemic. Many precautions have been initiated by both Central and state governments and followed by the citizens of the nation. The entire nation was unaware of what would be the consequences until it reached its peaks. Though the viral spread broke world-wide during December 2019, there was not much impact on a bigger country like India though it had its borders widespread with China.COVID-19 has emerged into the world during the period of December 2019 in China. According to World Health Organization (WHO) (World Health Organization, n.d.), the origin of the disease is from the wet market of Wuhan City, China. COVID-19 is one of the corona virus which has its base in China (Wu et al., 2020). The disease was subsequently named as COVID-19. COVID-19 spread within China and to the rest of the

DOI: 10.4018/978-1-7998-8318-0.ch016

world.WHO initially did not warn the entire world as they were not sure about the spread and implications of the disease. Later, on 30 January 2020, the World Health Organization declared the outbreak a public health emergency of international concern.

While countries around the world continue to mobilize to contain the spread of COVID-19, mental health experts say, we can't lose sight of equally alarming issue-the long term mental health impact the pandemic is going to leave on society. The pandemic breakout has left a traumatic impact on everyone across the world. One thing that we know about trauma is that while it's happening we do whatever we can do to survive. We bear down and just get through it, which is what we're all trying to do right now. We are still very much in the traumatic phase. Unlikely, there is a residual stress, depression certainly due to financial pressures. Learning how to re-engage with the world in this new normal is going to be a hassle task for a lot of people and those are exactly the kinds of risk factors that led to increasing rates of depression, anxiety and even suicide. The rapid spread of pandemic gave little chance to prepare for, or even process all that has happened in terms of job loss and the complete uprooting of the everyday life and relationships.

2. DISRUPTION OF EDUCATION SYSTEM WORLDWIDE DUE TO LOCKDOWN AND CORONAVIRUS

It is the largest disruption of global education system in history. When the pandemic first hit around, 1 and half billion students around the world were sent home as schools, universities and other educational institutions closed their doors. Though some have reopened but class isn't in session for half of the world's students. The lucky ones were able to pursue their studies online but for many even in developed countries that's not an option. The world was already struggling to come to grips with educational inequality, now the United Nations warns that the COVID-19 (ICMR, n.d.) crisis is making it worse by disproportionately affecting the most vulnerable, and threatening to wipe out decades of progress for this generation and possibly beyond.

At the height of the pandemic, educational institutions in about 190 countries closed their doors. For students and many parents the foundational problems are still there. A lot of parents don't understand technology, a lot of families don't have internet or digital technology at home, but the universities are much better prepared by which the professors and other university educators were able to use the cloud and other messenger services like email that is all in accordance with data protection laws which was not the case before. The Corona crisis has made students realize where they stand in their technical skills and what needs to be done for up skilling..

In most classrooms students and professors still use analog tools and learning materials (). These traditional teaching techniques didn't work during the lockdown period, so the educational institutions spent money in digitalizing the classrooms. It is definitely the case where teachers and parents are investing their own private resources to digitalize the daily academic learning. All these have also been criteria for the stress development among students. They are aware of how slowly things are moving and want to see a change, there's room for improvement in the digitalization of certain academies to say the least, there's not enough resources for all the students, it was not enough to make proper home schooling possible but some students think that they are better prepared in case of another lockdown. At the beginning it was new for all of us, the system, the computers, working at home, and now they think that it would work better than first time but the classroom that can switch effortlessly from the college to each student's room it remains a distant goal.

According to UNESCO report 157 crore students across the world will have been affected because of the pandemic, the on-going situation and closure of education institutions both higher and the school education institutions across the world. Students have long been prone to stress, anxiety and depression. According to the CDC (Centres for Disease Control and Prevention), 3 out of 4 Americans between the ages of 18 and 24 report poor mental health tied to the pandemic.

3. PANDEMIC IMPACT ON STUDENT EDUCATION

Many students are not well right now. Drop a pin on the map of the world and the story is much the same. Heightened isolation, depression, anxiety, mental health crisis, courtesy of a college experience stripped almost entirely of a campus life, tradition and structure on top of a pandemic (Baum et al., 1981). Many students say that the college feels like a very far away concept right now. About 41% of student age groups started going to therapy due to their worsening mental stability. More than 5, 60,000 undergraduate students didn't enrol to the academic year of 2020 when compared to 2019. Even in normal times, those who report mental health struggles are nearly twice as likely to drop out of school as their peers. There are stressors for those studying remotely. 1 in 4 Americans between the ages of 18 and 24 reported having seriously considered suicide in the last 30 days, according to the CDC and nearly 70% of college presidents now say that the student mental health ranks among their top concerns.

Researchers say that before the pandemic 65% of college students were wrestling with anxiety, 30% with mental health conditions, 10% with thoughts of suicide. The on-going global crisis is being combined with an over reliance on technology and social media, are making today's young people more anxious than ever before and COVID isolation has made it worse. We are tribal people as human beings. We need a tribe and college is a place where many students find their tribes. 80% of students around the country say that COVID has negatively impacted their mental and spiritual health, and their career aspirations.

Students say deliberately that the main problem is they feel like they're in all of this alone emotionally and academically. But in fact this pandemic to every student is almost the same, everyone are travelling through the same boat and if they face a storm then the passengers experience it collectively.

4. CORONAVIRUS OUTBREAK: STUDENT MENTAL HEALTH

It's been over a year now that the world has been dealing with COVID-19 and its devastating ramifications. Some have lost loved ones, others their jobs. All we have done is staying at our homes in the lockdown and abled to live our lives very used to. And for many of us, this has taken a massive tour on our mental health (Sneha, 2020). About 55.1% of students consider themselves very happy and positive persons who are in control. But, in the pandemic first time in their lives about 44% of students felt lonely, anxious, depressed and stressed. Some of them have even experienced some nervous breakdowns and panic attacks. This made everyone wonder who else is feeling like this. Several studies have found that the young adults are especially prone to mental health effects due to this rampant. A survey on the mental health issues caused due to the COVID-19 epidemic says that the social isolation which was necessary to enforce on the people so as to protect ourselves and themselves from the infection at a certain point, made people crave for seeing the people again, shaking hands and hugging each other as a greeting. The motivation was not often there especially when it comes to doing something for their respective Universities/Institutions at times of cultural activities, discussions in clubs etc. Children of age 11 to 15 are petrified as they perceive that this might become an enduring and never ending situation. Financially even the elite groups have faced pathetic situations during the times of COVID-19 and still experiencing them. Mentally it just does a lot to the people. It's so true from the researches that young bloods are especially the ones experiencing increased level of anxiety and depression during the pandemic (Gaidhane et al., 2020) (Cooke et al., 2020).

Students who suffer from a generalized anxiety disorder have shut down because they don't know what to do. They are not prepared for this. They didn't run a rehearsal of the situation in their head as they generally do before taking their exams. It's exhausting to not worry because it is practically impossible to not worry. For those who are already suffering from anxiety and depression, the pandemic could be exasperating the symptoms. It is also now a high time to feel concerned for those who have never suffered from any mental health issues in the past and are now facing a life disruption without fully digesting the long-term impact that this pandemic may have on them. A large body of scientific studies depict that there is a close relationship between indicators such as unemployment, mental health and suicide. Research out of the **Great Recession** of 2008 reveals that a 1% increase in unemployment was accompanied by a 1% increase in the rates of suicides in the US. This explains that there is a peak and valley response in our nervous system when faced with trauma. Right now we are at a peak in which we are engaging with the immediate threat of the pandemic. We go into survival mode which can be highly taxing on a person's mental health and in fact it doesn't actually let us think through, or process what's happening to us. It's only the interest which is getting and keeping us alive, and keeping us safe.

The pandemic has been a different experience for people of different age groups; young people who are turning their careers, building momentum in their careers may have experienced this more of a negative impact, in terms of the disruption as well as the financial insecurities in the future. According to the ILO (International Labour Organization) about 114 million jobs were lost in 2020 world-wide, and it's especially young people who haven't been able to establish a career, who are worried about their prospects in the job market. But it's not all gloom and doom, experts say that once we are aligning about the rise of mental health struggles during the pandemic is we are able to talk about it more openly. This rampant can be seen as an opportunity to raise awareness on mental health as well as increased access to mental health care services for vulnerable groups as well as just the fact that we all have been through dark light in the past year, it's very hard to continue to deny that physical and mental health aren't equally predominant.

5. THE OBJECTIVE OF THE STUDY

During the current COVID-19 epidemic, this study was done to better understand and assess students' perceptions of academic stress and coping methods based on their emotional intelligence. Because of the abrupt lockout, education has been disrupted, making it difficult for educational institutions, educators, students, and their families, and negatively impacting their psychological, emotional, social, and financial well being. The study's goals are to investigate students' experiences of academic stress as a result of online education. To investigate how students deal with academic stress by employing emotional intelligence coping methods. To investigate the impact of academic stress and emotional intelligence on male and female students, as well as other factors such as academic failure, present online education, and current home environment.

Also, the presented research aimed at understanding the mental health of the students during first and second wave of coronavirus (covid 19) pandemic. More than 27 million Indian undergraduate students are home schooling. The study conducted on Telangana and Andhra Prasad student community to understand and assess the academic load, online schooling, fear of loneliness and depression, procrastinate (delay or postpone) about their personal goals in life, perceive the risk of contagion during this period of COVID-19 pandemic, perceive the condition of social isolation imposed during this period of COVID-19 pandemic. All of the constructs were assessed by collection of self-report in anonymous surveys. The results of this study showed that academic workload, separation from school, and fears of contagion had negative effects on college students' health via perceived stress. Our results enrich the literature on stress and health and offer novel practical implications for all circles of the society to ensure students' health under the context of the COVID-19 epidemic.

5.1 Methodology

The process of data collection was through forms and conducted an interview surveys with 227 students at a private university in the India to understand the effects of online education during pandemic on their mental health and well-being. The data were analyzed through quantitative and qualitative methods.

5.1.1 Data Collection

The questionnaire consists of both quantitative and qualitative aspects of the students mental model. The survey was performed by sharing the form and the total number of students participated are 227 (Male: 147, Female: 80). The following questions and the responses are listed below:

1. On a scale of 1 to 7 (7 being the happiest), how happy do you consider yourself? Scale of 1 to 7 (7 being happiest). The responses were out of 227, 27 students express that they are happiest with a scale of 7. The highest number students are on average with 4 and 5 scale i.e, 99. 57 students express less being happy in this pandemic period with scale of 1,2,3.
2. Compared to your colleagues or friends, do you consider yourself happier than most of them? We even gathered how happy are their compared with the friends during the pandemic situcation. In which we got 50-50 saying 50 happy and 50 sad compared with their friends.
3. How often do you feel lonely during this pandemic? The answers are captured using a scale of values from Always to Never. The following were the percentage of loneliness expressed by the students. Always- 8.8%, Often- 20.3%, Sometimes-44.1%, Rarely-18.9%, Never-7.9%.
4. How often do you feel depressed due to pandemic situation? The answers are captured using a scale of values from Always to Never. The following were the percentage of depression experienced by the students due to pandemic and lockdown. Always-6.2% Often- 17.6%, Sometimes- 38.3%, Rarely-18.9%, Never- 18.9%. Apart from the feel of depression we tried to identify the scenarios through which they are expressing depression by asking the following questions.
    a. Have you recently been suffering from headache or pressure in your head? The response was 50-50 wherein the students are suffering with headache. The following were the percentages of students expressed headache or pressure in head due pandemic (Yes: 46.7%, No: 53.3%).

b. Have you recently lost much sleep through worry? The response were worth of concern with 52% of students expressed sleeping disorder and 48% of students responded as No(Yes: 52%, No: 48%).

The emotional aspects of the students are captured using the below questionnaire.

5. Have you recently been worried about your family or close friends? The students expressed worry about their family and friends during the ongoing pandemic. 72.7% students have fear about their family and close friends. (Yes: 72.7%, No: 27.3%)
6. How do you perceive the condition of social isolation imposed during this period of COVID-19 pandemic? We got mixed response for the social isolation with 5.7% as extremely stressful and 18.1% students expressed no stress. (Not at all stressful: 18.1%, Somewhat stressful: 33%, Moderately stressful: 28.6%, Very stressful: 14.5%, Extremely stressful: 5.7%)
7. What are the usual psychological or emotional effects of stress you've noticed at yourself?

From the collected data, Figure 1 show the students Psychological or emotional stress. By observing the above figure, we can analyse that majority of students are irritable or angry, which states that stress levels are high due to COVID-19 pandemic.

Figure 1. Students psychological or emotional stress

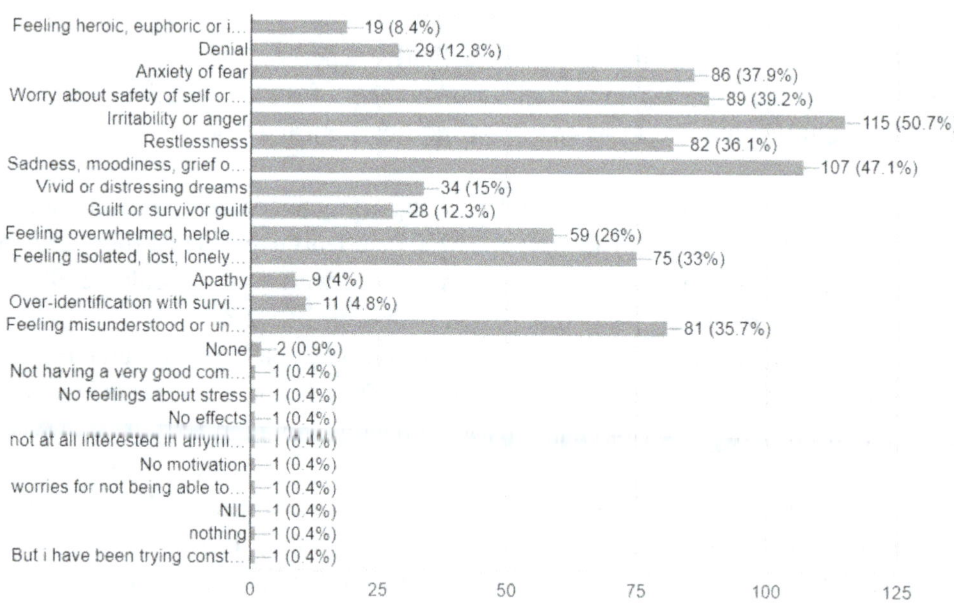

8. What are the usual behavioral effects of stress you've noticed at yourself? The data collected should that majority of students are losing their efficiency and effectiveness, these being highest at 48% and 41% of change in activity levels. 39.6% of irritability, outburst of anger and frequent arguments. It concludes that due to ongoing pandemic the students are facing lot of stress.

Figure 2. Usual behavioral effects of stress among students

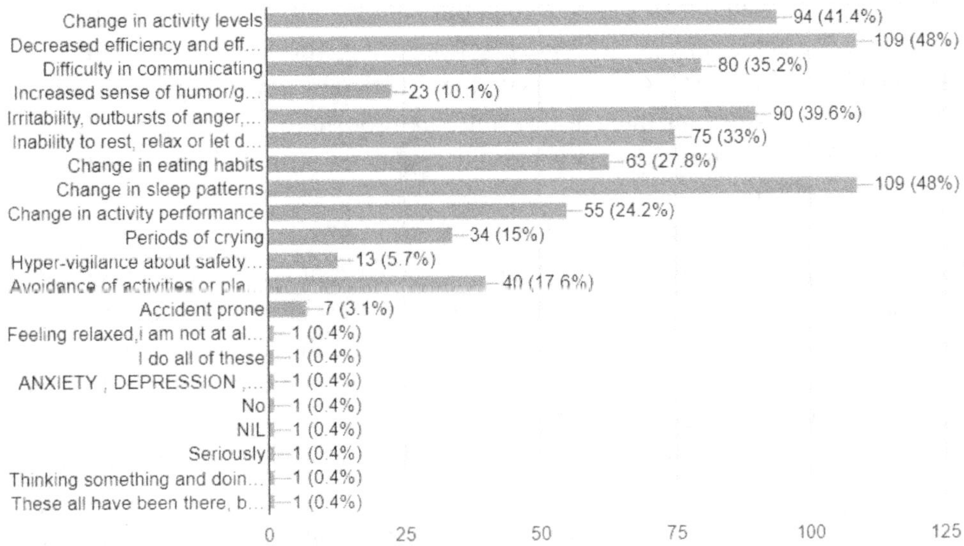

Apart from this we even collected the data from the students based on the ongoing first wave and second wave stress. Around 78.4% students are feared and feeling stressful due to second wave of covid 19.

9. Which was more stressful to you (First wave/ Second wave)? (First wave: 36.6%, Second wave: 78.4%)

Most of the students reported stress due to second wave of covid-19 pandemic.

6. FINDING AND DISCUSSION

COVID-19 raised the degree of anxiety and worry about their own health and the health of their loved ones, according to the great majority of students. Over a third of those who expressed concern raised questions for their families and relatives who were more vulnerable, such as older adults, those with pre-existing health problems, and those who work in environments where they are more likely to be exposed to COVID-19, such as essential and health care workers. Due to different kinds of distraction, the great majority of students had difficulties concentrating on academic work. Almost half of them stated that their house is a distracting atmosphere and that it is a better place to relax than study. At home, participants said they were more likely to be interrupted by family members and domestic tasks. Lack of responsibility, as well as social media, the internet, and video games, all had an impact on students' focus. Due to the absence of interaction and extended concentration to a computer screen, several students said that online classes were prone to distraction. Additionally, repetitive living patterns were highlighted by several as having a detrimental impact on academic task focus. The COVID 19 epidemic disrupted the sleep habits of the majority of pupils, with over one-third describing the disturbances as serious. Half of the students who experienced some disturbance said they stayed up later or woke up later than before the COVID-19 outbreak.

Education system around the world has stumbled apart due to the pandemic and the emergency measures taken by the governing authorities in view of public safety. This abrupt displacement has impacted the educationalists, educators, parents and students around the world drastically in the initial phases. However, the educationalists have tried to cope with the situation by amending the ways of teaching and learning methodologies transforming them into virtual classroom networks. This has made it possible for the students to continue their learning journey and stay connected with their mentors and peers through virtual platforms (Bao, 2020).

Nevertheless the virtual classroom system has got its own setbacks. Though students had a momentary interest in the virtual learning patterns, with time they are becoming victims to several physical and mental health issues such as irritation, anger, anxiety, fear, depression which is further moving towards extremities like suicidal tendencies. This is situation prevailing among the younger generation - the future of the world, is no less than the pandemic itself and is a matter of concern holding a top priority (Sneha & Aluvala, 2021) (Unger & Meiran, 2020).

The educationalists have found the virtual education methodology the better way to continue the students' learning journey in the interest of the future of their nations. Henceforth, many educational institutions have started several student-help sessions and motivational webinars as part of their academic structure to address the physical and mental health concerns of their students and help them out of their emotional traumas. The mentors are also doing their share of contribution by motivating the students periodically towards a happy and healthy lifestyle habits.

7. CONCLUSION

The results of this study showed that academic workload, separation from college, and fears of contagion had negative effects on college students' health via perceived stress. Our findings show that students prefer self-management, which should be encouraged in future research. Self-management of mental health disorders has showed some potential thanks to digital technology and telehealth apps. To conclude, with all the motivation, care and support that are available from family, friends and faculty, students should be able to realize the positivity within them and replenish their strength whenever they tend to fall weak both physically and emotionally, to move towards their career goals and accomplish them with discipline, dedication and diction.

REFERENCES

Bao, W. (2020). COVID-19 and online teaching in higher education: A case study of Peking University. *Human Behavior and Emerging Technologies*, *2*(2), 113–115. doi:10.1002/hbe2.191 PMID:32510042

Baum, A., Singer, J. E., & Baum, C. S. (1981). Stress and the environment. *The Journal of Social Issues*, *37*(1), 4–35. doi:10.1111/j.1540-4560.1981.tb01056.x

Cooke, Eirich, Racine, & Madigan. (2020). Prevalence of posttraumatic and general psychological stress during COVID-19: A rapid review and meta-analysis. *Psychiatry Research*, *292*. https://www.sciencedirect.com/science/article/pii/S0165178120326445 doi:10.1016/j.psychres.2020.113347

Gaidhane, Khatib, Zahiruddin, Gaidhane, Telrandhe, & Godhiwal. (2020). Depression, Anxiety and Stress among the general population in the time of COVID-19 lockdown: A cross-sectional study protocol. *IJRPS, 11*(1), 360-364.

ICMR. (n.d.). *Important COVID-19 Guidelines*. http://pbhealth.gov.in/Coronaguidelines.html

Sneha. (2020). *Recognizing stress based on social interactions in social networks*. Lambert Academic Publishing.

Sneha, Y., & Aluvala, S. (2021). Impact of Experimental Learning on Graduates Success in Engineering Education. *Journal Of Engineering EducationTransformations, 34*(0), 666–669. doi:10.16920/jeet/2021/v34i0/157240

Unger, S., & Meiran, W. R. (2020). Student attitudes towards online education during theCOVID-19 viral outbreak of 2020: Distance learning in a time of social distance. *International Journal of Technology in Education and Science, 4*(4), 256–266. doi:10.46328/ijtes.v4i4.107

World Health Organization. (n.d.). https://www.worldometers.info/coronavirus/

Wu, J. T., Leung, K., Bushman, M., Kishore, N., Niehus, R., de Salazar, P. M., Cowling, B. J., Lipsitch, M., & Leung, G. M. (2020). Estimating clinical severity of covid-19 from the transmission dynamics in Wuhan, China. *Nature Medicine, 26*(4), 506–510. doi:10.103841591-020-0822-7 PMID:32284616

Chapter 17
Web-Based Learning for Adolescents: Innovation and Challenges

Cyncia Matsika
Midlands State University, Zimbabwe

Munyaradzi Zhou
https://orcid.org/0000-0002-2184-0290
Midlands State University, Zimbabwe

Tinashe Gwendolyn Zhou
Midlands State University, Zimbabwe

Wilfreda I. Chawarura
Midlands State University, Zimbabwe

ABSTRACT

This chapter assesses the innovations and challenges of web-based learning for adolescents. The technology acceptance model (TAM) was used to establish the determinants for the challenges and innovations of web-based learning for adolescents. Challenges coined include virtual engagement, adaptability, motivation, technical issues, time management, learning styles, destructions, communication skills, and the digital divide. Web-based learning is an emerging and innovative approach to teaching and learning that meets the evolving needs of adolescent learning. Adolescents should embrace the transformation from traditional brick-and-mortar classroom practices to a web-based environment. Stakeholders must continuously overcome the challenges of web-based learning. Future research can concentrate on developing well-defined web-based course content that supports the holistic learning needs of adolescents and the development of a web-based learning model for adolescents.

DOI: 10.4018/978-1-7998-8318-0.ch017

INTRODUCTION

Once upon a time, teaching and learning were highly didactic and evocative as it was comprised mainly of face-to-face lectures and group discussions in a brick-and-mortar setup. According to Freire (2013), students waited like empty vessels to be filled by their teachers. Disruptions like the COVID-19 pandemic and technological advancements have modified teaching and learning worldwide and exposed numerous innovations and challenges of education systems. (Chiemeke & Mike Imafidor, 2020; Valverde-Berrocoso et al., 2020). Web-based learning refers to the impartation of knowledge utilizing Information Communication Technology (ICT)s and the internet. It is an emerging teaching and learning approach that allows the construction of educational materials, storing them in digital repositories, and reusability of educational materials in different contexts. Various studies have examined several benefits of platforms like Google for education to expose an array of valuable tools and resources for adolescents' learning. However, several challenges of web-based learning faced by adolescents need exploration.

The purpose of this chapter is to delineate web-based learning for adolescents through the evaluation of web-based education that utilizes inordinate forms of video streaming and many advanced functionalities in educational software to conduct teaching and learning. Pertinent innovations and challenges of web-based learning for adolescents are studied. The chapter will also explore online challenges like virtual engagement, adaptability issues, motivation, technical issues, time management, learning styles, destructions, communication skills, and the digital divide posed by web-based learning to adolescents. Information Communication Technologies and web-based innovative technologies that support learning like Web 2.0, Web 3.0, and augmented and virtual teaching and learning are discussed. The overarching research question is "What are the innovations and challenges of web-based learning for adolescents?"

BACKGROUND

Web-based learning refers to the replacement of face to face teaching and learning between student and teacher by using the utmost form of online education via a web browser to promote interactive learning that leads to more productive learning (Bower & Torrington, 2020; Chiemeke & Mike Imafidor, 2020; Means et al., 2009; Tsai & Machado, 2002). Web-based learning also entails various educational interventions that use the Internet (Cook, 2020; Mahalakshmi & Radha, 2020). The World Health Organization (WHO) defines adolescents as people between 10 and 19 years of age. The digital changes the education systems experienced globally have allowed the incorporation of new teaching and learning methods.

Web-based learning is an ideal solution to satisfy the demands of modern teaching and learning. It proffers a professional, flexible, sustainable, and economical approach to teaching and learning. Integration of formal curricula with just-in-time knowledge benefits the education ecosystem (Marzano & Ochoa-Siguencia, 2017). The research only focused on low-qualified adults and did not consider adolescents. This research study focuses on the challenges faced by adolescents during web-based learning.

Exposure to technological gadgets like smartphones, computers, and tablets ushered in a digital generation of young children (Ali, 2018). In his findings, Doctor. Ali revealed that university students who grew up in the digital era (those born between 1982 and 1991) embraced the power of digital connectedness at a faster rate. However, this particular study did not focus on adolescents. Young people use technological gadgets from very tender ages. Vrasidas (2015) reiterates that the availability of ICT resources is not enough since other supportive factors like teacher readiness are crucial for implementation. Adolescents use technical gadgets for social media communication like WhatsApp, Facebook, and Twitter.

Ms and Toro (2016) discussed the characteristics, disadvantages, and advantages of the current web-based learning environments and the factors that influence the acceptance of these technologies in higher education in general. This study is distinct as it aims at narrowing down the examination of web-based learning to adolescents. The factors that impact the effectiveness of web-based learning in an educational context, the excellence of the teacher, and the educational purpose (Hatcher et al., 2013). For web-based learning to be successful, the teacher must also play a pivotal role.

Most researchers identified the challenges of online teaching and did not focus on the innovations and challenges faced by adolescents (Mansureh et al., 2017). This study focuses on the challenges of web-based learning by adolescents and further focuses on emerging learning environments and technologies. Yusof and Saadon (2012) coined that web-based learning has some effects on students' learning. They advocated for more studies to enquire about the best mode of web-based technology in teaching and learning. The research study is on the effects of web-based learning on students studying grammar proficiency at tertiary institutions. Their conclusions are not generalizable to adolescents. They did not narrow their study down to the innovations and challenges that adolescents face during web-based learning.

Technology offers pupils the digital competencies needed in the industry and various learning and socialization opportunities (Gottschalk, 2019). The author recommended that an investigation on the benefits of technology use for adolescents in the teaching and learning activities needs further inquiry (Gottschalk, 2019). This notion contradicts other scholars who believe it is a dangerous fallacy to conclude that the digital generation knows all about information technology. Young people born during the digital era may lack the theoretical knowledge base necessary for particular industrial work despite being techno-savvy (Ali, 2018). Misconceptions on the subject may create employability issues for young people in the job market.

Inal and Korkmaz (2019) remarked that emerging technologies cause changes and constructive developments in all aspects of life. However, they limited their research to the foreign language field and did not consider web-based learning in particular.

Web-based learning presents numerous benefits as it enables class attendance from home and enables students to engage in emerging forms of teaching and learning (Ms & Toro, 2016). Augmented and Virtual Reality (AVR) provides learners with a stimulating learning environment. AVR boosts excitement to learn, increases class participation and improves understanding of concepts (Cabero-Almenara et al., 2019). Rana (2014) recommended applying new e-learning methodologies to capture the learners' interests and preferences based on the holistic needs of a particular group of learners.

A quality product should be simple and easy to use and perform its purpose effortlessly. A good product ought to be usable and beneficial for learners to accept. The Technology Acceptance Model (TAM) (see Fig. 1.) focuses on these elements. The book chapter applied Davi's (1989) TAM model that influences the intention by the user to adopt emerging technologies. TAM consists of two factors, perceived ease of use and perceived usefulness of particular emerging technology (Braun, 2013; Cabero-Almenara et al., 2019; Charness & Boot, 2015; Davis, 2013; Dehghani et al., 2020; Divakarla & Kumari, 2010; Gamage, 2020; Rafique et al., 2020; Valverde-Berrocoso et al., 2020). Learners who perceive web-based learning as complex are likely to resist using it vice-versa.

Technological Acceptance Model

According to Davis (1989), the Technological Acceptance Model (TAM) examines the influence and intention of an individual to use an emerging technology, herein referred to as web-based learning. TAM measures two factors, the perceived ease of use and perceived usefulness. Adolescents who perceive web-based learning as difficult are likely to resist using the new technology, while those who consider it easy to use will be willing and happy to adopt it.

Figure 1. Technology Acceptance Model (TAM)
Source: (Ma & Liu, 2011)

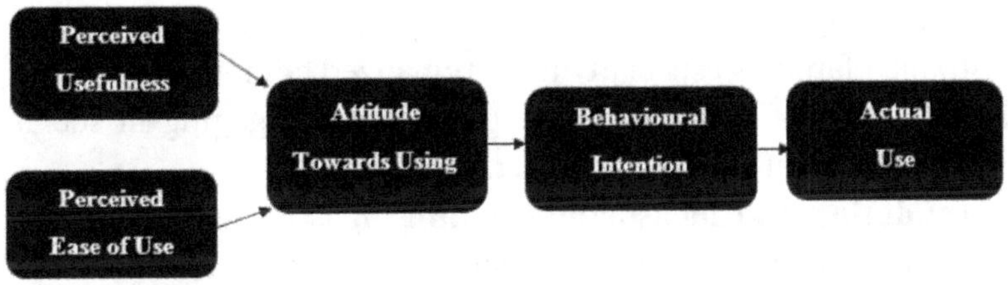

Matsika and Zhou (2021) applied the TAM to analyze the factors that affect the adoption and use of Augmented and Virtual Reality technology in higher and tertiary education. TAM in this research study is applied to expose the usefulness and ease of use items of web-based learning by adolescents in the results and discussion section.

MAIN FOCUS OF THE CHAPTER

The COVID-19 pandemic has disrupted face to face teaching and learning as schools have been mandated to temporarily close not only in Zimbabwe but the world over (Abad-Segura et al., 2020; Bao, 2020; Bloom et al., 2020; Chiemeke & Mike Imafidor, 2020; Dhawan, 2020; Mahalakshmi & Radha, 2020; Murphy, 2020; Schleicher, 2020; Toquero, 2020). The closure of schools for a prolonged period results in high economic and social costs. According to UNESCO (2021), disparities already exacerbated in the education sector because of the disruptions include:

- **Learning interruption**: Temporary closure of schools interrupts the normal learning activities, consequently depriving adolescents of the opportunity for development and growth. The demerits of interrupted learning are disproportionate for underprivileged adolescents who tend to have fewer educational opportunities beyond school.
- **Stress and confusion for teachers**: When schools close indefinitely, teachers' obligations become unclear as they wouldn't know how to maintain connections with pupils to support teaching and learning. Transitions to web-based learning platforms are confusing and frustrating as it is an emerging technology.

- **Parents unpreparedness for web-based and homeschooling**: Most parents are not qualified or prepared to home school adolescents from home, they may not have the proper resources or expertise.
- **Difficulties creating, maintaining, and improving web-based learning**: When formal schools close, the demand for web-based learning increases. Unplanned transformation of teaching and learning from the traditional classrooms to homes in a hurried manner presents various technical and human challenges.
- **Associated economic costs**: Internet connection, relevant ICT gadgets and the associated infrastructure is costly to implement and maintain. Parents will also need to physically monitor the learners as some of them need help on how to navigate the web-based platform.
- **Rise in dropout rates**: It is difficult to ensure that the adolescent child return and stay in school after lock down especially when economic pressures force them to work and generate income for financially constrained families especially in marginalized communities.
- **Social isolation**: Closure of schools deprive pupils from engaging in social activities, as such many adolescents miss out of on social contact.
- **Challenges in validating and measuring learning**: It is difficult to administer formative and summative assessments and evaluations especially high-stake examinations when schools are closed. The web-based platforms are not yet viable to conduct high stake examinations as fairness issues, digital divide concerns and other academic ethical concerns are yet to be addressed. These aforementioned disruptions may trigger disengagement by the learners.

Web-based learning is a panacea to mitigating the disruptions being encountered in teaching and learning. Despite the availability of numerous web-based research studies, very little if any research has paid attention to the challenges being encountered by adolescents during web-based learning. The main objective of this research study therefore, is to unearth the challenges encountered by adolescents during web-based learning. The challenges are discussed in the ensuing sections. Four different urban schools with grade 7 classes aged between 10 and 13 years of age participated in the research study. 180 adolescents across these classes, both male and female responded to the questionnaires and interviews.

METHODOLOGY

Questionnaires and interviews were administered on adolescents at four different urban schools in Zimbabwe's Northern Central District who are between 10 and 13 years of age, hence focus groups were used to collect the data. Their respective teachers were asked to help in the data collection to promote trust as adolescents normally do not warm up to strangers. The Technology Acceptance Model framework was then used determine the various challenges that adolescents encounter during web-based learning and the innovations and progress of web-based learning. A systematic literature review was also conducted in this research study to expose that previously conducted studies on web-based learning did not narrow down to the factors that affect adolescence during online teaching and learning. A qualitative research method was used as it made it possible for follow up questions to be asked on obtained answers which created very valuable conversations around web-based learning issues. Because of the prevailing lockdown and COVID-19 restrictions, other methods like observation technique, were difficult to be successfully conduct.

RESULTS AND DISCUSSION

An overview of the experiences in the use of web-based learning is given in the Figure. 2. The results show that developing countries are still experiencing basic ICT problems which requires basic ICT solutions hence there is need to promote basic ICT solutions like provision and expansion of ICT infrastructure, awareness on the use of ICTs and training of users so that no one is left behind in the process of digitization.

Figure 2. Survey results on experiences in the use of web-based learning

Research Question	Very Difficult	Difficult	Neutral	Easy	Very Easy	N
1. Was it easy to adapt to web-based learning from the traditional face to face?	94	23	35	16	12	180
2. Do you have reliable internet connection?	111	0	0	0	69	180
3. How long have you had a digital tool?	112	0	0	0	67	179
4. How do you manage the time required for web-based learning?	153	25	0	0	0	178
5. Do you prefer web-based learning to face to face learning?	87	19	45	17	12	180
6. Are you familiar with Microsoft Office packages?	82	69	2	14	13	180
7. Is the feedback on web-based assignments helpful ?	75	54	26	18	7	180
8. Do you find it easy to communicate during web-based learning?	68	62	21	16	13	180

The application of the TAM on challenges and innovations related to web-based learning were unearthed. In addition, challenges of Web-based learning informed through literature review are presented in this section. The use of a structured approach (TAM) to analyse the challenges and innovations concerning web-based learning by adolescents helped to inform and expose the challenges that adolescents encounter during web-based learning.

Challenges Encountered During Web-Based Learning Informed Using TAM

The analysis on the challenges encountered by adolescents during web-based learning is outlined in the following sections in relation to the TAM elements.

Usefulness items according to the TAM:

1. Using web-based technology will facilitate easy adaptation for the learner and allow the achievement of learning tasks more swiftly.
2. Using web-based technology will enhance the learning experience and allow scheduling of tasks by the learner.
3. Using web-based technology will enhance the learning performance.
4. Using web-based technology will increase productivity.
5. Learners will find web-based learning as a useful learning tool.

Ease of Use Items:

6. Learning to use web-based technology will be easy for the learners.
7. Learners will find it effortless to get the web-based platform to perform intended tasks.
8. Learners interaction with the web-based platform will be apprehensible and flexible.

Challenges of Web-based Learning Based on Literature Review

Several issues like considering the different learning styles and the background of the learners should be addressed when designing a web-based platform for learners (Weay et al., 2015). The Internet can represent an important source of supplemental information, but it cannot be used for all types of learner, in all contexts, and always the same way (Marzano & Ochoa-Siguencia, 2017). The literature reviewed identified the merits and challenges of web-based learning, which include:

1. Adaptability Issues

There is lack of adequate web-based learning strategies as training for the learners and teachers is not properly conducted (Ms & Toro, 2016). From the interviews that were conducted by the researchers, 52% of the learners found it difficult to adapt from the traditional face to face classroom setup to web-based learning (see Figure 2). The sudden change-over posed adaptability issues to the adolescents as they found it difficult to focus. The learners should embrace web-based learning open-mindedly.

2. Technical Issues

62% of the learners did not possess a reliable high speed internet connection that is able to download and upload the learning materials (see Figure 2). As such, they encountered problems during virtual lessons, the internet disconnected and for those who used data bundles, they depleted even during the online lecture. A few learners did not have access to the internet or did not have a digital device to use which created a digital division between students who had access to the necessary technology and those that did not. The speed of the internet connection plays a pivotal role during web-based learning as it determines how quickly one is able to login to attend lectures and whether or not one can miss lessons. The video and picture quality also gets affected by the internet speed.

3. Computer Education Issue

Some learners were also not knowledgeable about how to use Microsoft packages like Microsoft Word and Microsoft power point, 46% of the learners admitted to having lack of experience with these packages. (see Figure 2). They used their technical gadgets for gamification and social media mostly but not for learning purposes, it was their first time to do web-based learning after forced closure of face to face learning by the COVID-19 pandemic. Learners found it challenging to do tasks like creating and submitting work to meet deadlines.

4. Time Management Issue

A massive 86% of the learners admitted to struggling with managing their time (see Figure 2). Web-based learning is a new phenomenon to them and juggling between their social activities and web-based learning is a challenge. Despite the flexibility of web-based learning, the majority of the learners could not effectively balance the time needed for learning.

5. Self- Motivation Issue

Several challenges encountered by the learners during web-based lessons demoralized them as they stopped to see the light at the end of the tunnel. 48% couldn't wait to revert back to face to face learning, they missed the physical classroom interactions. (see Figure 2).

6. Distraction Issues

One would be tempted to imagine that learning from home is completely amazing, however, because of the presence of parents and siblings, learners had to be confined in a room for the duration of the lessons. Home setup (electrical gadgets like the microwave, blender, radio, television and even the sound of depleting electricity tokens that need to be recharged) are completely distracting as they produce noise that echoes back during live lessons.

7. Learning Styles Issues

Most students were exposed to only the face to face method of learning; web-based learning was completely brand sparking new to them. Web-based learning presented completely new styles of learning like video tutorials, virtual lessons and power point presentations. Students have different learning abilities; some take less time than others to adapt to new methods of learning as exhibited by the first research question where only 7% of the learners found it easy to adapt to web-based learning (see Figure 2). Learners who take longer to adapt tend to lose confidence, concentrate less and even fail to create assignments and generally perform less than fast adapters.

8. Communication Issues

38% of the learners found it very difficult to communicate during web-based learning (see Figure 2). They had to familiarise themselves with icons that helped them to raise their hands during virtual classrooms, mute their microphones to avoid any distractions from possible background noise. Even writing their assignments was not convincing enough to their teachers as concepts behind their work was not clear and could not be clearly understood by their teachers. Some students admitted to being shy, not able to explicitly express their feelings and not well versed with some application software, as such, these attributes affected their level of effective communication among their peers and teachers alike.

9. Virtual Engagement Issues

Web-based lectures were being delivered using different methods like live sessions, recorded video tutorials, power point presentations, live chats on google classroom and messages. These methods where not engaging enough like the traditional face to face method as 42% of the students could not understand the feedback from their teachers (see Figure 2). Learners also admitted that they did not do any follow ups with their teachers to clear any possible doubts.

10. Feedback Issue

Feedback was not individualized to learners as the teachers broadcasted the solutions to assignments. Learners also admitted to misunderstanding the feedback sometimes and still did not make any effort to clarify issues with their respective teachers.

Innovations

Several innovations that support web-based learning for adolescents are now available online as open source software. The 2020 typology of free web-based learning technologies studied Web 2.0 tools, reviewed and classified freely available Web-based learning technologies and provided a list of 226 technologies which were further arranged into types and clusters that can be used via a web browser to conduct teaching and learning. The learning technologies promote interactive and results oriented learning They gauged the trends of web-based technologies that were available from 2015 to 2020. The tools were clustered into text based tools (discussion forums, note-taking and document creation); image based tools (image creation and editing, drawing and painting, online white boarding, diagramming, mind-mapping, mapping, word clouds); audio tools (audio creation and editing, video tools, video creation and editing, video streaming); Multimodal production tools (digital pin-boards, presentations, lesson authoring); digital story telling tools (online book creation, comic strip creation, animated videos); website creation tools (wikis, individual creation tools, blogs); knowledge organisation and sharing tools (file sharing, social bookmarking, aggregators, republishing, timeline creators); data analysis tools (conducting surveys, online spreadsheets; infographics); 3D modelling tools (3D model repositories, 3D model creation); Coding tools; assessment tools; social networking systems; learning management systems and web-conferencing tools (Bower & Torrington, 2020; Mahalakshmi & Radha, 2020).

According to Lal (2011) a highly powerful integration of graphics and semantic data called Scalable Vector Graphics (SVG) is a possible tool to support web-based learning. This is also known as Web 3.0 as shown in Figure 3.

Innovations like immersive 3-D online environments and 3-D social networking systems that have the potential to permeate the AVR world and gamification platforms and amalgamate them with the Web will have rewarding impact on education. Web 3.0 technologies offer benefits of 3D Labs, 3Dwikis, AI based search engines, Virtual environments such as the Avatar, Semantic Digital Libraries that may produce increased benefits in the delivery of effective web-based learning to students (Lal, 2011). The technological world is transforming in a way that benefits the education systems and teaching and learning institutions should embrace this evolution and position themselves strategically.

SOLUTIONS AND RECOMMENDATIONS

The solutions and recommendations include availability of ICT infrastructure and tools, endowing ICT skills to all users, harnessing disruptive technologies (Artificial Intelligence, Machine Learning, Virtual Reality and so on), acquiring soft skills like self-management, and improving pedagogical and andragogic skills.

Figure 3. Web-based learning using Web 3.0

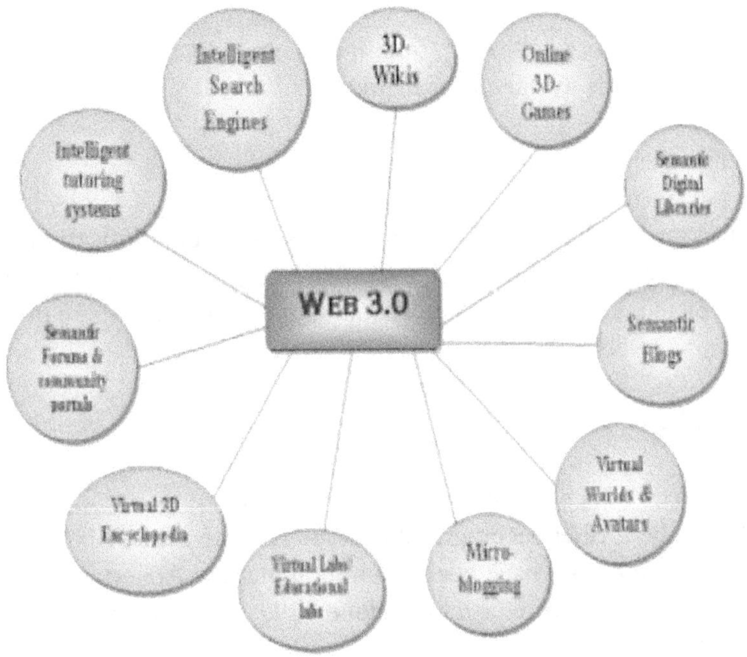

1. How to overcome the Adaptability challenges?

At this moment in time, adaptive learning is augmenting technologies like Virtual and Augmented Reality (AVR) and utilizing Artificial Intelligence (AI) to tailor make learning content to meet individualized learning needs. AVR technologies and their accompanying applications increase productivity in education as they motivate learners, foster engagement and critical thinking plus promote knowledge transfer (Curcio et al., 2017).

2. How to overcome the Technical challenges?

Parents should invest on a reliable high-speed internet connection at their home to promote web-based learning for their children. They should also purchase smart gadgets and keep details of their technical support team from their Internet Service Provider (ISP) for easy and timeous troubleshooting of any possible connection or technical issues.

3. How to overcome the Computer Education challenge?

Learners should have access to online support from the I.T department at their respective schools for easy technical support. They should also be trained on how to use the web-based platform and application software that will be used to do assignments prior to engaging on any online learning activities.

4. How to overcome the Time Management challenge?

For good results from web-based learning, time management is of paramount importance. Factors to be monitored for effective time management include:

a. Avoiding Distractions

Entertainment platforms like WhatsApp, twitter, Facebook and Gamification should be avoided at all cost during learning times as they distract the learner and affect the learning process negatively as work might not be completed. Learners need to schedule time for breaks where possible so that they do not miss any lessons or overwhelm themselves.

b. Creating a To-Do List

Learners are encouraged to keep a diary where they can diarize their daily activities and timelines every day. Large tasks can be further decomposed into sub tasks to achieve better results. Learners can use the diary to check if all activities have been run to completion. Adhering to the list improves time management to a greater extent as every task to be done will be written down.

c. Seek Help

Learners should also seek assistance and guidance where necessary from their parents at home during web-based learning.

d. Avoid Multitasking

Learners are encouraged to complete their tasks in chronological order, one after the other and not concurrently. Doing multiple tasks at the same time requires a lot of experience and concurrent controls as this can affect productivity.

5. How to overcome the Self- Motivation challenge?
   a. Learner Involvement

Learners should always show up for all learning sessions. Missing on some lessons would create gaps and confusion as some concepts are the building blocks to the next concepts. Learners should always be up to date and in touch with all the learning events. WhatsApp group platforms may be created for purposes of information sharing.

b. Schedule Learning Time

Effective learning calls for a lot of scheduling; learners should schedule their learning times so that they do not miss any. Breaks are also necessary to avoid over working and after the breaks, learners should not lose their learning enthusiasm.

c. Keeping a Positive Attitude

Learners are encouraged to love, adapt and enjoy web-based learning and maintain a positive attitude towards their learning activities. Attitude will take learners places.

6. How to overcome the Destruction challenge?

Since web-based learning takes place from home, learners should inform their family members of their learning times to avoid any possible destructions. The learning environment should be restricted during learning times. This is when scheduling activities including breaks come in handy as the learners should also not shut themselves from family times.

7. How to overcome the Learning Style challenge?

Learners should know the learning style that work best for them. Some learners understand better when they write notes down, some through group discussions, others through visual presentations or video lectures. Learners should follow the learning style that best work for them to enhance their individual learning experience.

8. How to overcome the Communication challenge?

Web-based learning enhances the learning experience (Huang et al., 2016). It facilitates interaction and communication with others. Communication ensures that there is feedback and diffusion of ideas. Learners need to communicate effectively across the board for attainment of better results.

9. How to overcome the Virtual Engagement challenge?

Communication with teachers is necessary to iron out any ambiguities. Teachers are there to guide the learners.

10. How to overcome the Feedback challenge?

As alluded before in the communication section, learners are encouraged to approach their teachers for personalised feedback on performance. Feedback also improves the learning patterns.

FUTURE RESEARCH DIRECTIONS

Future research may concentrate on examining how the web-based innovative tools like the integrated Scalable Vector Graphics can be incorporated into web-based learning to reap their full benefits in education.

CONCLUSION

Web-based learning is an emerging innovative approach to learning which can digitally empower the education system. Challenges faced by adolescents include adaptability issues, technical or digital divide issues, time mismanagement, demotivation, distractions, complex learning styles, communication issues, virtual engagement pitfalls, and poor feedback from educators. Web-based learning is a panacea to the disruptions currently faced by teaching and learning institutions around the globe which students and educators should embrace. Scalable Vector Graphics (SVG) and the 2020 typology of free web based learning technologies can be utilized by educators and learners for undisrupted teaching and learning even during pandemics like the COVID-19.

REFERENCES

Ali, W. (2018). Influence of Evolving Technology in Emerging Online Lives of the Digital Native University Students. *Asia Pacific Journal of Contemporary Education and Communication Technology*, *4*(2), 141–155. doi:10.25275/apjcectv4i2edu15

Bower, M., & Torrington, J. (2020). Typology of free Web-based learning technologies. *Educause*. doi:10.13140/RG.2.2.11064.16647

Cabero-Almenara, J., Fernández-Batanero, J. M., & Barroso-Osuna, J. (2019). Adoption of augmented reality technology by university students. *Heliyon*, *5*(5), e01597. Advance online publication. doi:10.1016/j.heliyon.2019.e01597 PMID:31193247

Chiemeke, S., & Mike Imafidor, O. (2020). Web-based Learning In Periods of Crisis: Reflections on the Impact of Covid-19. *International Journal of Computer Science and Information Technologies*, *12*(3), 33–46. doi:10.5121/ijcsit.2020.12303

Cook, D. A. (2020). Web-based learning: Pros, cons and controversies. *Clinical Medicine. Journal of the Royal College of Physicians of London*, *7*(1), 37–42. doi:10.7861/clinmedicine.7-1-37 PMID:17348573

Curcio, I. D. D., Dipace, A., & Norlund, A. (2017). Virtual realities and education. *Research on Education and Media*, *8*(2), 60–68. doi:10.1515/rem-2016-0019

Freire, P. (2013). Pedagogy of the oppressed. In The Applied Theatre Reader. doi:10.4324/9780203891315-58

Gottschalk, F. (2019). *Impacts of Technology Use on Children: Exploring Literature on the Brain, Cognition and Well-Being*. OECD Education Working Papers, No. 195. OECD Publishing.

Hatcher, M., Henson, J., & LaRosa, P. (2013). Determinants of Success in an Online Management Information Systems Course: The Significance of Grade Point Average As a Performance Predictor. *International Journal of Business Strategy*, *13*(4), 129–132. doi:10.18374/IJBS-13-4.7

Huang, T. C., Chen, C. C., & Chou, Y. W. (2016). Animating eco-education: To see, feel, and discover in an augmented reality-based experiential learning environment. *Computers & Education*, *96*, 72–82. doi:10.1016/j.compedu.2016.02.008

Inal, M., & Korkmaz, Ö. (2019). The effect of web based blended learning on students' academic achievement and attitudes towards English course. *Education and Information Technologies, 24*(4), 2603–2619. doi:10.100710639-019-09890-7

Lal, M. (2011). LACLO 3 - Web 3.0 in Education & Research. *International Journal of Information Technology, 3*(2), 973–5658.

Mahalakshmi, K., & Radha, R. (2020). Covid 19: A Massive Exposure Towards Web Based Learning. *Journal of Xidian University, 14*(4). Advance online publication. doi:10.37896/jxu14.4/266

Mansureh, K., Lipschuetz, A., & Santiague, L. (2017). Issues and Challenges for Teaching Successful Online Courses in Higher Education: A Literature Review. *Journal of Educational Technology Systems.*

Marzano, G., & Ochoa-Siguencia, L. (2017). Challenges of Web-Based Participatory Learning. *Society. Integration. Education. Proceedings of the International Scientific Conference, 2*(May), 458. 10.17770ie2017vol2.2395

Means, B., Toyama, Y., Murphy, R., Bakia, M., & Jones, K. (2009). Evaluation of Evidence-Based Practices in Online Learning. *Structure (London, England), 66.* www.ed.gov/about/offices/list/opepd/ppss/reports.html

Ms, P., & Toro, U. (2016). *A Review of Literature on Knowledge Management using ICT in Higher Education.* Academic Press.

Rana, H., Rajiv, R., & Lal, M. (2014). E-learning: Issues and Challenges. *International Journal of Computers and Applications, 97*(5), 20–24. doi:10.5120/17004-7154

Tsai, S., & Machado, P. (2002). E-learning, Online Learning, Web-based Learning, or Distance Learning. *Unveiling the Ambiguity in Current Terminology., 2001*, 3–5.

Valverde-Berrocoso, J., del Carmen Garrido-Arroyo, M., Burgos-Videla, C., & Morales-Cevallos, M. B. (2020). Trends in educational research about e-Learning: A systematic literature review (2009-2018). *Sustainability (Switzerland), 12*(12), 5153. Advance online publication. doi:10.3390u12125153

Vrasidas, C. (2015). The rhetoric of reform and teachers' use of ICT. *British Journal of Educational Technology, 46*(2), 370–380. doi:10.1111/bjet.12149

Weay, A. L., Abdullah, S. H., Masood, M., & Muniandy, B. (2015). Design issues and challenges for web-based history learning system for adolescents in Malaysia. *ACM International Conference Proceeding Series,* 50–54. 10.1145/2846439.2846443

Yusof, N. A., & Saadon, N. (2012). The Effects of Web-based Language Learning on University Students' Grammar Proficiency. *Procedia - Social and Behavioral Sciences, 67*(November), 402–408. doi:10.1016/j.sbspro.2012.11.344

KEY TERMS AND DEFINITIONS

Augmented and Virtual Reality: Refers to technologies that simulate a real-life environment.

Digital Divide: Alludes to the gap between those who do not have access to information and communication technologies and those who do.

Distance Learning: Refers to the physical separation between a student and a teaching during teaching and learning.

E-Learning: Refers to learning that is conducted electronically.

Educational Software: Refers to different programs that are used for educational purposes.

Educational Technology: Refers to various digital technologies that are used for teaching and learning.

Online Learning: Refers to teaching and learning that is conducted over the internet.

TAM: Alludes to the Technology Acceptance Model.

Teaching and Learning: Refers to the impartation and acquisition of knowledge.

Video Streaming: Delivering a video file over the internet.

Web-Based Learning: Learning via the internet.

Chapter 18
Artificial Intelligence and Blockchain as Disruptive Technologies in Adolescent Lives

Munyaradzi Zhou
https://orcid.org/0000-0002-2184-0290
Midlands State University, Zimbabwe

Cyncia Matsika
Midlands State University, Zimbabwe

Tinashe Gwendolyn Zhou
Midlands State University, Zimbabwe

Wilfreda I. Chawarura
Midlands State University, Zimbabwe

ABSTRACT

COVID-19 and future pandemics drastically change the way of life globally. Research has predominantly focused on the use and integration of disruptive technologies in industry and commerce. Little of the recent studies focused on the implementation of artificial intelligence and blockchain technologies in educational applications. The chapter focuses on how these can be implemented, from development, deployment, use, and maintenance of applications. A computer program's lifespan is usually spent during its use. The qualitative case study was administered using a digital learning platform that provides interactive learning for primary and secondary learners. The disruptive technologies inform new teaching methodologies and the development of student-centered algorithms for learning. Further research includes privacy issues in the implementation of disruptive technologies and data-sharing governance issues and evaluating the effectiveness of artificial intelligence and blockchain-based learning platforms.

DOI: 10.4018/978-1-7998-8318-0.ch018

INTRODUCTION

Artificial Intelligence (AI) and Blockchain technologies are applied in adolescent's lives to create new possibilities or frontiers. Blockchain technology is viewed as a myriad integrated platform of distributed databases that supports data collection, aggregation, and analysis. The highly secure platform supports audit trails that support the accountability and responsibility of users. Blockchain supports the learning process of adolescents in educational platforms by enforcing accountability. Researchers mainly focused on the integration of the disruptive technologies (Artificial Intelligence and Blockchain technologies), their pros and cons, and their use but not specifically for education in adolescents and how it is implemented (from development, deployment, use, and maintenance) (Anderson et al., 2018; Fedorova & Skobleva, 2020; Sgantzos & Grigg, 2019; Zovko & Gudlin, 2020). The book focuses specifically on the application of disruptive technologies in primary education. The application of AI and Blockchain concepts in education is essential to endow skills that are essential to create or add value in pupils. This can be in form of unearthing new teaching methodologies and synchronization of recent data sources. The technologies have the potential of unearthing new insights and innovations (get more use out of the data). AI algorithms crawl on the web which consists of unstructured, semi-structured, and structured data. Unstructured data such as videos and images on Digital Learning Platforms (DLPs) are indexed to contextualize information on these objects in form of text, thus the creation of value. The unstructured data can be labeled automatically and certified (Anderson et al., 2018).

AI and Blockchain technology makes sharing of data seamless through a matchmaking of contents and this is done in a trusted manner (Ahmad et al., 2021). It is essential to cluster subjects offered, schools, districts, and regions the adolescents are studying. The aggregation of related content results in informing new knowledge domains which are student-centric. The interactive digital learning platform can be blended with various technologies such as supporting drill and practice, tutorials, and simulation activities. Students can view and interact with virtual features or items and receive customized instruction or feedback. Learning using entertaining educational games results in practicing skills and learner productivity through engagement. The use of Artificial Intelligence coupled with data science algorithms on big data supports analytics of student performance, and adaptive learning (student-centered) and offers customized learning activities and resources to address the unique needs of the student (Guan et al., 2020). The collection and analysis of data (data discovery) is enhanced by the cloud model, globalized-networked information communication devices.

Data discovery in DLPs is attained through the partnership of stakeholders such as schools' authorities and other educators, students, electronic learning platform developers, researchers, policymakers, community members, and students' guardians with the Ministry of Primary and Secondary Education (MoPSE). The MoPSE has special units such as the Curriculum Development Unit and the Technical Services (CDTS) that focus on quality delivery of education. It focuses on the development, validation, rolling-out and continuous improvement of the curriculum through stakeholder involvement. Stakeholders involved include educationists, industry and commerce experts and captains, local and international educational institutions (primary to tertiary level) (MoPSEC, 2021). CDTS is responsible for validating and giving a 'proxy' to publish curriculum material to the public domain. Digital Learning Applications in Zimbabwe include eNhava, Ruzivo, The Learning Passport, and MoPSE e-Library. The rolling out of these applications is certified by the CDTS (Watt, 2018; ZARNet, 2021). The CDTS also supports training activities and awareness campaigns on products on offer. The involvement of stakeholders and teamwork is an essential component of requirements engineering or analysis in any system developed.

The DLPs (systems) supported are mainly web-based. Android and iOS applications are still few. The Unstructured Supplementary Service Data (USSD) applications are not supported.

User requirements analysis is critical during the system lifecycle. AI and Blockchain technologies are blended during requirements engineering to come up with highly engaging and interactive solutions. Semantic analysis during the development and use of systems results in the value of the applications deployed. Standardization processes include checking the semantics of the data, linking related data (data cleaning to make sure data is under the same domain). Quality data increases the level of competence in obtaining critical skills by learners. Data managers, assessors, students, lecturers, and academics are responsible for the certification of the blockchain to support accountability. Value-driven applications complement in addressing the challenges posed in the education system due to COVID-19.

The challenges of COVID-19 include banning gatherings in public and private places. Schools are no exception, hence there is a need for calls to coming up with educational applications which imitate the teacher and fulfill continual learning until students finish their syllabi (Dhawan, 2020; Teräs et al., 2020). The rise of the virtual learning trend has to be supported now and in the future for long-life learning. The chapter focuses on AI and blockchain as disruptive technologies in adolescents' lives in primary education. It fulfills the following objectives:

- To identify how AI and Blockchain technologies support data sharing and interactive learning.
- To ascertain user requirements analysis elements adhered to by independent developers of educational applications.
- To inform how artificial intelligence and blockchain technologies are used in adolescents' education.
- To come up with processes involved in developing educational applications for primary education whilst merging disruptive technologies.

BACKGROUND

Artificial Intelligence and blockchain transform the way of life and seamlessly results in enhanced decision making and innovativeness. Due to advancements in Information Communication Technologies (ICTs) and the emergency of the COVID-19 pandemic, there is a need to integrate disruptive technologies in the DLPs. In 2020, students failed to attend face-face lessons, giving a relative attendance rate of less than 10%. The low pass rate resulted from the implementation of lockdowns in compliance with World Health Organization (WHO) regulations and national government policies. Reactive and transformative decisions address the current status quo, future pandemics and increase the intellectual capacity of learners.

The Grade 7 results published by the Zimbabwe Schools Examinations Council (ZIMSEC) show that the pass rate declined sharply by 9.79 percent from the 2019 record of 46.9 percent. Eighty-eight schools have recorded a 0 percent pass rate. The primary school learners are in the adolescent age range of 11 to 13 years (NewZimbabwe, 2021). The Grade 7 results published by the Zimbabwe Schools Examinations Council (ZIMSEC) show that the pass rate declined sharply by 9.79 percent from the 2019 record of 46.9 percent. Eighty-eight schools have recorded a 0 percent pass rate. The primary school learners are in the adolescent age range of 11 to 13 years (UNICEF, 2019). The adoption and use of DLPs that embed AI and blockchain technologies improve the pass rates of Grade 7 students. Intelligent learning dashboards complement the coverage of the entire syllabus by learners.

Disruption is the transformation brought by the advent of new technologies, inventions, processes in the marketplace (Girasa, 2020; Makridakis et al., 2018; Zovko & Gudlin, 2020). Blockchain technology is a highly asymmetric encrypted platform that supports independent distributed transacting in the digital landscape that supports personal privacy, trust among parties, and permanent recording of transactions (Makridakis et al., 2018; Zovko & Gudlin, 2020). The technology reinforces trust and enables each party to retain control of their data. Timestamps are used to authenticate entries. AI and blockchain technologies are an 'element'. AI is the use of computer systems as intelligent agents to learn or perceive the environment and take actions for meaningful decision making (Zovko & Gudlin, 2020). The development of DLPs that are interactive aids the learner in the completion of the syllabi.

Figure 1 shows the processes of Blockchain.

Figure 1. Blockchain process
Source: JRC "Blockchain, Now and Tomorrow" 2019

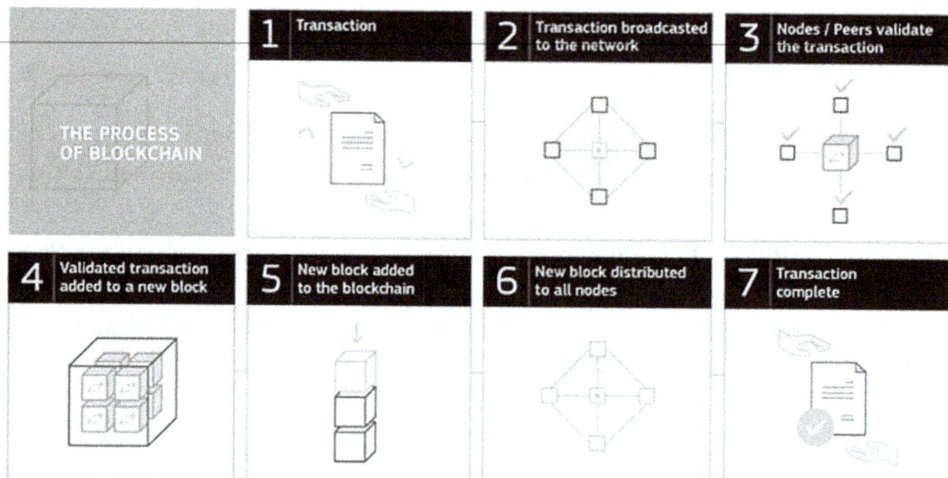

The learner, teacher, and other academics commit a transaction. Transactions processes include accessing the DLPs, accessing learning materials, writing evaluation and assessment materials, and the uploading of learning materials. The transactions are broadcasted to the network and this includes schools and DLPs databases. Teachers, CDTS, DLPs managers, authorities or experts in schools, and the DLPs ecosystem validate the transactions. The validated transactions are added to the new block which can be treated as a knowledge base. The new block is added to the blockchain and AI algorithms are applied to create semantics that supports life-long learning. New blocks are distributed to all nodes and the transaction is completed.

MAIN FOCUS OF THE CHAPTER

The chapter mainly focuses on dovetailing the elements aforementioned in the objectives listing.

AI and Blockchain Technology in Education

The integration of disruptive technologies, AI and Blockchain releases great potential in different sectors including education as it supports seamless data sharing and improvement in the quality of life (Atlam et al., 2020; Grech & Camilleri, 2017). Blockchains can secure the availability and seamless access to integrity-protected large data sets, from various domains, thus facilitating the effective and more profound learning of AI algorithms. The challenges of cybercrimes which include offenses against the confidentiality, integrity, and availability of computer data and systems, computer-related offenses, and content-related offenses, and the types of cybercrimes included within these categories call for the use of blockchain technology (Naz et al., 2019; Yli-Huumo et al., 2016). Blockchain and distributed ledger technologies provide novel opportunities for protecting educational data collection, linkages, processing, and storage through decentralized identity and other tamper-proof privacy mechanisms (almost impossible to falsify). These systems can allow users greater sovereignty through tools that enable them to own and control their data. Artificial intelligence provides further possibilities for enhancing system and user security, enriching data sets, and supporting improved analytical models (Kim & Park, 2020). The process of Blockchain is identified and elements are related to stakeholders such as schools, administrators, MoPSEC, CDTS, individuals, and so on.

Advanced machine learning and artificial intelligence (AI) algorithms support the creation of student-centered learning applications which are interactive and focuses on the attainment of learning outcomes. The applications are also capable of giving feedback to the users even though they cannot wholly replace the teacher. This is achieved through data cleaning, inferring, analysis, and classification to make predictions. The data is collected or recorded through the use of Blockchain platforms and they support the seamless flow of information (Sharma et al., 2020).

The successful implementation of interactive learning systems results from harnessing data from multiple sources such as individual profiles, school profiles, and subject-centered profiles. Student profiling unearths opportunities including scrutinizing performance per student or given module. Adaptive learning and assessments result in the development of customized profiles and training materials for each learner based on their aptitude, ability, experience, and preferred mode of learning. This is attained using learning analytics. Appropriate pedagogical technological models that focus on competency-based skills, academic performance (through profiling), and increasing learners' engagement are essential in implementing pedagogical skills during the development of applications. This will ultimately result in improving the completion of courses by students as the academic progression is earmarked. AI algorithms help in assessing if the learning objectives have been attained at any given moment thus thriving to achieve the learning outcomes. Intelligent Chatbots through predictive classification algorithms can detect students who could probably quit and remedial actions are made during the academic journey. Big data is facilitating the implementation of the algorithms which support student profiling and learning analytics (Guan et al., 2020).

User Requirements Analysis Elements Adhered to by Independent Developers of Educational Applications

Digital Learning platforms are being developed without proper user requirements analysis and integration with pedagogical elements (conceptualisation, construction, and dialogue) in the final product (Hadjerrouit, 2007; Rodríguez, 2013). This adversely affects how interactive the systems are to the learners and the educators. Developers and authorities responsible for rolling out educational applications must be endowed with User Centered Design (UCD) knowledge and skills set so that the end product is used interactively and supports the desired technological artefacts (Masood & Thigambaram, 2015). Teachers in schools and psychologists are a good skills set since they understand how adolescents think and interact hence their engagement with educators, guardians, developers, product owners, curriculum developers ultimately result in digital learning applications which support user interaction. Pilot studies and post-implementation testing to check for usability, usefulness, and how desirable the artefact is, is critical (Unterkalmsteiner, 2015). During system use, continuous monitoring is essential for the applications to remain student-centered thus it supports the needs, capabilities, expectations, and experiences of the learners. Pedagogical elements in the applications must thrive to attain continuous improvement and case-based solutions (Dhawan, 2020).

Artificial Intelligence and Blockchain Technologies in Adolescents' Education

Feasibility studies for the application of AI and Blockchain in delivering digital learning applications are essential as this ultimately results in quality learning processes for adolescents. The use of timestamps in blockchains and chronological order of records supports progressive tracking of learning processes on DLPs. This is further enhanced through the property of Blockchain of making records tamper-proof and their recordation to be made at distributed nodes (schools, students, curriculum developers, and other stakeholders can transact at any point) (Atienza-Mendez & Bayyou, 2019). The data can be shared seamlessly across nodes or different learning platforms (knowledge bases) in a trusted manner without switching accounts. To enhance student engagements, their evaluation and assessment results are certified or given badges in affiliation with organisations they perceive as role models. Employers can easily access these results in processes such as recruitment (Sun et al., 2018). Blockchain applications can be applied in different categories, namely, learning objectives and competencies, copyrights and certificates administration, evaluation and assessment of the learner's professional ability, safeguarding collaborative learning environment, competitions management, enhancing learners' interactions in interactive digital platforms, and supporting life-long learning (Atienza-Mendez & Bayyou, 2019; Samah et al., 2019; Siraj & Taggart, 2014). Undoubtedly, Blockchain offers trust and security, authentication (access control) procedures, and generally easy administration of the DLPs in different facets hence blending it with AI is a great asset (Samah et al., 2019). AI supports and endows lifelong learning, training, and skills development hence Ministry of Education of other sectors must support the development and deployment of AI-based applications (OECD Publishing, 2019).

Processes Involved in Developing Educational Applications for Primary Education Whilst Merging Disruptive Technologies

Stakeholders' involvement is critical in coming up with student-centered applications. Technology and the educational applications supports acquiring of necessary skills by students. The stakeholders like teachers and parents are very key since they determine the adoption and use of the technologies. School authorities might train both students and teachers in the use of the DLPs. The platforms can be complemented using blended learning (Grimus, 2020). DLPs need a strong theoretical background based on purposeful, systemic research, and a sound pedagogy hence a pragmatic innovative approach needs to be adopted (Serdyukov, 2017). Efforts to come up with Intelligent DLPs that can at least 50% replace human intervention is critical since it supports great continued education of the students. The partnership of stakeholders such as developers, curriculum developers, educators, developers, and owners of the DLPs is critical for the continuous improvement of the products (Crittenden et al., 2019). The products need to support evaluation and assessment to be seamless, effective, and efficient. In addition, it must support inclusive education to cater to the marginalised and students with special needs. Training and education of students, parents, and instructors and continuous feedback in the rolled-out systems are key. Teamwork among the stakeholders aforementioned, industry, and education experts can holistically support continued and engaged education (UK Department for Education, 2019). The chapter further contributes to the methodological gap.

Methodological Gap

A systematic literature review has been applied in studying the use of disruptive technologies in education (Pulist, 2020). Different case studies have been applied to the application of BCT in higher education (Fedorova & Skobleva, 2020). This book chapter applied qualitative interviews to collect data in the software development team (technical team in applications development) in the interactive DLPs. A single case study was used as it resulted in theoretical saturation in the development process of the digital learning platform and how they are applying AI and Blockchain concepts (Boddy, 2016). The digital learning platforms targeted provide curriculum for primary and secondary school learners. In this research, the primary level digital platform is assessed on how it is developed using disruptive technologies and hinges on the targeted sample of adolescents.

SOLUTIONS AND RECOMMENDATIONS

Solutions in this book chapter include new teaching models, segmentation of learners' needs including performance analysis, and suggesting areas for improvement. The availability of an online artificial-teacher who can recommend to a live teacher in case semantics and interaction is derailed. The live teachers will be available more on a call-center-like basis.

The Blockchain features are not yet implemented in the educational applications hence there is a need for the call for the roadmap to modernise and keep in track with what is existing. This includes a facility to help the students to analyse performance on the platform and suggest areas of improvement. For example, after data modelling and analysis the system can suggest that the candidate is not strong in Mathematics and other areas hence it suggests improvement in those areas. Artificial Intelligence features are existing in the applications but they are not fully-fledged, the concepts grows from point-to-point. It is critical to implement AI tools to help learners in the learning process. The learning environment is in form of gaming (learning games). The gaming mode provides an interactive environment for learners to progressively learn according to their competency and practice. The engaging platform supports students to access either the new or old curriculum. The application is web-based and it needs to be complemented with mobile application versions (iOS and Android) and Unstructured Supplementary Service Data (USSD) platforms. The use of USSD supports offline learning. Mobile Applications are still under consideration (they need to be redefined since the content is in gigabytes). A smart virtual assistant gaming mode is supported in the English language and needs customization to support major indigenous languages in Zimbabwe, namely, Shona and Ndebele.

The educational applications have features including supporting study groups which are both private and public; and checking of language use to deter offensive language (vulgar words) use. The platforms support the points system and in the future, there is a need to have stores to redeem points in monetary value as an extension of the existing ways of point systems/redeeming. The facility will encourage or motivate students to participate as it rewards points redeemable in monetary value (Shrestha et al., 2020). Data sharing, monetary incentivising in monetary terms, and accessing copyrighted material for a time frame specified further supports learning (Xuan et al., 2020).

The Artificial Intelligence bot can crawl on users' sessions to identify content that is mostly accessed, discussed, and the ML algorithms synchronise content to be customized based on the needs of the learner and this ultimately results in improvement of content and value-addition in many dimensions. An appealing Chabot can also hint or encourage or motivate students to participate as the environment is points-rewarding enabled. Engagement is essential; there is a need for appreciation on both ends of stakeholders (students/teachers). During application testing and implementation, it is essential to satisfy the needs of systems users including owners and system users. Continuous improvement is essential, through stakeholders' involvement (who proffer solutions and suggestions), more quality content can always be customized (OECD, 2016). Deployment of applications needs all stakeholders support. The educator must be more welcoming to transit from the use of hardcopies–textbooks to electronic materials and learning. DLPs are a good customized environment imitating the use of textbooks. Schools are populated on the DLPs. Private schools, make it a culture that all students must be registered and the notion must be applied in public schools. DLPs owners donate tablets mainly in government schools. Private schools have a good pool of donors of ICT equipment including laptops and tablets. Organisations are donating tablets even for the underprivileged. In addition, the donations include main servers to schools with a capacity of a minimum of 15 PCs in the laboratories. Small/Minor/Mirror servers are also being donated so that the students can access learning material offline (without an Internet connection) and this can be complemented with the use of USSD. The CDTS and the DLPs owners provide awareness campaigns on the use of the platforms. Furthermore, the CDTS must initiate the implementation and teaching of the DLPs in teachers training colleges curriculum.

FUTURE RESEARCH DIRECTIONS

The book chapter mainly focused on how disruptive technologies (AI and Blockchain) are applied to improve academic performance in adolescents' lives. The processes in the use of disruptive technologies needs to be further studied. These include authorisation procedures in the use of blockchains in education applications, the concept of User experience (UX), and User Interface (UI) (the design, evaluation, usability testing, and implementation) in adolescents as they use these interactive applications. The concepts of examining how data aggregation, as well as the impact of data collection, storage, analysis, use, and sharing, on privacy and security is essential during the lifetime of the applications (Dutta et al., 2020). Privacy issues to do with the disruptive technologies and data sharing governance issues and protections are also essential.

CONCLUSION

The research supports the application of disruptive technologies in education and coming up with ICT-based teaching methodologies in form of virtual assistants, and or algorithms for learning. ICT-based teaching methodologies include personalised lessons, instructions, capstones for each learner or group based on learning pedagogies. The learning dashboards must consist of real-time sentiment analysis Chatbots. The Chatbots need to be continually updated by developers through refinement of the vocabulary after thorough requirements analysis from the system users and during the lifespan of the applications. The application of Natural Language Processing (NLP) for more engaged, interactive platforms needs to be complemented with switching to live teachers for more hands-on and personalised engagement. AI is used to introduce new concepts which are student-centered whilst interacting with Blockchains. The application of Blockchain technology for the performance of adolescents is still in its infancy. Education of adolescents is engaged well through DLPs which include gaming learning, and training simulations. The DLPs must support inclusive education and target marginalised groups such as adolescents and women. This book chapter contributes to Sustainable Development Goal 4 (SDG4) which focuses on the attainment of quality, lifelong and inclusive education. The Digital Learning Applications must support performance through the analysis of student behaviour, patterns of use, the optimisation of caching strategies, pre and post-assessments, conducting of lessons, augmented virtual reality practical laboratory sessions, test preparation sessions, PrepEngine dashboard. The platform must be artificial super intelligent and target units like individual learners, schools, and subjects.

REFERENCES

Ahmad, R. W., Salah, K., Jayaraman, R., Yaqoob, I., Ellahham, S. & Omar, M. (2021). The role of blockchain technology in telehealth and telemedicine. *International Journal of Medical Informatics, 148*(November). doi:10.1016/j.ijmedinf.2021.104399

Anderson, J., Rainie, L., & Luchsinger, A. (2018). Artificial intelligence and the future Humans. *Pew Research Centre, 169*(1), 57–88. doi:10.4000/rei.8727

Atienza-Mendez, C., & Bayyou, D. G. (2019). Blockchain Technology Applications in Government. *IJCAT -. International Journal of Computers and Technology, 6*(11), 68–74. doi:10.4018/978-1-7998-3632-2.ch003

Atlam, H. F., Azad, M. A., Alzahrani, A. G., & Wills, G. (2020). A review of Blockchain in Internet of Things and AI. *Big Data and Cognitive Computing, 4*(4), 1–27. doi:10.3390/bdcc4040028

Boddy, C. R. (2016). Sample size for qualitative research. *Qualitative Market Research, 19*(4), 426–432. doi:10.1108/QMR-06-2016-0053

Crittenden, W. F., Biel, I. K., & Lovely, W. A. III. (2019). Embracing Digitalization: Student Learning and New Technologies. *Journal of Marketing Education, 41*(1), 5–14. doi:10.1177/0273475318820895

Dhawan, S. (2020). Online Learning: A Panacea in the Time of COVID-19 Crisis. *Journal of Educational Technology Systems, 49*(1), 5–22. doi:10.1177/0047239520934018

Dutta, P., Choi, T. M., Somani, S., & Butala, R. (2020). Blockchain technology in supply chain operations: Applications, challenges and research opportunities. *Transportation Research Part E, Logistics and Transportation Review*, *142*(May), 102067. Advance online publication. doi:10.1016/j.tre.2020.102067 PMID:33013183

Fedorova, E. P., & Skobleva, E. I. (2020). Application of blockchain technology in higher education. *European Journal of Contemporary Education*, *9*(3), 552–571. doi:10.13187/ejced.2020.3.552

Girasa, R. (2020). Artificial Intelligence as a Disruptive Technology: Economic Transformation and Government Regulation. In *Artificial Intelligence as a Disruptive Technology*. Palgrave Macmillan. doi:10.1007/978-3-030-35975-1

Grech, A. & Camilleri, A. F. (2017). *Blockchain in Education*. doi:10.4018/978-1-5225-9478-9.ch014

Grimus, M. (2020). Emerging Technologies: Impacting Learning. *Pedagogy and Curriculum Development*, (February), 127–151. Advance online publication. doi:10.1007/978-981-15-0618-5_8

Guan, C., Mou, J., & Jiang, Z. (2020). Artificial intelligence innovation in education: A twenty-year data-driven historical analysis. *International Journal of Innovation Studies*, *4*(4), 134–147. doi:10.1016/j.ijis.2020.09.001

Hadjerrouit, S. (2007). E-learning as a means of social inclusion: The case of syrian refugees in Germany. *Interdisciplinary Journal of Knowledge and Learning Objects*, 3.

Kim, J., & Park, N. (2020). Blockchain-based data-preserving AI learning environment model for AI cybersecurity systems in IoT service environments. *Applied Sciences (Switzerland)*, *10*(14), 4718. Advance online publication. doi:10.3390/app10144718

Makridakis, S., Polemitis, A., Giaglis, G., & Louca, S. (2018). Blockchain: The next Breakthrough in the Rapid Progress of AI. In *Artificial Intelligence - Emerging Trends and Applications* (pp. 197–219). http://www.intechopen.com/books/trends-in-telecommunications-technologies/gps-total-electron-content-tec-prediction-at-ionosphere-layer-over-the-equatorial-region%0AInTec%0Ahttp://www.asociatiamhc.ro/wp-content/uploads/2013/11/Guide-to-Hydropower.pdf

Masood, M., & Thigambaram, M. (2015). The Usability of Mobile Applications for Pre-schoolers. *Procedia: Social and Behavioral Sciences*, *197*(February), 1818–1826. doi:10.1016/j.sbspro.2015.07.241

MoPSEC. (2021). *Curriculum development and technical services department*. http://mopse.co.zw/department/curriculum-development-and-technical-services-department

Naz, M., Al-zahrani, F. A., Khalid, R., Javaid, N., Qamar, A. M., Afzal, M. K., & Shafiq, M. (2019). A Secure Data Sharing Platform Using Blockchain and Interplanetary File System. *Sustainability (Switzerland)*, *11*(24), 1–24. doi:10.3390u11247054

NewZimbabwe. (2021, February 7). Teachers Group Blames Govt For Low Grade 7 Pass Rate. *NewZimbabwe*. https://www.newzimbabwe.com/teachers-group-blames-govt-for-low-grade-7-pass-rate/

OECD. (2016). *Innovating Education and Educating for Innovation: The Power of Digital Technologies and Skills*. OECD Publishing. doi:10.1787/9789264265097-

OECD Publishing. (2019). Artificial Intelligence in Society. In *Artificial Intelligence in Society*. Author.

Pulist, S. K. (2020). Blockchain Technology Applications in Government. *IEEE Technical Committee on Learning Technology*, *21*(1), 16–18. doi:10.4018/978-1-7998-3632-2.ch003

Rodríguez, R. M. (2013). Implementation process of technology in Education: The case of Blackboard 9.1 in the University of Manchester. *Actualidades Investigativas en Educación*, *13*(3). Advance online publication. doi:10.15517/aie.v13i3.12043

Samah, A., Marwah, A., Saira, G., & Ali, A. (2019). applied sciences Blockchain-Based Applications in Education. *Applied Sciences (Basel, Switzerland)*, *9*(2400), 1–18.

Serdyukov, P. (2017). Innovation in education: What works, what doesn't, and what to do about it? *Journal of Research in Innovative Teaching & Learning*, *10*(1), 4–33. doi:10.1108/JRIT-10-2016-0007

Sgantzos, K., & Grigg, I. (2019). Artificial intelligence implementations on the blockchain. Use cases and future applications. *Future Internet*, *11*(8), 1–15. doi:10.3390/fi11080170

Sharma, A., Bahl, S., Bagha, A. K., Javaid, M., Shukla, D. K., & Haleem, A. (2020). Blockchain technology and its applications to combat COVID-19 pandemic. *Research on Biomedical Engineering*. Advance online publication. doi:10.100742600-020-00106-3

Shrestha, A. K., Vassileva, J., & Deters, R. (2020). A Blockchain Platform for User Data Sharing Ensuring User Control and Incentives. *Frontiers in Blockchain*, *3*(October), 1–22. doi:10.3389/fbloc.2020.497985

Siraj, I. & Taggart, B. (2014). *Exploring Effective Pedagogy in Primary Schools : Evidence from Research*. Academic Press.

Sun, H., Wang, X., & Wang, X. (2018). Application of blockchain technology in online education. *International Journal of Emerging Technologies in Learning*, *13*(10), 252–259. doi:10.3991/ijet.v13i10.9455

Teräs, M., Suoranta, J., Teräs, H., & Curcher, M. (2020). Post-Covid-19 Education and Education Technology 'Solutionism': A Seller's Market. *Postdigital Science and Education*, *2*(3), 863–878. doi:10.100742438-020-00164-x

UK Department for Education. (2019). *Realising the potential of technology in education: A strategy for education providers and the technology industry*. https://www.gov.uk/government/publications/realising-the-potential-of-technology-in-education

UNICEF. (2019). *Investing in a safe, healthy and productive transition from childhood to adulthood is critical*. https://data.unicef.org/topic/adolescents/overview/#:~:text=Adolescents

Unterkalmsteiner, M. (2015). *Coordinating Requirements Engineering and Software Testing*. Blekinge Institute of Technology.

Watt, E. (2018). *Digital learning service is helping a million children at school and home in Zimbabwe*. https://theirworld.org/news/digital-learning-service-helps-1m-children-in-zimbabwe

Xuan, S., Zheng, L., Chung, I., Wang, W., Man, D., Du, X., Yang, W., & Guizani, M. (2020). An incentive mechanism for data sharing based on blockchain with smart contracts. *Computers & Electrical Engineering*, *83*, 106587. Advance online publication. doi:10.1016/j.compeleceng.2020.106587

Yli-Huumo, J., Choi, S., Park, S., & Smolander, K. (2016). Where is current research on Blockchain technology? - A systematic review. PLoS ONE, 11(10). doi:10.1371/journal.pone.0163477

ZARNet. (2021). *eNhava - Reinventing Education.* http://www.zarnet.ac.zw/index.php/enhava/

Zovko, V., & Gudlin, M. (2020). Artificial Intelligence as a Disruptive Technology. *International Conference - The Future of Education.* 10.1007/978-3-030-35975-1

KEY TERMS AND DEFINITIONS

Academic Performance: Is the attainment of a goals according to defined module outlines parameters.

Applications: Refers to computer programs to attain specific tasks.

Artificial Intelligence: Is the process of solving problems using computers in a human-like nature or even in a much smarter way than humans in a field of enquiry.

Blockchain Technology: Is a distributed tamper-proof transacting digital platform.

Curriculum: Is the domain of knowledge to be acquired in a course of study.

Data: Is unprocessed information in form of text, media and so on.

Digital Learning: Is an effective non-face-to-face learning through the use of digital technologies.

Disruptive Technologies: Are emerging technologies that transforms in a better the lives of people in an unforeseen way. In this book chapter it refers to artificial intelligence and blockchain technology.

Requirements Analysis: Is the definition of system users' expectations for example through a story.

Virtual Assistants: Are technology-based platforms trying to replace humans.

Chapter 19
Fake News Polarization for Sentiment Analysis

Chirag Visani
Marwadi University, India

Vishal Sorathiya
https://orcid.org/0000-0002-6121-5940
Marwadi University, India

Sunil Lavadiya
Marwadi University, India

ABSTRACT

The popularity of the internet has increased the use of e-commerce websites and news channels. Fake news has been around for many years, and with the arrival of social media and modern-day news at its peak, easy access to e-platform and exponential growth of the knowledge available on social media networks has made it intricate to differentiate between right and wrong information, which has caused large effects on the offline society already. A crucial goal in improving the trustworthiness of data in online social networks is to spot fake news so the detection of spam news becomes important. For sentiment mining, the authors specialise in leveraging Facebook, Twitter, and Whatsapp, the most prominent microblogging platforms. They illustrate how to assemble a corpus automatically for sentiment analysis and opinion mining. They create a sentiment classifier using the corpus that can classify between fake, real, and neutral opinions in a document.

INTRODUCTION

Knowledge sharing has been an important aspect in this 21st century and this has become more effective with the usage of the common platform we all know the Internet. Internet users use many communication tools and the most widely used is microblogging platforms are Facebook, Twitter, and WhatsApp which contains arbitrarily large text posts (Zhang, Dong, and Yu 2018). This Corpus can be in form of

DOI: 10.4018/978-1-7998-8318-0.ch019

real emotions, fake emotions, or objective texts.(Parikh and Atrey 2018). The Corpus, which is available online, promises a once-in-a-lifetime opportunity for NLP and related areas. One of the fields with a lot of real-world applications is sentiment analysis.Sentiment analysis can be used by a user to research products or services before purchasing a product or buying a service. Marketers can use this to research public opinion of their company and products. Last but not least organizations can also use this to gather critical feedback about problems in newly released products. The popularity of the internet has increased the use of e-commerce websites and news as well (Lin et al. 2014). So it can be helpful to other users to make decisions making. It can be very useful to make business competitive. But many users or organizations post spam news to promote or defame a brand or specific product or people. Spam news is Very common nowadays on e-commerce websites. Many business organizations hire people to post fake news on their behalf (Jiang, Cao, and Chen 2013). As a factor, spam news recognition has grown increasingly extremely important in recent years. Online user reviews for both products and companies have had a major impact on others' basic buying decisions. Despite the difficulty with which the surveys are available and the considerable consequences for merchants, there is a growing driving factor to manage the audits, which are often benefit-driven (Mukherjee, Liu, and Glance 2012). Entrepreneurship on the internet Audit spam is rapidly replacing websites that provide client audits.- Bitter or positive reviews that aren't worthy; surveys that are written despite the fact that the analysts have never used the business. As an outcome, auditing spam identification has become increasingly important in recent years. Today, plenty of researchers are taking aim at them (Jindal, Liu, and Lim 2010). Spam surveys can be categorized into three parts.The detailed comparative study of the previously available techniques is shown in Table 1 and Table 2.

- Untruthful News or Reviews: Those undeserving positive news that is presented on advance a particular brand or item regardless of the possibility that those brand or item not meriting that as well as a negative survey that is presented on slander a particular brand or item regardless of the possibility that those brand or item are great.
- News or reviews on brands only: Those surveys that do not give any remark concerning an item or administration however give clear remark for brand or vendor. This can be helpful yet it does not give any insights about item or administrations so we can consider as spam.
- Non-reviews on news: That news that contains advertisements, question-answer, or other content irrelevant to the product or services.

In this examination range, there are fundamentally two methodologies ready to manage this issue. One is supervised learning and the second one is unsupervised learning. Both have a few points of interest and downsides. In supervised learning analysts have great precision yet there is a required great measure of space information or data about setting (e.g. what is being looked into, examples of the analysts, and so on). So this technique is not by any means helpful for individuals without this sort of area learning. In an unsupervised approach, this sort of data about the setting is not required. So these sorts of research can be exceptionally valuable for everybody (Ramesh Babu Durai and Duraisamy 2011). Be that as it may, there is one primary downside of precision. This kind of methodology does not give great exactness.

Challenges and Issues

The major challenges and issues of the problem:

- High complexity: There is an immeasurable number of item audits on the web so recognition of survey spam is extremely mind-boggling. It is altogether different from email and webspam locations. Since for such spam, it is anything but difficult to discover preparing information for model building and to assess the subsequent models yet it is not such case with survey spam recognition.
- Consideration of sentiment of news: Now and again it is additionally conceivable that news is composed in a negative tone however it is an honest survey so we need to likewise consider the sorts of supposition of reviews.
- Low accuracy of fake news detectors: Numerous scientists have executed distinctive strategies to identify fake audits. However, all past techniques have less exactness in recognizing fake or spam surveys. We must enhance this precision.

In this chapter, we try to improve the accuracy of spam news and review detection. The main aim of this chapter is to improve the overall accuracy of fake news as well as spam review.

EXISTING SOLUTIONS

The summarized comparative analysis is presented in Table 1 along with its limitations, used methods, advantages, and parameters. The observed parameters from all these methods in terms of accuracy, threshold, precision, recall, and a decision is shown in Table 2.

PROPOSED FRAMEWORK

This research work is divided in to three main part. The primary part is in charge of getting and pre-preparing the survey from the input side or the web. Before arranging the surveys, the second stage is an information phonetic assessment, which does a semantic inquiry on the information from the previous segment. The third segment performs survey characterization of the audits into two class's spam or non-spam surveys. The below Figureure portray a diagram of the proposed structure.

a. Data Linguistic Analysis Component

The data language evaluation phase conducts a diversity of linguistic tasks on new data gathered from the preceding section. It excecuted like you see in *figure 1*,

1. **Text or Sentence Filter**: This component is basically used to cleave the entire data into individual syntax. Often. these data start with a capital letter, which is the equivalent of a syntax split sign. If words such as Pvt., Ltd.,etc comes before the "." Symbol, it's best to put it at the ending of the sentence. "/" as well as "?" are examples of diacritical marks. Syntax splitting is also implemented using "," and "."
2. **Tokenization**: This section organises the content into very simple entities, such as numbers, accentuation, and words.

3. **Part-of-speech (POS)**: Not all of the words in source content information are relevant for tilt analysis. Every token is provided a part-of-speech (POS) label, which might be a descriptive word, verb, or qualifier.

Table 1. Comparison with the existing solution

Sr No	Paper Name	Advantages	Limitation	Used Method	Parameters
1	Chinese Review Spam Classification Using Machine Learning Method (Xi 2012)	Good accuracy (around 93%) with SVM method.	They don't explore the effect of review spam detection for another product dataset.	SVM	Accuracy
2	Review Spam Detector with Rating Consistency Check (Sharma and Lin 2013)	-They use an algorithm that is not context-dependent. -Easy to deploy. -Result of this is very effective even though they do not use context.	When the system is unable to decide about review then it flags a potential list of evidence to determine spam or not spam review.	Unsupervised Method	Accuracy
3	Using Unanticipated Rules for Find Abnormal Review Patterns (Jindal, Liu, and Lim 2010)	They formulate the problem of review spam detection as finding unexpected rules. The technique is domain-independent.	Gives low accuracy for spam detection.	Unsupervised methods	Accuracy
4	Towards Online Review Spam Detection (Lin et al. 2014)	Good accuracy result with SVM, Avg accuracy logistic regression, and low accuracy unsupervised method.	Result accuracy is not better than previous research of others.	SVM, Logistic Regression (Supervised Method), and Unsupervised Methods for each experiment	Accuracy
5	Detecting Product Review Spammers using Activity Model (Jiang, Cao, and Chen 2013)	They improve the prediction model that using the behavioral approach to detect review spammers who try to manipulate review ratings on some target product group.	They get average accuracy of detection of review spammer through the use of user behavior.	Unsupervised methods	Precision, recall, F-measure
6	Chinese Spammer Review Detection System Development and Construction (Xu et al. 2013)	Chinese Spam Design and Implementation news Detection System.	It's poor than other methods for large datasets.	Naïve Bayes(Supervised Method)	Accuracy
7	Topic-Based Mixture Modeling and Semi-Supervised Training Improve Twitter Sentiment Analysis (Xiang and Zhou 2014)	-Used their algorithm of a large amount of low-cost un-annotated data to improve accuracy	Needed to improve the accuracy in distinct domains	Semi-supervised learning (topic-based method)	Accuracy,
8	Sentiment Analysis and Opinion Mining (Social media as Twitter) (Da Silva, Coletta, and Hruschka 2016)	Good accuracy in SSL approach	SSL depends on a small initial sample of labeled data Unsupervised approach results are worse compare to SSL	SSL(wrapper-based, topic-based), unsupervised approach	Accuracy, Precision, Recall, threshold

Table 2. Parameters observed

Ref.	Accuracy	Recall	Precision	Threshold	Decision
(Xi 2012)	Yes	No	No	No	No
(Sharma and Lin 2013)	Yes	No	No	No	No
(Jindal, Liu, and Lim 2010)	No	No	No	No	No
(Lin et al. 2014)	Yes	No	No	No	No
(Jiang, Cao, and Chen 2013)	Yes	No	No	No	No
(Xu et al. 2013)	Yes	No	No	Yes	No
(Xiang and Zhou 2014)	Yes	No	No	No	No
(Da Silva, Coletta, and Hruschka 2016)	No	Yes	Yes	Yes	No

4. **Removing words**: This procedure removes or filters English stopwords from a document by eliminating every token from the stopword list that equals a stopword, or simply put, it removes popular words with no significance.

b. Fake News Classification Component

This is our most important component in the proposed framework. This component classifies the given text into the Fake or original class. This component works as follows:

Application of unsupervised method: In the first place we check the survey content against taking after four criteria for grouping of audit into spam or non-spam These criteria are given below:

1. **Questions in news or review:** The news which contains the question is typically utilized with the end goal of investigating. So the survey related to the question is for the most part not being considered an honest review.
2. **Product/company comparisons:** Numerous Competitor organizations compose undeserving negative surveys for the contender's item, with an accentuation to guide the analysts to their particular items/administrations. So these sorts of surveys are additionally not being considered honest news. For example, A news channel or paper writes: "Product X does not work. I bought Product B and now I am happy with this product".
3. **News written in capital letters:** News that starts with a capital letter is considered spam or false. because most of these surveys are used to gain attention or for ads. news with all capital letters should be ignored. For example: "THIS PERSON IS NOT GOOD FOR OUR COUNTRY".
4. **URL links in News:** The analysts in some cases leave a connection to their site or it is a commercial. In the vast majority of the cases, the analyst composes the post to leave the connection and the survey is futile. So, these sorts of surveys are additionally not being considered honest news.

If the survey falls flat against these criteria then we can name that the audit can be a "spam or fake" survey. At that point, we apply the SVM managed procedure to the rest of the non-spam audit to order it encourage into "spam or fake" and "no spam or real " class.

Figure 1. Proposed diagram of sentimetal data analysis

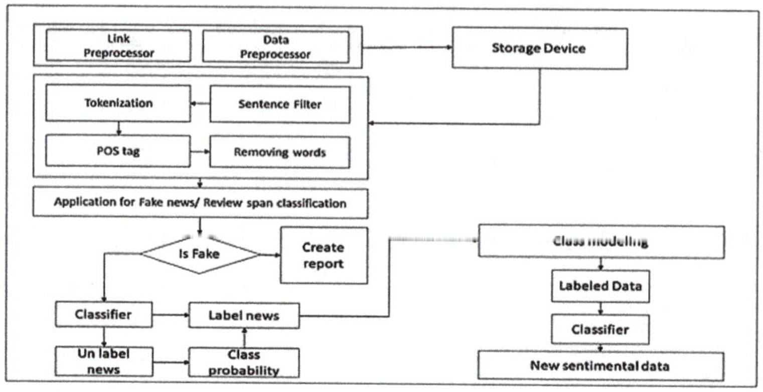

c. Application of the Supervised Method

Next, we apply directed systems to the rest of the audit to characterize spam or non-spam. Building a classifier can be used to implement regulated systems. Illustrations which may be physically labeled or user names are used to create this classifier.For the most part utilized managed calculations are Support Vector Machines (SVM), Naive Bayes classifier, and Maximum Entropy. SVMs are more fitting for audit spam orders. The proposed framework will utilize an SVM classifier for the arrangement as research has demonstrated that SVM performs better when information is substantial. In an SVM-based order, two arrangements of archives are required: preparing and a test set. A preparation set is utilized to prepare the programmed classifier to take in the qualities of reports, and a test set is utilized to arrange inconspicuous records. Machine Learning begins with gathering preparing datasets. following steps are used to prepared a classifier while preparing the information.

Once you select any supervised allocated address at that time the selection process is done .so now we can familiarize with how grammer or syntax are represented. A lot of frequently acclimated appearance in allocation are how many Term attendance and their frequency are present . These appearances cover n-grams or uni-grams and they are presence or not. So basically these appearances accept on the whole and propitious acclimated in classification.

The bag-of-words approach is one strategy of translating argument into appearance. An aspect of the affection set is anniversary chat, and an anniversary certificate is represented by a set of absolute numbers. In the document-gram model, the area anniversary aspect of the set represents the abundance of that chat and may be explicitly described as:

Take, for example, f1...fm, which is a predefined set of m occurrences that can appear in a document; examples also include chat "diseases" or the bigram "particularly unwell." Take, for example, ni(k), which is the number of times fi appears in certificate d. Then we may say that the certificate agent d: = (n1(k), n2(k),...nm(k)) represents the anniversary certificated.

An argument is represented in one of the models as an irregular aggregation of words, behindhand grammar, and even text structure, resulting in the lack of logical abstractions in the input data processing. To abate compounds loss, the N-gram archetype is utilised, in which anniversary sub-sequence of breadth n is specified as a feature, rather as anniversary conversation. Since this approach increases the complexity of the process, it is unable to account for ongoing dependencies while preserving conversation adjustment inside N-words. So a size 1 N-gram is labeled a "unigram," a size 2 is called a "bigram," a size 3 is called a "trigram," and a size 4 or more is usually called a "N-gram." Below is an N-gram set of an example news speech.:

"I did not like the blue pen."
Bigrams: ["I did", "not like", "the blue pen"]
Trigrams: ["I did not", "like the blue pen"]
Unigrams: ["I", "did", "not", "like", "the", "blue"," pen"]

d. Text Classification Component

A classifier is established using tagged news(tweet), and it is then used to assess the profitability of each unlabeled bargain (this primary footfall occurs alone once). Next that, a sample of tweets with a stylish anticipation college rather than an aplomb start are selected. They're then included in the labelled tweets dataset, which is then used to create an affair model, from which anniversary tweets' affair distributions are recorded. Then, clusters that support the affair distributions are recognized, and an accurate effect paradigmatic for the each cluster is generated. The unlabeled tweets are categorized using the consistent effect admixture archetypal. An acknowledged action occurs until a certain number of samples have been completed or no further tweets have been found to be the answer to the set of labelled tweets.

As apparent from a simulation study, SVM gives the simplest accurateness as a neighborhood of all classifiers, so we'll use SVM as a capital algorithm for assessment spam allocation and affect the analysis.

This proposed framework for analyzing spam or fake news detection. We also explained how the anniversary basic framework will work. Our new access is predicated on the affiliation of the unsupervised and supervised methods. We accept SVM for supervised adjustment because it's best for this blazon of problems like analysis spam detection.

CONCLUSION

Detection of spam/fake news becomes a crucial analysis botheration nowadays. Now a days firm or group organizations in eCommerce are establishing advanced their efforts to acquisition the simplest arrangement for analysis fake news detection. Although, a number of the algorithms and methods accept been utilized in exposition fake apprehension gives satisfactory outputs, However, no algorithm or approach can guarantee that all obstacles will be fulfilled.The extended strategy is based on improving the success rate of spurious detection.There are an enormous charge in e-commerce for such applications because every alignment wants genuine news analysis arrangement to apperceive how users feel about their articles and services. We proposed access for audition spam in auberge news. The proposed access is predicated on accumulation methods from abstracts mining and argument mining. The proposed address anatomy and apparatus were presented and explained. It involves the afterward accomplish for audition spam: abstracts accretion and pre-processing abstract linguistic assay and assuredly analysis spam classification. it's chip access of supervised adjustment (SVM) and unsupervised method. it's congenital to access the accurateness of spam detection. It can yield advantages of both SVM supervised adjustment and unsupervised adjustment to advance accurateness results. Testing the acceptable arrangement was performed by application two abstracted datasets. These after-effects gave a bright appearance that the arrangement is in a position of audition spam analysis with over ~91% accuracies. During the appraisal, it had been accessible to ascertain that it's achievable and reliable to body arrangement able of classifying analysis argument into spam and non-spam chic automatically. However, it's bright that an abundant amount of opinions are difficult to allocate due to the complication of the animal language. The proposed system's limitation includes an area annex of opinions. The proposed arrangement can allocate alone auberge analysis into spam and non-spam class.

REFERENCES

Jiang, Cao, & Chen. (2013). Detecting Product Review Spammers Using Activity Model. *Proceeding of International Conference on Advanced Computer Science and Electronics Information (ICACSEI 2013)*, 650–653.

Jindal, N., Liu, B., & Lim, E. P. (2010). Finding Unusual Review Patterns Using Unexpected Rules. *International Conference on Information and Knowledge Management, Proceedings*. 10.1145/1871437.1871669

Lin, Y. (2014). Towards Online Anti-Opinion Spam: Spotting Fake Reviews from the Review Sequence. *ASONAM 2014 - Proceedings of the 2014 IEEE/ACM International Conference on Advances in Social Networks Analysis and Mining*, 261–64.

Mallet Official Page. (2010). http://mallet.cs.umass.edu/index.php/Main_Page

Mukherjee, A., Liu, B., & Glance, N. (2012). Spotting Fake Reviewer Groups in Consumer Reviews. *WWW'12 - Proceedings of the 21st Annual Conference on World Wide Web*, 191–200. 10.1145/2187836.2187863

Parikh & Atrey. (2018). Media-Rich Fake News Detection: A Survey. *2018 IEEE Conference on Multimedia Information Processing and Retrieval (MIPR)*, 436–41. https://ieeexplore.ieee.org/document/8397049/

Python Official Site. (2019). https://www.python.org/

Ramesh Babu Durai, C., & Duraisamy, V. (2011). Content Based Image Retrieval Using Novel Gaussian Fuzzy Feed Forward-Neural Network. *Journal of Computational Science*, *7*(7), 958–961.

Sharma, K., & Lin, K. I. (2013). Review Spam Detector with Rating Consistency Check. *Proceedings of the Annual Southeast Conference*, 34–39.

Silva, Da, Coletta, & Hruschka. (2016). A Survey and Comparative Study of Tweet Sentiment Analysis via Semi-Supervised Learning. *ACM Computing Surveys*, *49*(1).

Xi, Y. (2012). Chinese Review Spam Classification Using Machine Learning Method. *Proceedings - 2012 International Conference on Control Engineering and Communication Technology, ICCECT 2012*, 669–72.

Xiang, B., & Zhou, L. (2014). Improving Twitter Sentiment Analysis with Topic-Based Mixture Modeling and Semi-Supervised Training. *52nd Annual Meeting of the Association for Computational Linguistics, ACL 2014 - Proceedings of the Conference*, 434–39.

Xu, X. (2013). Design and Implementation of Chinese Spam Review Detection System. Lecture Notes in Computer Science (Including Subseries Lecture Notes in Artificial Intelligence and Lecture Notes in Bioinformatics), 508–18.

Zhang, J., Dong, B., & Yu, P. S. (2018). FAKEDETECTOR: Effective Fake News Detection with Deep Diffusive Neural Network. *2020 IEEE 36th International Conference on Data Engineering (ICDE)*, 1826–29. https://ieeexplore.ieee.org/document/9101457/

Compilation of References

Unger, S., & Meiran, W. R. (2020). Student attitudes towards online education during theCOVID-19 viral outbreak of 2020: Distance learning in a time of social distance. *International Journal of Technology in Education and Science*, *4*(4), 256–266. doi:10.46328/ijtes.v4i4.107

World Health Organization. (n.d.). https://www.worldometers.info/coronavirus/

Wu, J. T., Leung, K., Bushman, M., Kishore, N., Niehus, R., de Salazar, P. M., Cowling, B. J., Lipsitch, M., & Leung, G. M. (2020). Estimating clinical severity of covid-19 from the transmission dynamics in Wuhan, China. *Nature Medicine*, *26*(4), 506–510. doi:10.103841591-020-0822-7 PMID:32284616

ICMR. (n.d.). *Important COVID-19 Guidelines*. http://pbhealth.gov.in/Coronaguidelines.html

Baum, A., Singer, J. E., & Baum, C. S. (1981). Stress and the environment. *The Journal of Social Issues*, *37*(1), 4–35. doi:10.1111/j.1540-4560.1981.tb01056.x

Sneha. (2020). *Recognizing stress based on social interactions in social networks*. Lambert Academic Publishing.

Gaidhane, Khatib, Zahiruddin, Gaidhane, Telrandhe, & Godhiwal. (2020). Depression, Anxiety and Stress among the general population in the time of COVID-19 lockdown: A cross-sectional study protocol. *IJRPS, 11*(1), 360-364.

Cooke, Eirich, Racine, & Madigan. (2020). Prevalence of posttraumatic and general psychological stress during COVID-19: A rapid review and meta-analysis. *Psychiatry Research, 292*. https://www.sciencedirect.com/science/article/pii/S0165178120326445 doi:10.1016/j.psychres.2020.113347

Bao, W. (2020). COVID-19 and online teaching in higher education: A case study of Peking University. *Human Behavior and Emerging Technologies*, *2*(2), 113–115. doi:10.1002/hbe2.191 PMID:32510042

Sneha, Y., & Aluvala, S. (2021). Impact of Experimental Learning on Graduates Success in Engineering Education. *Journal Of Engineering EducationTransformations*, *34*(0), 666–669. doi:10.16920/jeet/2021/v34i0/157240

Abas, H., Yusoff, R. L., & Tukiman, Z. (2017). Apps: Platform Dakwah untuk Generasi Y dan Z. Dalam Hasanah Abd Khafiz, Zainab Ismail (penyunting). Prosiding Seminar Antarabangsa Dakwah ke 3, Hlm. 58-67. Penerbit Jabatan Pengajian Dakwah dan Kepimpinan Universiti Kebangsaan Malaysia: Bangi Selangor

Abbas, A., Gulzar, R., & Hussain, Z. (2019). Article. *Journal of Communication and Cultural Trends*, *1*(1), 26–44. doi:10.32350/jcct.11.02

Abd Rahim. (2017). Tahap Penggunaan Instagram dalam kalangan pelajar Universiti Putra Malaysia. *Journal of Business and Social Development, 5*(2).

Abdi, H. (2003). Factor Rotations in Factor Analysis. In M. Lewis-Beck, A. Bryman, & T. Futing (Eds.), *Encyclopedia of Social Sciences Research Methods*. Sage.

Abdulellah Alaboudi, A. (2020). *Impact of digital technology on Saudi students*. Academic Press.

Acharya, L., Jin, L., & Collins, W. (2018). College life is stressful today – Emerging stressors and depressive symptoms in college students. *Journal of American College Health*, *66*(7), 655–664. doi:10.1080/07448481.2018.1451869 PMID:29565759

Ackerman, C. (2020, February 2). + *Reviews*. https://positivepsychology.com/mindfulness-apps/

Adegboji, O. B., & Toyo, O. D. (2006). The impact of the Internet on research: The experience of Delta State University, Abraka, Nigeria. *Library Philosophy and Practice*, *8*(2), 1–8.

Agarwal, V. (2012). Harmful effects of media on children and adolescents Burden of care of principal caregivers of autistic children and adolescents visiting health facilities in Lucknow city.

Ahmad, R. W., Salah, K., Jayaraman, R., Yaqoob, I., Ellahham, S. & Omar, M. (2021). The role of blockchain technology in telehealth and telemedicine. *International Journal of Medical Informatics, 148*(November). doi:10.1016/j.ijmedinf.2021.104399

Ainin, S., Parveen, F., Moghavvemi, S., Jaafar, N. I., & Shuib, N. L. M. (2015). Factors influencing the use of social media by SMEs and its performance outcomes. *Industrial Management & Data Systems*, *115*(3), 570–588. doi:10.1108/IMDS-07-2014-0205

Akintolu, M., & Uleanya, C. (2021). Ensuring Sustainable Development Goal in Rural Africa through Adult Literacy Programme: A Case Study of Technology Usage in Developing Nations. *Universal Journal of Educational Research, 8*.

Akram, W., & Kumar, R. (2017). *A Study on positive and negative effects of social media on society*.

Al Nuaim. (2020). The Use of Virtual Classrooms in E-learning: a case study in King Abdulaziz University. Academic Press.

Al Zboon, M. S., Al Ghammaz, S. A. D., & Al Zboon, M. S. (2018). The Impact of the Use of YouTube and Facebook on Students' Academic Achievement in Geography Course at the University of Jordan for the Bachelor's Degree. *Modern Applied Science*, *12*(3), 164. doi:10.5539/mas.v12n3p164

Alabdulkareem, S. A. (2015). Exploring the use and the impacts of social media on teaching and learning science in Saudi. *Procedia: Social and Behavioral Sciences*, *182*, 213–224. doi:10.1016/j.sbspro.2015.04.758

Al-Darwish, A. I., & Choe, P. (2019). A framework of information security integrated with human factors. In *International Conference on Human-Computer Interaction* (pp. 217-229). Springer. 10.1007/978-3-030-22351-9_15

Alderson Broaddus University. (n.d.). *Common Issues for College Students*. https://ab.edu/counseling-services/common-issues-for-college-students/

Al-Dheleai, Y. M., & Tasir, Z. (2017). Using facebook for the purpose of students' interaction and its correlation with students' academic performance. *The Turkish Online Journal of Educational Technology*, *16*(4), 170–178.

Alexander, M. W., Truell, A. D., & Zhao, J. J. (2012). Expected advantages and disadvantages of online learning: Perceptions from college students who have not taken online courses. *Issues in Information Systems*, *13*(2), 193–200.

Ali, I. (2015). Big Data: Apa dan pengaruhnya pada perpustakaan? (What is Big Data and its Influence to Library). *Perpustakaan Nasional*, *22*(4).

Ali, W. (2018). Influence of Evolving Technology in Emerging Online Lives of the Digital Native University Students. *Asia Pacific Journal of Contemporary Education and Communication Technology*, *4*(2), 141–155. doi:10.25275/apjcectv4i2edu15

Alkhawaja, M. I., & Halim, M. S. B. A. (2019). Challenges of E-Learning System Adoption in Jordan Higher Education. *International Journal of Academic Research in Business & Social Sciences*, *9*(9), 487–494. doi:10.6007/IJARBSS/v9-i9/6317

AlMindeel, R., & Martins, J. T. (2020). Information security awareness in a developing country context: Insights from the government sector in Saudi Arabia. *Information Technology & People*, *34*(2), 770–788. doi:10.1108/ITP-06-2019-0269

Alnatheer, M. A. (2014). A conceptual model for understanding information security culture. *International Journal of Social Science and Humanity*, *4*(2), 104–107. doi:10.7763/IJSSH.2014.V4.327

Alruwais, N., Wills, G., & Wald, M. (2018). Advantages and challenges of using e-assessment. *International Journal of Information and Education Technology (IJIET)*, *8*(1), 34–37. doi:10.18178/ijiet.2018.8.1.1008

Alsubaie, M., Abbott, R., Dunn, B., Dickens, C., Keil, T. F., Henley, W., & Kuyken, W. (2017). Mechanisms of action in mindfulness-based cognitive therapy (mbct) and mindfulness-based stress reduction (mbsr) in people with physical and/or psychological conditions: A systematic review. *Clinical Psychology Review*, *55*, 74–91. doi:10.1016/j.cpr.2017.04.008 PMID:28501707

Altawalbeh, M. A. (2021). The Impact Of Using Social Sites On Academic Performance For Students At Jordanian Universities. *Turkish Journal of Computer and Mathematics Education*, *12*(7), 3264–3270.

Altman, N. S. (1992). An introduction to kernel and nearest-neighbor nonparametric regression. *The American Statistician*, *46*(3), 175–185.

American Addiction Centers. (2019). *Coping with College Stress: Exploring perceived stress and coping methods in college.* https://americanaddictioncenters.org/learn/college-coping-mechanisms/

American Addiction Centers. (2020, June 1). *Guide to Substance Abuse and Mental Health in College.* https://americanaddictioncenters.org/rehab-guide/college

American College Health Association. (2010). *American College Health Association-National College Health Assessment II: Reference Group Data Report Spring 2010.* Linthicum, MD: American College Health Association. https://www.acha.org/HealthyCampus/HealthyCampus/Student_Objectives.aspx

American College Health Association. (2019). *American College Health Association-National College Health Assessment II: Undergraduate Student Executive Summary Spring 2019.* American College Health Association.

American College Health Association. (n.d.). *Who We Are.* https://www.acha.org/ACHA/About/About_ACHA/ACHA/About/About_ACHA.aspx?hkey=9deff142-0898-4f21-8a69-f5e0621eb59c

American Institute of Stress. (2019, September 9). *Stress: An Epidemic Among College Students.* https://www.stress.org/stress-an-epidemic-among-college-students

American Psychiatric Association, DSM-5 Task Force. (2013). *Diagnostic and statistical manual of mental disorders: DSM-5* (5th ed.). American Psychiatric Publishing, Inc. doi:10.1176/appi.books.9780890425596

American Psychological Association. (2019). *Stress in America: Stress and Current Events. Stress in America™ Survey.* https://www.apa.org/news/press/releases/stress/2019/stress-america-2019.pdf

American Psychological Association. (n.d.). *Ego resilience.* https://dictionary.apa.org/ego-resiliency

Amichai-Hamburger, Y. (2008). Potential and promise of online volunteering. *Computers in Human Behavior*, *24*(2), 544–562. doi:10.1016/j.chb.2007.02.004

Amichai-Hamburger, Y., Kingsbury, M., & Schneider, B. H. (2013). Friendship: An old concept with a new meaning? *Computers in Human Behavior, 29*(1), 33–39. doi:10.1016/j.chb.2012.05.025

An Assessment of Zimbabwe Secondary School Teachers' Attitudes towards the Use of Smart Phones in the Classroom: A Case of Midlands Province, Zimbabwe. (2017). *International Journal of Social Sciences & Educational Studies, 4*(1). doi:10.23918/ijsses.v4i1p12

Anderson, M., & Jiang, J. (2018). *Teens' Social Media Habits and Experiences _ Pew Research Center.*

Anderson, M., Jiang, J., & Smith, A. (2018). *For media or other inquiries* (Vol. 28).

Anderson, C. A., Berkowitz, L., Donnerstein, E., Huesmann, L. R., Johnson, J. D., Linz, D., Malamuth, N. M., & Wart ella, E. (2003). The Influence of Media Violence on Youth. *Psychological Science in the Public Interest, 4*(3), 81–110. doi:10.1111/j.1529-1006.2003.pspi_1433.x PMID:26151870

Anderson, J., Rainie, L., & Luchsinger, A. (2018). Artificial intelligence and the future Humans. *Pew Research Centre, 169*(1), 57–88. doi:10.4000/rei.8727

Anderson, M., & Jiang, J. (2018). *Teens.* Social Media & Technology.

Andonia, M., Robua, V., Flynna, D., Abramb, S., Geachc, D., Jenkinsd, D., McCallumd, P., & Peacock, A. (2019). Blockchain technology in the energy sector: A systematic review of challenges and opportunities. *Renewable & Sustainable Energy Reviews, 100*, 143–174.

Andreoni, J., & Vesterlund, L. (2001). Which is the Fair Sex? Gender Differences in Altruism. *The Quarterly Journal of Economics, 116*(1), 293–312. doi:10.1162/003355301556419

Ani, U. D., He, H., & Tiwari, A. (2019). Human factor security: Evaluating the cybersecurity capacity of the industrial workforce. *Journal of Systems and Information Technology, 21*(1), 2–35. doi:10.1108/JSIT-02-2018-0028

Ansari, J. A. N., & Khan, N. A. (2020). Exploring the role of social media in collaborative learning the new domain of learning. *Smart Learning Environments, 7*(1), 9. Advance online publication. doi:10.118640561-020-00118-7

App Annie. (2021). *The App Analytics and App Data Industry Standard.* Retrieved 4 May 2021, from https://www.appannie.com/en/?sfdcid=7016F000002MS1c&utm_campaign=ww-logo-201910-1910-digital-2021-partnership&utm_content=report-&utm_medium=partnership&utm_source=digital-2021

Apple. (2018, July 5). *The App Store turns 10.* https://www.apple.com/newsroom/2018/07/app-store-turns-10/

Appleby, D. C. (2019). The soft skills college students need to succeed now and in the future. *Education Directorate, 5*(2). https://www.apa.org/ed/precollege/psn/2017/09/soft-skills

Aripin, N. (2016). "YouTube" dan Gneerasi Muda Islam: Satu Pendekatan Kelompok Fokus dalam Kalangan Pelajar Universiti. *Jurnal Komunikasi Malaysian Journal of Communication Jilid, 32*(1), 371–400.

Arora, N., & Singh, N. (2017). Factors Affecting the Academic Performance of College Students. iManag's. *Journal of Educational Technology, 14*(1), 47–53. doi:10.26634/jet.14.1.13586

Asha'ari. (2018). Pola Agama dalam penggunaan Instagram di kalangan mahasiswa. *Jurnal Pengajian Islam.*

Atienza-Mendez, C., & Bayyou, D. G. (2019). Blockchain Technology Applications in Government. *IJCAT -. International Journal of Computers and Technology, 6*(11), 68–74. doi:10.4018/978-1-7998-3632-2.ch003

Atlam, H. F., Azad, M. A., Alzahrani, A. G., & Wills, G. (2020). A review of Blockchain in Internet of Things and AI. *Big Data and Cognitive Computing, 4*(4), 1–27. doi:10.3390/bdcc4040028

Austermann, M. I., Thomasius, R., & Paschke, K. (2021). Assessing Problematic Social Media Use in Adolescents by Parental Ratings: Development and Validation of the Social Media Disorder Scale for Parents (SMDS-P). *Journal of Clinical Medicine*, *10*(4), 617.

Austin, W. A., & Totaro, M. W. (2011). Gender differences in the effects of internet usage on high school absenteeism. *Journal of Socio-Economics*, *40*(2), 192–198. doi:10.1016/j.socec.2010.12.017

Aydin, S. (2012). A review of research on Facebook as an educational environment. *Educational Technology Research and Development*, *60*(6), 1093–1106. doi:10.100711423-012-9260-7

Ayu, D. (2016). Motif dan kepuasan penggunaan Instagram. In *Kesenjangan Antara Motif Dan Kepuasan Penggunaan Media Sosial Instagram pada Mahasiswa Universitas Muhammadiyah Surakarta Angkatan 2013*. Universitas Muhamad-diyah Sukarta.

Azano, A. P., & Stewart, T. T. (2015). Exploring place and practicing justice: Preparing preservice teachers for success in rural schools. *Journal of Research in Rural Education*, *30*(9), 1–12. https://jrre.psu.edu/

Azizi, S. M., Soroush, A., & Khatony, A. (2019). The relationship between social networking addiction and academic performance in Iranian students of medical sciences: A cross-sectional study. *BMC Psychology*, *7*(1), 1–8. doi:10.118640359-019-0305-0 PMID:31053171

Baird, S., Seager, J., Sabarwal, S., Guglielmi, S., & Sultan, M. (2020). *Adolescence in the Time of COVID-19*. Academic Press.

Balakrishnan, V. (2015). Cyberbullying among young adults in Malaysia: The roles of gender, age and Internet frequency. *Computers in Human Behavior*, *46*, 149–157. doi:10.1016/j.chb.2015.01.021

Balleys, C., Millerand, F., Thoër, C., & Duque, N. (2020). Searching for oneself on YouTube: Teenage peer socialization and social recognition processes. *Social Media + Society*, *6*(2), 2056305120909474. doi:10.1177/2056305120909474

Bamber, M. D., & Schneider, J. K. (2020, January 7). College students' perceptions of mindfulness-based interventions: A narrative review of the qualitative research. *Current Psychology (New Brunswick, N.J.)*. Advance online publication. doi:10.100712144-019-00592-4

Bandura, A. (1986). *Social foundations of thought and action: A social cognitive theory*. Prentice Hall.

Bandura, A. (1994). Social cognitive theory of mass communication. In J. Bryant & D. Zillmann (Eds.), *Media effects: Advances in theory and research* (pp. 61–90). Lawrence Erlbaum.

Bandura, A. (2006). Adolescent Development from an Agentic Perspective. In F. Pajares & T. Urdan (Eds.), *Self-Efficacy Beliefs of Adolescents* (pp. 1–43). Information Age.

Bar-On, M. E., Broughton, D. D., Buttross, S., Corrigan, S., Gedissman, A., González De Rivas, M. R., Rich, M., Shifrin, D. L., Brody, M., Wilcox, B., Hogan, M., Holroyd, H. J., Reid, L., Sherry, S. N., Strasburger, V., & Stone, J. (2001). Children, adolescents, and television. *Pediatrics*, *107*(2), 423–426. doi:10.1542/peds.107.2.423 PMID:11158483

Bashman, L. (2020). *Online learning punishes the poor*. Independent Online. Available athttps://www.iol.co.za/news/opinion/online-learning-punishes-the-poor-46963674

Beauchemin, J. D. (2018). Solution-focused wellness: A randomized controlled trial of college students. *Health & Social Work, 43*(2), 94-100. http://dx.doi.org.tcsedsystem.idm.oclc.org/10.1093/hsw/hly007

Beckman, K., Bennett, S., & Lockyer, L. (2019). Reproduction and transformation of students' technology practice: The tale of two distinctive secondary student cases. *British Journal of Educational Technology*, *39*(6), 346. doi:10.1111/bjet.12736

Beech, R., & Schoeppe, A. (1974). Development of Value Systems in Adolescents. *Developmental Psychology, 10*(5), 644–656. doi:10.1037/h0037016

Bekalu, M. A., McCloud, R. F., & Viswanath, K. (2019). Association of Social Media Use With Social Well-Being, Positive Mental Health, and Self-Rated Health: Disentangling Routine Use From Emotional Connection to Use. *Health Education and Behavior, 46*(2_suppl), 69–80.

Bener, A., Al-Mahdi, H. S., Ali, A. I., Al-Nufal, M., Vachhani, P. J., & Tewfik, I. (2011). Obesity and low vision as a result of excessive Internet use and television viewing. *International Journal of Food Sciences and Nutrition, 62*(1), 60–62. doi:10.3109/09637486.2010.495711 PMID:20645888

Bertapelli, F., Pitetti, K., Agiovlasitis, S., & Guerra-Junior, G. (2016). Overweight and obesity in children and adolescents with Down syndrome-prevalence, determinants, consequences, and interventions: A literature review. *Research in Developmental Disabilities, 57*, 181–192. doi:10.1016/j.ridd.2016.06.018 PMID:27448331

Best, P., Manktelow, R., & Taylor, B. (2014). Online communication, social media and adolescent wellbeing: A systematic narrative review. *Children and Youth Services Review, 41*, 27–36.

Beyens, I., Pouwels, J. L., van Driel, I. I., Keijsers, L., & Valkenburg, P. M. (2020). The effect of social media on well-being differs from adolescent to adolescent. *Scientific Reports, 10*(1).

Bhardwaj, A., Avasthi, V., & Goundar, S. (2017). Impact of social networking on Indian youth: A survey. *I.J. of Electronics and Information Engineering, 7*(1), 41–51.

Bhikkhu, T. (2018, March 5). *The Buddha's Original Teachings on Mindful*ness. https://tricycle.org/trikedaily/satipatthana-sutta-mindfulness/

Bianchi, S. M., & Robinson, J. (1997). What did you do today? Children's use of time, family composition, and acquisition of social capital. *Journal of Marriage and Family, 59*(2), 332–344. doi:10.2307/353474

Bich Diep, P., Minh Phuong, V., Dang Chinh, N., Thi Hong Diem, N., & Bao Giang, K. (2021). Health Science Students' Use of Social Media for Educational Purposes: A Sample from a Medical University in Hanoi, Vietnam. *Health Services Insights, 14*. Advance online publication. doi:10.1177/11786329211013549 PMID:34017181

Bishop, K. K. (2016). The relationship between retention and college counseling for high-risk students. *Journal of College Counseling, 19*(3), 205–217. doi:10.1002/jocc.12044

Bishwas, P.C., Hakim, M.U., Alam, M.S. (2020). Online Class and Its Psychological Impact on Satisfaction of University Students in Bangladesh during COVID – 19 Pandemic. *Ioer International Multidisciplinary Research Journal, 2*(4), 101 – 109.

Bittles, A. H., Petterson, B. A., Sullivan, S. G., Hussain, R., Glasson, E. J., & Montgomery, P. D. (2002). The influence of intellectual disability on life expectancy. *The Journals of Gerontology. Series A, Biological Sciences and Medical Sciences, 57*(7), M470–M472. doi:10.1093/gerona/57.7.M470 PMID:12084811

Blanchet, P., Bebin, M., Bruet, S., Cooper, G. M., Thompson, M. L., Duban-Bedu, B., Gerard, B., Piton, A., Suckno, S., Deshpande, C., Clowes, V., Vogt, J., Turnpenny, P., Williamson, M. P., & Alembik, Y. (2017). Clinical Sequencing Exploratory Research Study Consortium; Deciphering Developmental Disorders Consortium, Glasgow E, McNeill A. MYT1L mutations cause intellectual disability and variable obesity by dysregulating gene expression and development of the neuroendocrine hypothalamus. *PLOS Genetics, 13*(8), e1006957. doi:10.1371/journal.pgen.1006957 PMID:28859103

Blue Whale. (2021). *The truth behind an online 'suicide challenge'*. BBC News. Retrieved 3 October 2021, from https://www-bbc-com.cdn.ampproject.org/v/s/www.bbc.com/news/blogs-trending-46505722.amp?amp_js_v=a6&amp_gsa=1&usqp=mq331AQKKAFQArABIIACAw%3D%3D#aoh=16332251619975&referrer=https%3A%2F%2Fwww.google.com&amp_tf=From%20%251%24s&ampshare=https%3A%2F%2Fwww.bbc.com%2Fnews%2Fblogs-trending-46505722

Blumer, H. (2012). Symbolic interactionism. *Contemporary Sociological Theory, 62.*

Blumler, J. G., & Katz, E. (1974). The uses of mass communications: Current perspectives on gratifications research. *Sage (Atlanta, Ga.).*

Boddy, C. R. (2016). Sample size for qualitative research. *Qualitative Market Research, 19*(4), 426–432. doi:10.1108/QMR-06-2016-0053

Boer, M., Stevens, G. W. J. M., Finkenauer, C., de Looze, M. E., & van den Eijnden, R. J. J. M. (2021). Social media use intensity, social media use problems, and mental health among adolescents: Investigating directionality and mediating processes. *Computers in Human Behavior, 116*, 106645.

Boer, M., van den Eijnden, R. J. J. M., Boniel-Nissim, M., Wong, S. L., Inchley, J. C., Badura, P., Craig, W. M., Gobina, I., Kleszczewska, D., Klanšček, H. J., & Stevens, G. W. J. M. (2020). Adolescents' Intense and Problematic Social Media Use and Their Well-Being in 29 Countries. *The Journal of Adolescent Health, 66*(6), S89–S99.

Bohon, L. M., Cotter, K. A., Kravitz, R. L., Cello, P. C. Jr, Fernandez, Y., & Garcia, E. (2016). The Theory of planned behavior as it predicts potential intention to seek mental health services for depression among college students. *Journal of American College Health, 64*(8), 593–603. doi:10.1080/07448481.2016.1207646 PMID:27386898

Boitard, R., Pourazad, M. T., Nasiopoulos, P., & Slevinsky, J. (2015). Demystifying High-Dynamic-Range Technology: A new evolution in digital media. *IEEE Consumer Electronics Magazine, 4*(4), 72–86. doi:10.1109/MCE.2015.2463294

Bonilla-Molina, L. (2020). Covid-19 on Route of the Fourth Industrial Revolution. *Postdigital Science and Education*, 1-7.

Bóo, S. J. M., Childs, F. J., Cooney, S., Datta, B., Dufour, G., Jones, P. B., & Galante, J. (2020). A follow-up study to a randomized control trial to investigate the perceived impact of mindfulness on academic performance in university students. *Counselling & Psychotherapy Research, 20*(2), 286–301. doi:10.1002/capr.12282

Boote, D. N., & Beile, P. (2005). Scholars before researchers: On the centrality of the dissertation literature review in chapter preparation. *Educational Chapter, 34*(6), 3-15.

Bourdieu, P., & Passeron, J. C. (1977). Reproduction in Education, Society and Culture. *Sage (Atlanta, Ga.).*

Bower, M., & Torrington, J. (2020). Typology of free Web-based learning technologies. *Educause*. doi:10.13140/RG.2.2.11064.16647

Boyd, D. M., & Ellison, N. B. (2008). Social network sites: Definition, history, and scholarship. *Journal of Computer-Mediated Communication, 13*(1), 210–230. doi:10.1111/j.1083-6101.2007.00393.x

Braider, L., La Lima, C., Crimarco, N., Hollander, B., Reid-Russell, A., Kane, J., & Greenwald, B. (2019). Characterization of psychiatrically hospitalized college students. *Journal of American College Health, 67*(7), 654–660. doi:10.1080/07448481.2018.1500471 PMID:30365917

Brechwald, W. A., & Prinstein, M. J. (2011). Beyond homophily: A decade of advances in understanding peer influence processes. *Journal of Research on Adolescence, 21*(1), 166–179. doi:10.1111/j.1532-7795.2010.00721.x PMID:23730122

Brown, B. B. (2004). Adolescents' relationships with peers. In R. M. Lerner & L. Steinberg (Eds.), Handbook of Adolescent Psychology (2nd ed., pp. 363–394). John Wiley & Sons.

Burke, P. (2004). Identities and Social Structure: The 2003 Coley-Mead Award Address. *Social Psychology Quarterly, 67*(1), 5–15. doi:10.1177/019027250406700103

Burns, M. E., Houser, M. L., & Farris, K. L. B. (2018). Theory of planned behavior in the classroom: An examination of the instructor confirmation-interaction model. *Higher Education: The International Journal of Higher Education Research, 75*(6), 1091–1108. doi:10.100710734-017-0187-0

Bushman, B. J., & Anderson, C. A. (2001). Media Violence and the American Public: Scientific Facts Versus Media Misinformation. *The American Psychologist, 56*(6-7), 477–489. doi:10.1037/0003-066X.56.6-7.477 PMID:11413871

Buzzetto-More, N. (2014). An Examination of Undergraduate Student's Perceptions and Predilections of the Use of YouTube in the Teaching and Learning Process. *Interdisciplinary Journal of E-Skills and Lifelong Learning, 10*, 17–32. doi:10.28945/1965

Cabero-Almenara, J., Fernández-Batanero, J. M., & Barroso-Osuna, J. (2019). Adoption of augmented reality technology by university students. *Heliyon, 5*(5), e01597. Advance online publication. doi:10.1016/j.heliyon.2019.e01597 PMID:31193247

Çankayaa, S., & Odabas, H. F. (2009). Parental controls on children's computer and Internet use, World Conference on Educational Sciences 2009. *Procedia: Social and Behavioral Sciences, 1*, 1105–1109.

Carlo, G., McGinley, M., Hayes, R., Batenhorst, C., & Wilkinson, J. (2007). Parenting styles or practices? Parenting, sympathy, and prosocial behaviors among adolescents. *The Journal of Genetic Psychology, 168*(2), 147–176. doi:10.3200/GNTP.168.2.147-176 PMID:17936970

Carmack, H. J., & Heiss, S. N. (2018). Using the theory of planned behavior to predict college students' intent to use linkedin for job searches and professional networking. *Communication Studies, 69*(2), 145–160. doi:10.1080/10510974.2018.1424003

Carter, S. (2018). *Social media may be making you overspend—and it's not just because of the ads* [Blog]. Retrieved from https://www.cnbc.com/2018/03/15/social-media-may-make-you-overspend-and-its-not-just-because-of-ads.html

Cash, H., Rae, C. D., Steel, A. H., & Winkler, A. (2012). Internet Addiction A Brief Summary of Research and Practice. *Current Psychiatry, 8*(8), 292–298. Retrieved from https://www.ingentaconnect.com/content/ben/cpsr/2012/00000008/00000004/art00010

Caton, S., & Chapman, M. (2017). *The Use of Social Media and People with Intellectual Disability*. Academic Press.

Cawley, G. C., & Talbot, N. L. C. (2010). On Over-fitting in Model Selection and Subsequent Selection Bias in Performance Evaluation. *Journal of Machine Learning Research, 11*(70), 2079–2107.

Cenameri, M., & Taçi (Luga), P. (2015). The Impact of Technology on Adolescent Behavior. *Academic Journal of Interdisciplinary Studies*. doi:10.5901/ajis.2015.v4n2s2p288

Cervantes, R. C., & Palacio, U. M. (2020). Estimation of obesity levels based on computational intelligence. *Informatics in Medicine Unlocked, 21*, 100472. doi:10.1016/j.imu.2020.100472

Cetinkaya, L. (2017). An educational technology tool that developed in the natural flow of life among students: WhatsApp. *International Journal of Progressive Education, 13*(2), 29–47.

Chai, H.-Y., Niu, G.-F., Lian, S.-L., Chu, X.-W., & Sun, X.-J. (2019). Why social network site use fails to promote well-being? The roles of social overload and fear of missing out. *Computers in Human Behavior, 100*, 85–92. doi:10.1016/j.chb.2019.05.005

Chan, C. D., Cor, D., & Band, M. P. (2018). Privilege and Oppression in Counselor Education: An Intersectionality Framework. *Counselor Education Faculty Publications and Presentations*., 50. https://archives.pdx.edu/ds/psu/24227

Chang, R. (2002). *Healthy computing*. Retrieved September 20, 2006, From http://www.slais.ubc.ca/courses/libr500/01-02- wt2/www/R_Chang/index.html

Chapman, L. (2002). Russian roulette or Pandora's box: use of Internet as a research tool. In *VALA 2002. 11th Biennial Conference and Exhibition*. Victorian Association for Library Automation Inc. Retrieved 24 December, 2006, from http://www.vala.org.au/vala2002/2002pdf/18Chpmn.pdf

Chenlin, Y., ChangJiang, L., Jingzhou, H., & Jinfeng, W. (2018). Academic stress and evaluation of a mindfulness training intervention program. *NeuroQuantology: An Interdisciplinary Journal of Neuroscience and Quantum Physics*, *16*(5).

Chetty, Qigui, Gcora, Josie, Wenwei, & Fang. (2018). Bridging the digital divide: measuring digital literacy. *Economics: The Open-Access, Open-Assessment E-Journal, 12*(23), 1-20.

Chiemeke, S., & Mike Imafidor, O. (2020). Web-based Learning In Periods of Crisis: Reflections on the Impact of Covid-19. *International Journal of Computer Science and Information Technologies*, *12*(3), 33–46. doi:10.5121/ijcsit.2020.12303

Cholissodin & Riyandani. (2016). *Analisis Big Data (Teori & Aplikasi) "Big Data vs Big Information vs Big Knowledge"*. Versi 1.01

Cichosz, P. (2015). *Data Mining Algorithms: Explained Using R*. Wiley. doi:10.1002/9781118950951

Cipolletta, S., Malighetti, C., Cenedese, C., & Spoto, A. (2020). How can adolescents benefit from the use of social networks? The igeneration on instagram. *International Journal of Environmental Research and Public Health*, *17*(19), 1–15.

Clarke, M. (2003). *Healthy computing. It's in your hands*. Retrieved September 20, 2006, from http://www.slais.ubc.ca/courses/libr500/02-03-wt2/www/M_Clarke/Webpages/index.htm

Clark, L. S. (2011). Parental mediation theory for the digital age. *Communication Theory*, *21*(4), 323–343. doi:10.1111/j.1468-2885.2011.01391.x

CNN. (n.d.). http://edition.cnn.com/2010/WORLD/asiapcf/04/01/korea.parents.starved.baby/index.html

Cole, N. N., Nonterah, C. W., Utsey, S. O., Hook, J. N., Hubbard, R. R., Opare-Henaku, A., & Fischer, N. L. (2015). Predictor and moderator effects of ego resilience and mindfulness on the relationship etween academic stress and psychological well-being in a sample of Ghanaian college students. *The Journal of Black Psychology*, *41*(4), 340–357. doi:10.1177/0095798414537939

Collins, P. H. (2019). *Intersectionality as critical social theory*. Duke University Press.

Colmenarejo, G. (2020). Machine Learning Models to Predict Childhood and Adolescent Obesity: A Review. *Nutrients*, *12*(8), 2466. doi:10.3390/nu12082466 PMID:32824342

Colwill, C. (2009). Human factors in information security: The insider threat–Who can you trust these days? *Information Security Technical Report*, *14*(4), 186–196. doi:10.1016/j.istr.2010.04.004

Cook, D. A. (2020). Web-based learning: Pros, cons and controversies. *Clinical Medicine. Journal of the Royal College of Physicians of London*, *7*(1), 37–42. doi:10.7861/clinmedicine.7-1-37 PMID:17348573

Cooper, G., Barkatsas, T., & Strathdee, R. (2016). The Theory of Planned Behaviour (TPB) in Educational Research using Structural Equation Modelling (SEM). In T. Barkatsas & A. Bertram (Eds.), *Global Learning in the 21st Century. Global Education in the 21st Century Series*. doi:10.1007/978-94-6300-761-0_9

Cortes, C., & Vapnik, V. (1995). Support-Vector Networks. *Machine Learning*, *20*(3), 273–297. doi:10.1007/BF00994018

Coury, D. L., Anagnostou, E., Manning-Courtney, P., Reynolds, A., Cole, L., McCoy, R., Whitaker, A., & Perrin, J. M. (2012). Use of psychotropic medication in children and adolescents with autism spectrum disorders. *Pediatrics*, *130*(Suppl 2), S69–S76. doi:10.1542/peds.2012-0900D PMID:23118256

Cowan, N. (2014). Working Memory Underpins Cognitive Development, Learning, and Education. *Educational Psychology Review*, *26*(2), 197–223. doi:10.100710648-013-9246-y PMID:25346585

Coyne, S. M., Warburton, W. A., Essig, L. W., & Stockdale, L. A. (2018). Violent video games, externalizing behavior, and prosocial behavior: A five-year longitudinal study during adolescence. *Developmental Psychology*, *54*(10), 1868–1880. doi:10.1037/dev0000574 PMID:30234338

Coyne, S. M., Young, B., Rogers, A. A., Zurcher, J. D., Stockdale, L., Booth, M., ... Booth, M. (2020). Does time spent using social media impact mental health? An eight year longitudinal study. *Computers in Human Behavior*, *104*(1), 1–9. doi:10.1016/j.chb.2019.106160

Crittenden, W. F., Biel, I. K., & Lovely, W. A. III. (2019). Embracing Digitalization: Student Learning and New Technologies. *Journal of Marketing Education*, *41*(1), 5–14. doi:10.1177/0273475318820895

Crone, E. A., & Konijn, E. A. (2018). Media use and brain development during adolescence. *Nature Communications*, *9*(1), 10. doi:10.103841467-018-03126-x PMID:29467362

Curcio, I. D. D., Dipace, A., & Norlund, A. (2017). Virtual realities and education. *Research on Education and Media*, *8*(2), 60–68. doi:10.1515/rem-2016-0019

D'Alberton, F., & Scardovi, A. (2021). Children Exposed to Pornographic Images on the Internet: General and Specific Aspects in a Psychoanalytic Perspective. *The Psychoanalytic Study of the Child*, *74*(1), 131–144. doi:10.1080/00797308.2020.1859277

D'Esposito, J. E., & Gardner, R. M. (1999). University Students' Perceptions of the Internet: An Exploratory Study. *Journal of Academic Librarianship*, *25*(6), 456–461. doi:10.1016/S0099-1333(99)00078-6

Daniel, S. W. T., Carin, L., Dzau, V., & Wong, T. Y. (2020). Digital technology and COVID-19. *Nature Medicine, VOL*, *26*, 458–464.

Daraha, K. (2013). The Effect of the Internet Use on High School Students: A Case Study of Pattani Province of Thailand. *Procedia: Social and Behavioral Sciences*, *91*. Advance online publication. doi:10.1016/j.sbspro.2013.08.422

Dean, J. (2014, April 7). *Mindfulness Meditation: 8 Quick Exercises That Fit into Your Day*. https://www.spring.org.uk/2014/04/mindfulness-meditation-8-quick-exercises-that-easily-fit-into-your-day.php

Debell, M., & Chapman, C. (2004). *Computer and internet use by children and adolescents in 2001: Statistical analysis report*. National Center for Education Statistics. doi:10.1037/e609922011-001

Decouflé, P., & Autry, A. (2002). Increased mortality in children and adolescents with developmental disabilities. *Paediatric and Perinatal Epidemiology*, *16*(4), 375–382. doi:10.1046/j.1365-3016.2002.00430.x PMID:12445156

Delen, E., & Bulut, O. (2011). 'The relationship between students' exposure to technology and their achievement in science and math'. *The Turkish Online Journal of Educational Technology*, *10*(3), 311–317.

Dellavalle, R. R. (2014). Dermatology on Instagram. *Dermatologi Online Journal, 20*(7).

Deloitte. (2018). Preparing tomorrow's workforce for the Fourth Industrial Revolution. For business: A framework for action. *Global Business Coalition for Education*, 1–58. https://www2.deloitte.com/global/en/pages/about-deloitte/articles/gx-preparing-tomorrow-workforce-for-the-fourth-industrial-revolution.html?id=gx:2sm:3tw:4GBCed2018:6abt:20180919220600:SocialImpact-Report&utm_source=tw&utm_campaign=GBCed2018&utm_content=

Department of Basic Education. (2004). *White paper on E-Education: Transforming learning and teaching through information and communication technologies.* Available: http://www.education.gpg.gov.za/Legislation/White%20paper%207%20on%20e-Education.pdf

Department of Basic Education. (2017). *Rural Education Draft Policy*. Retrieved from http:// pmg-assets.s3-website-eu-west-1.amazonaws.com/180126draftruraledupolicy.pdf

Department of Education South Africa. (2019, August 30). *National Education infrastructure Management System Standard Report August 2019*. Author.

Dewberry, C., & Jackson, D. J. R. (2018). An application of the theory of planned behavior to student retention. *Journal of Vocational Behavior*, *107*, 100–100. doi:10.1016/j.jvb.2018.03.005

Dhawan, S. (2020). Online Learning: A Panacea in the Time of COVID-19 Crisis. *Journal of Educational Technology Systems*, *49*(1), 5–22. doi:10.1177/0047239520934018

Di Pietro, G., Biagi, F., Costa, P., Karpiński, Z., & Mazza, J. (2020). *The likely impact of COVID-19 on education: Reflections based on the existing literature and recent international datasets* (Vol. 30275). Publications Office of the European Union.

Dienlin, T., & Johannes, N. (2020). The impact of digital technology use on adolescent well-being. *Dialogues in Clinical Neuroscience*, *22*(2), 135–142. doi:10.31887/DCNS.2020.22.2/tdienlin PMID:32699513

Dontre, A. J. (2020). The influence of technology on academic distraction: A review. *Human Behavior and Emerging Technologies*, (August), 1–12. doi:10.1002/hbe2.229

Doorsamy, W., Paul, B. S., & Marwala, T. (2020). *The Disruptive Fourth Industrial Revolution: Technology, Society and Beyond*. Springer Nature Switzerland AG. doi:10.1007/978-3-030-48230-5

Dounavi, K., & Tsoumani, O. (2019). Mobile Health Applications in Weight Management: A Systematic Literature Review. *American Journal of Preventive Medicine*, *56*(6), 894–903. doi:10.1016/j.amepre.2018.12.005 PMID:31003801

Du Plessis, P. (2014). Problems and complexities in rural schools: Challenges of education and social development. *Mediterranean Journal of Social Sciences*, *5*(20), 1109–1117. doi:10.5901/mjss.2014.v5n20p1109

Du Preez, P., & le Grange, L. (2020). *The COVID-19 pandemic, online teaching/learning, the digital divide and epistemological access*. Unpublished paper.

Dufva, T., & Dufva, M. (2019). Grasping the future of the digital society. *Futures*, *107*, 17–28. doi:10.1016/j.futures.2018.11.001

Dugan, T. M., Mukhopadhyay, S., Carroll, A., & Downs, S. (2015). Machine Learning Techniques for Prediction of Early Childhood Obesity. *Applied Clinical Informatics*, *6*(3), 506–520. doi:10.4338/ACI-2015-03-RA-0036 PMID:26448795

Dumford, A. D., & Miller, A. L. (2018). Online learning in higher education: Exploring advantages and disadvantages for engagement. *Journal of Computing in Higher Education*, *30*(3), 452–465. doi:10.100712528-018-9179-z

Dunstan, J., Aguirre, M., Bastías, M., Nau, C., Glass, T. A., & Tobar, F. (2020). Predicting nationwide obesity from food sales using machine learning. *Health Informatics Journal, 26*(1), 652–663. doi:10.1177/1460458219845959 PMID:31106648

Dutta, P., Choi, T. M., Somani, S., & Butala, R. (2020). Blockchain technology in supply chain operations: Applications, challenges and research opportunities. *Transportation Research Part E, Logistics and Transportation Review, 142*(May), 102067. Advance online publication. doi:10.1016/j.tre.2020.102067 PMID:33013183

Duverger, P., & Steffes, E. (2012). Using YouTube Videos as a Primer to Affect Academic Content Retention. *Instutute of Education Science, 23*(2), 51–66.

Dvořáková, K., Kishida, M., Li, J., Elavsky, S., Broderick, P., Agrusti, M., & Greenberg, M. (2017). Promoting healthy transition to college through mindfulness training with first-year college students: Pilot randomized controlled trial. *Journal of American College Health, 65*(4), 259–267. doi:10.1080/07448481.2017.1278605 PMID:28076182

Dworkin, S. L. (2012). Sample size policy for qualitative studies using in-depth interviews. *Archives of Sexual Behavior, 41*(6), 1319–1320. doi:10.100710508-012-0016-6 PMID:22968493

EAB. (2020). *Students Spend More Time at Work Than in Class.* https://eab.com/insights/daily-briefing/student-affairs/students-spend-more-time-at-work-than-in-class/

Ebehard, B., Podio, M., Alonso, A. P., Radovica, E., Avotina, L., Peiseniece, L., Sendon, M. C., Lozano, A. G., & Sole-Pla, J. (2017). Smart work: The transformation of the labour market due to the fourth industrial revolution (I4.0). *International Journal of Business and Economic Sciences Applied Research, 10*(3), 47–66.

Eijnden, R., Lemmens, J., & Valkenburg, P. (2016). The social media disorder scale: Validity and psychometric properties. *Computers in Human Behavior, 61*, 478–487. doi:10.1016/j.chb.2016.03.038

Eisenberg, N., Miller, P. A., Shell, R., McNalley, S., & Shea, C. (1991). Prosocial development in adolescence: A longitudinal study. *Developmental Psychology, 27*(5), 849–857. doi:10.1037/0012-1649.27.5.849

Ekanem, J. E. (2015). Problems and Prospects of Education Resource Centers in Nigeria. *Journal of Education and Practice, 6*(30), 66–70.

Embi. (2011). *Aplikasi Web 2.0 Dalam Pengajaran dan Pembelajaran.* Pusat Pembangunan Akademik Universiti Kebangsaan Malaysia.

Encyclopedia Britannica. (2015, July 15). *Smṛtyupasthāna Buddhist philosophy.* https://www.britannica.com/topic/smrtyupasthana

Epstein, B. (2015). Providing Psychological Counseling in Community Colleges: Even Greater Challenges and Fewer Resources. *Journal of College Student Psychotherapy, 29*(4), 289–295. doi:10.1080/87568225.2015.1074020

Erikson, E. H. (1994). *Identity: Youth and Crisis* (7th ed.). WW Norton & Company.

Erreygers, S., Vandebosch, H., Vranjes, I., Baillien, E., & De Witte, H. (2017). Nice or naughty? The role of emotions and digital media use in explaining adolescents' online prosocial and antisocial behavior. *Media Psychology, 20*(3), 374–400. doi:10.1080/15213269.2016.1200990

Erstad, O. (n.d.). *Identity, Community, and Learning Lives in the Digital Age.* Academic Press.

Esposito, G., van Bavel, R., Baranowski, T., & Duch-Brown, N. (2016). Applying the model of goal-directed behavior, including descriptive norms, to physical activity intentions: A contribution to improving the theory of planned behavior. *Psychological Reports, 119*(1), 5–26. doi:10.1177/0033294116649576 PMID:27229344

Ess, C. (2017). Digital Media Ethics. Oxford Research Encyclopedia of Communication, 2017(Sep), 1–47. doi:10.1093/acrefore/9780190228613.013.508

Eye Health: Technology's Effect on Eye's. (2021). Retrieved 6 May 2021, from https://guardionhealth.com/technologys-effect-eyes/#:~:text=One%20of%20the%20most%20common,to%20a%20painful%20tension%20headache

Faraj, M., & Al-Abdallat, S. (2018). The Impact of YouTube and Facebook on the Achievement of Jordan University Students in English Language Course. *The Arab Journal for Quality Assurance in Higher Education, 11*(34), 3–24.

Farhana, Z., Tanni, S. A., Shabnam, S., & Chowdhury, S. A. (2020). *Secondary Education during Lockdown Situation Due to Covid-19 Pandemic in Bangladesh: Teachers' Response on Online Classes*. Academic Press.

Fedorova, E. P., & Skobleva, E. I. (2020). Application of blockchain technology in higher education. *European Journal of Contemporary Education*, *9*(3), 552–571. doi:10.13187/ejced.2020.3.552

Felver, J., Morton, M., & Clawson, A. (2018). Mindfulness-based stress reduction reduces psychological distress in college students. *College Student Journal*, *52*(3), 291–298.

Fernandes, B., Biswas, U. N., Mansukhani, R. T., Casarín, A. V., & Essau, C. A. (2020). The impact of COVID-19 lockdown on internet use and escapism in adolescents. *Revista de psicología clínica con niños y adolescentes, 7*(3), 59-65.

Fernandes, B., Biswas, U. N., Mansukhani, R. T., Casarín, A. V., & Essau, C. A. (2020). The Impact of COVID-19 Lockdown on Internet Use and Escapism in Adolescents. *Revista de Psicología Clínica con Niños y Adolescentes, 7*(3), 59-65.

Fine, G. A. (2004). Adolescence as Cultural Toolkit: High School Debate and the Repertoires of Childhood and Adulthood. *The Sociological Quarterly*, *45*(1), 1–20. doi:10.1111/j.1533-8525.2004.tb02395.x

Fischer, F. (2003). Public Analysis Handbook. Academic Press.

Fishbein, M., & Ajzen, I. (2010). *Predicting and changing behavior: The reasoned action approach*. Psychology Press.

Flett, J., Conner, T., Riordan, B., Patterson, T., & Hayne, H. (2020). App-based mindfulness meditation for psychological distress and adjustment to college in incoming university students: A pragmatic, randomized, waitlist-controlled trial. *Psychology & Health*, *35*(1), 1–26. doi:10.1080/08870446.2019.1711089 PMID:31204484

Fobian, A. D., Avis, K., & Schwebel, D. C. (2016). Impact of media use on adolescent sleep efficiency. *Journal of Developmental and Behavioral Pediatrics*, *37*(1), 9–14. doi:10.1097/DBP.0000000000000239 PMID:26651090

Fortin, A. (2018, February 13). *How mindfulness can benefit college students*. https://www.collegian.psu.edu/lifestyles/health/article_58131944-104c-11e8-9e6d-9fd2ac28debc.html

Francis, P. C., & Horn, A. S. (2017). Mental health issues and counseling services in us higher education: An overview of recent research and recommended practices. *Higher Education Policy*, *30*(2), 263–277. doi:10.1057/s41307-016-0036-2

Freire, P. (2013). Pedagogy of the oppressed. In The Applied Theatre Reader. doi:10.4324/9780203891315-58

Galante, J., Dufour, G., Vainre, M., Wagner, A. P., Stochl, J., Benton, A., & Jones, P. B. (2018). A mindfulness-based intervention to increase resilience to stress in university students (the mindful student study): A pragmatic randomized controlled trial. *The Lancet. Public Health*, *3*(2), 81. doi:10.1016/S2468-2667(17)30231-1 PMID:29422189

Gamson, W. A., Croteau, D., Hoynes, W., & Sasson, T. (1992). Media Images and the Social Construction of Reality. *Annual Review of Sociology*, *18*(1), 373–393. doi:10.1146/annurev.so.18.080192.002105

Ganguly, N., & Kumaraguru, P. (2019). The positive and negative effects of social media in India. *Communications of the ACM*, *62*(11), 96–99.

Gao, J., Zheng, P., Jia, Y., Chen, H., Mao, Y., Chen, S., Wang, Y., Fu, H., & Dai, J. (2020). Mental health problems and social media exposure during COVID-19 outbreak. *PLoS One*, *15*(4), 1–10. doi:10.1371/journal.pone.0231924 PMID:32298385

Garriott, P. O., Hudyma, A., Keene, C., & Santiago, D. (2015). Social cognitive predictors of first- and non-first-generation college students' academic and life satisfaction. *Journal of Counseling Psychology, 62*(2), 253-263. http://dx.doi.org.tcsedsystem.idm.oclc.org/10.1037/cou0000066

Gentile, D. A., Anderson, C. A., Yukawa, S., Ihori, N., Saleem, M., Ming, L. K., Shibuya, A., Liau, A. K., Khoo, A., Bushman, B. J., Rowell Huesmann, L., & Sakamoto, A. (2009). The effects of prosocial video games on prosocial behaviors: International evidence from correlational, longitudinal, and experimental studies. *Personality and Social Psychology Bulletin*, *35*(6), 752–763. doi:10.1177/0146167209333045 PMID:19321812

Gentile, D. A., Nathanson, A. I., Rasmussen, E. E., Reimer, R. A., & Walsh, D. A. (2012). Do you see what I see? Parent and child reports of parental monitoring of media. *Family Relations*, *61*(3), 470–487. doi:10.1111/j.1741-3729.2012.00709.x

George, M. J., Russell, M. A., Piontak, J. R., & Odgers, C. L. (2018). Concurrent and Subsequent Associations Between Daily Digital Technology Use and High-Risk Adolescents' Mental Health Symptoms. *Child Development*, *89*(1), 78–88. doi:10.1111/cdev.12819 PMID:28466466

Gerben, C. (2018). Factoring in Student Technology Use in Literacy Education: A Review of *The Class: Living and Learning in the Digital Age. Journal of Adolescent & Adult Literacy*, *61*(5), 593–595. doi:10.1002/jaal.725

Gerbner, G., & Gross, L. (1976). Living with television: The violence profile. *Journal of Communication*, *26*(76), 173–199. doi:10.1111/j.1460-2466.1976.tb01397.x PMID:932235

Ghasemi, V., Rabiei, K., Davoodi, S., & Rabiei, H. (2017). Mobile Internet Usage among Adolescents and Young Adults in Iran: A Sociological Survey. *Tarih Kültür ve Sanat Arastirmalari Dergisi*, *6*(1), 863. doi:10.7596/taksad.v6i1.785

Ghazali. (2015). Tindak Balas Pengguna YouTube Terhadap Kes Buli dalam Kalangan Remaja di Malaysia. *Sains Humanika, 6*(1), 9-17.

Ghazali, S., Mapjabil, J., Nor, A. M., Samat, N., & Jas, L. S. J. (2012). Dufusi Ruangan Budaya Transeksualisme dan Imaginasi Geografi Pelajar Lelaki Berpenampilan Silang di Universiti Tempatan Malaysia. *Journal of Social Sciences and Humanities*, *7*(1), 252–266.

Giedd, J. N. (2012). The digital revolution and adolescent brain evolution. *The Journal of Adolescent Health*, *51*(2), 101–105. doi:10.1016/j.jadohealth.2012.06.002 PMID:22824439

Girasa, R. (2020). Artificial Intelligence as a Disruptive Technology: Economic Transformation and Government Regulation. In *Artificial Intelligence as a Disruptive Technology*. Palgrave Macmillan. doi:10.1007/978-3-030-35975-1

Gnilka, P., Ashby, J., Matheny, K., Chung, Y., & Chang, Y. (2015). Comparison of coping, stress, and life satisfaction between Taiwanese and US college students. *Journal of Mental Health Counseling*, *37*(3), 234–249. doi:10.17744/mehc.37.3.04

Goggin, M. (1990). *Implementation, Theory, and Practice. Toward a Third Generation*. Forasmann and Company.

Goldberg, S. B., & Raymond, P. (2019). Mindfulness-based cognitive therapy for the treatment of current depressive symptoms: A meta-analysis. *Cognitive Behaviour Therapy*, *48*(6), 445–462. doi:10.1080/16506073.2018.1556330 PMID:30732534

Gómez-García, M., Hossein-Mohand, H., Trujillo-Torres, J. M., & Hossein-Mohand, H. (2020). The Training and Use of ICT in Teaching Perceptions of Melilla's (Spain) Mathematics Teachers. *Mathematics*, *8*(10), 1641. doi:10.3390/math8101641

González, M. R., Gasco, J., & Llopis, J. (2016). Facebook and academic performance: A positive outcome. *Anthropologist*, *23*(1–2), 59–67. doi:10.1080/09720073.2016.11891924

Gottschalk, F. (2019). *Impacts of technology use on children: Exploring literature on the brain, cognition and well-being*. Academic Press.

Gottschalk, F. (2019). *Impacts of Technology Use on Children: Exploring Literature on the Brain, Cognition and Well-Being*. OECD Education Working Papers, No. 195. OECD Publishing.

Grech, A. & Camilleri, A. F. (2017). *Blockchain in Education*. doi:10.4018/978-1-5225-9478-9.ch014

Greenhow, C., & Robelia, B. (2009). Informal learning and identity formation in online social networks. *Learning, Media and Technology*, *34*(2), 119–140. doi:10.1080/17439880902923580

Greitemeyer, T. (2009). Effects of songs with prosocial lyrics on prosocial thoughts, affect, and behavior. *Journal of Experimental Social Psychology*, *45*(1), 186–190. doi:10.1016/j.jesp.2008.08.003

Greitemeyer, T. (2011). Effects of prosocial media on social behavior when and why does media exposure affect helping and aggression? *Current Directions in Psychological Science*, *20*(4), 251–255. doi:10.1177/0963721411415229

Greitemeyer, T., Osswald, S., & Brauer, M. (2010). Playing prosocial video games increases empathy and decreases schadenfreude. *Emotion (Washington, D.C.)*, *10*(6), 796–802. doi:10.1037/a0020194 PMID:21171755

Grimus, M. (2020). Emerging Technologies: Impacting Learning. *Pedagogy and Curriculum Development*, (February), 127–151. Advance online publication. doi:10.1007/978-981-15-0618-5_8

Grindstaff, L., & Turow, J. (2006). Video Cultures: Television Sociology in the "New TV" Age. *Annual Review of Sociology*, *32*(1), 103–125. doi:10.1146/annurev.soc.32.061604.143122

Gross, E. F., Juvonen, J., & Gable, S. L. (2002). Internet use and well-being in adolescence. *The Journal of Social Issues*, *58*(1), 75–90. doi:10.1111/1540-4560.00249

Grossman, R., Giovanni, S., John, E., Nitin, A., & Huan, L. (2010). *Ensemble Methods in Data Mining: Improving Accuracy Through Combining Predictions*. Synthesis Lectures on Data Mining and Knowledge Discovery. *Morgan & Claypool.*, *2*(1), 1–126. doi:10.2200/S00240ED1V01Y200912DMK002

Guan, C., Mou, J., & Jiang, Z. (2020). Artificial intelligence innovation in education: A twenty-year data-driven historical analysis. *International Journal of Innovation Studies*, *4*(4), 134–147. doi:10.1016/j.ijis.2020.09.001

Guan, J., & Huck, J. (2012). Children in the digital age: exploring issues of cybersecurity. In *Proceedings of the 2012 iConference* (pp. 506-507). 10.1145/2132176.2132266

Gunnlaugsson, G., Whitehead, T. A., Baboudóttir, F. N., Baldé, A., Jandi, Z., Boiro, H., & Einarsdóttir, J. (2020). Use of Digital Technology among Adolescents Attending Schools in Bissau, Guinea-Bissau. *International Journal of Environmental Research and Public Health*, *17*(23), 8937. doi:10.3390/ijerph17238937 PMID:33271847

Guo, S. S., Wu, W., Chumlea, W. C., & Roche, A. F. (2002). Predicting overweight and obesity in adulthood from body mass index values in childhood and adolescence. *The American Journal of Clinical Nutrition*, *76*(3), 653–658. doi:10.1093/ajcn/76.3.653 PMID:12198014

Gutgesell, M. E., & Payne, N. (2004). Issues of adolescent psychological development in the 21st century. *Pediatrics, 25*(3), 79–85. PMID:14993515

Hadjerrouit, S. (2007). E-learning as a means of social inclusion: The case of syrian refugees in Germany. *Interdisciplinary Journal of Knowledge and Learning Objects*, 3.

Hagger, M. S., Chan, D. K. C., Protogerou, C., & Chatzisarantis, N. L. D. (2016). Using meta-analytic path analysis to test theoretical predictions in health behavior: An illustration based on meta-analyses of the theory of planned behavior. *Preventive Medicine, 89*, 154–161. doi:10.1016/j.ypmed.2016.05.020 PMID:27238207

Haithcox-Dennis, M., Brinkley, J., Richman, A., DeWeese, A., & Byrd, J. L. (2012). Mhealth on campus: Assessing undergraduates attitudes and utilization of mobile health applications. *Global Journal of Health Education and Promotion, 15*(1).

Hamad, S., Abu Hamra, E., Diab, R., Abu Hamad, B., Jones, N., & Małachowska, A. (2020). *Exploring the Impacts of Covid-19 on Adolescents in the Gaza Strip.* ODI.

Hanafi & Rahim. (2017). Penggunaan media sosial dan faktor-faktor yang mempengaruhi terbentuknya gerakan sosial pemimpin pelajar universiti di Bandar Pekan Baru, Riau, Indonesia. *Journal of Social Sciences and Humanities.*

Hanekom, P. (2020). Covid-19 exposes South Africa's digital literacy divide. *Mail & Guardian.* Available at: https://mg.co.za/opinion/2020-09-08-covid-19-exposes-south-africas-digital-literacy-divide/

Hardy, S. A., Carlo, G., & Roesch, S. C. (2010). Links between adolescents' expected parental reactions and prosocial behavioral tendencies: The mediating role of prosocial values. *Journal of Youth and Adolescence, 39*(1), 84–95. doi:10.100710964-008-9383-7 PMID:20091219

Hargittai, E., & Dobransky, K. (2017). Old dogs, new clicks: Digital inequality in skills and uses among older adults. *Canadian Journal of Communication, 42*(2). Advance online publication. doi:10.22230/cjc.2017v42n2a3176

Harrison, K. (2006). Scope of Self: Toward a Model of Television's Effects on Self-Complexity in Adolescence. *Communication Theory, 16*(2), 251–279. doi:10.1111/j.1468-2885.2006.00270.x

Haski-Leventhal, D. (2009). Altruism and volunteerism: The perceptions of altruism in four disciplines and their impact on the study of volunteerism. *Journal for the Theory of Social Behaviour, 39*(3), 271–299. doi:10.1111/j.1468-5914.2009.00405.x

Hastungkara, D. P., & Triastuti, E. (2019). Application Of E-Learning And Artificial Intelligence In Education Systems In Indonesia. *Anglo-Saxon, 10*(2), 117–133.

Hatch, K. E. (2011). *Determining the Effects of Technology on Children Senior Honors Projects.* University of Rhode Island. https://digitalcommons.uri.edu/cgi/viewcontent.cgi?article=1212&context=srhonorsprog

Hatcher, M., Henson, J., & LaRosa, P. (2013). Determinants of Success in an Online Management Information Systems Course: The Significance of Grade Point Average As a Performance Predictor. *International Journal of Business Strategy, 13*(4), 129–132. doi:10.18374/IJBS-13-4.7

Healy, S., Pacanowski, C. R., & Williams, E. (2019). Weight management interventions for youth with autism spectrum disorder: A systematic review. *International Journal of Obesity, 43*(1), 1–12. doi:10.103841366-018-0233-8 PMID:30305689

Hedding, D. W., Greve, M., Breetzke, G. D., Nel, W., & Van Vuuren, B. J. (2020). COVID-19 and the academe in South Africa: Not business as usual. *South African Journal of Science, 116*(7-8), 1–3. doi:10.17159ajs.2020/8298

Hellström, C., Nilsson, K. W., Leppert, J., & Åslund, C. (2015). Effects of adolescent online gaming time and motives on depressive, musculoskeletal, and psychosomatic symptoms. *Upsala Journal of Medical Sciences*, *120*(4), 263.

Hennessy, M., Byrne, M., Laws, R., Mc Sharry, J., O'Malley, G., & Heary, C. (2019). Childhood obesity prevention: Priority areas for future research and barriers and facilitators to knowledge translation, coproduced using the nominal group technique. *Translational Behavioral Medicine*, *9*(4), 759–767. doi:10.1093/tbm/iby074 PMID:30011024

Hennessy, T., Doris, L., & Brian, M. (2016). The Digital Divide in Farming: A Problem of Access or Engagement? *Applied Economic Perspectives and Policy*, *38*(3), 474–491. doi:10.1093/aepp/ppw015

Herath, T., & Rao, H. R. (2009). Protection motivation and deterrence: A framework for security policy compliance in organisations. *European Journal of Information Systems*, *18*(2), 106–125. doi:10.1057/ejis.2009.6

Herselman, M. E., Botha, A., Maremi, K. J., Dlamini, S. B., & Marais, M. A. (2020). *Mobile technology affecting teaching and learning in rural schools*. Academic Press.

Hilli, C. (2020). Distance teaching in small rural primary schools: A participatory action research project. *Educational Action Research*, *28*(1), 38–52. doi:10.1080/09650792.2018.1526695

Hofschire, L. J., & Greenberg, B. S. (2002). Media's impact on adolescents' body dissatisfaction. In J. D. Brown, J. R. Steele, & K. Walsh-Childers (Eds.), *Sexual teens, sexual media: Investigating media's influence on adolescent sexuality* (pp. 125–149). Lawrence Erlbaum Associates.

Hofstee, E. (2006). *Constructing a good dissertation.* Johannesburg: EPE.

Hohlfeld, T. N., Ritzhaupt, A. D., & Barron, A. E. (2010). Connecting schools, community, and family with ICT: Four-year trends related to school level and SES of public schools in Florida. *Computers & Education*, *55*, 391–405. doi:10.1016/j.compedu.2010.02.004

Holinka, C. (2015). Stress, emotional intelligence, and life satisfaction in college students. *College Student Journal*, *49*(2), 300–311.

Hollingworth, S., Mansaray, A., Allen, K., & Rose, A. (2011). Parents' perspectives on technology and children's learning in the home: Social class and the role of the habitus. *Journal of Computer Assisted Learning*, *27*, 347–360. doi:10.1111/j.1365- 2729.2011.00431.x

How Celebrities Influence Teens and Why it Matters. (2021). *Newport Academy*. Retrieved 2 October 2021, from https://www.google.com/url?q=https://www.newportacademy.com/resources/empowering-teens/celebrities-influence/&usg=AOvVaw3t-vK9SiJBcWeUwduiNSfS

Hsieh, K., Rimmer, J. H., & Heller, T. (2014). Obesity and associated factors in adults with intellectual disability. *Journal of Intellectual Disability Research*, *58*(9), 851–863. doi:10.1111/jir.12100 PMID:24256433

Huang, T. C., Chen, C. C., & Chou, Y. W. (2016). Animating eco-education: To see, feel, and discover in an augmented reality-based experiential learning environment. *Computers & Education*, *96*, 72–82. doi:10.1016/j.compedu.2016.02.008

Hubert, H. B., Feinleib, M., McNamara, P. M., & Castelli, W. P. (1983). Obesity as an independent risk factor for cardiovascular disease: A 26-year follow-up of participants in the Framingham Heart Study. *Circulation*, *67*(5), 968–977. doi:10.1161/01.CIR.67.5.968 PMID:6219830

Huberty, J., Green, J., Lee, C., Glissmann, C., Larkey, L., & Puzia, M. (2019). Efficacy of the mindfulness meditation mobile app "calm" to reduce stress among college students: Randomized controlled trial. *Journal of Medical Internet Research*, *21*(6), e14273. Advance online publication. doi:10.2196/14273 PMID:31237569

Hughes, N., & Burke, J. (2018). Sleeping with the frenemy: How restricting 'bedroom use'of smartphones impacts happiness and wellbeing. *Computers in Human Behavior, 85*, 236–244. doi:10.1016/j.chb.2018.03.047

Hu, Q., Dinev, T., Hart, P., & Cooke, D. (2012). Managing employee compliance with information security policies: The critical role of top management and organisational culture. *Decision Sciences, 43*(4), 615–660. doi:10.1111/j.1540-5915.2012.00361.x

Hutchins, M. (2021). *Social Media Use Linked To Anxiety, Depression Among Teens, New Study Finds.* Retrieved 6 May 2021, from https://www.wbur.org/hereandnow/2020/01/09/social-media-anxiety-depression-teens

Hu, Y., Mankikonda, L., & Kambhamati, S. (2014). What We Instagram: A First Analysis of Instagram Photo Content User Types. *Proceedings of the Eight International.*

Immordino-Yang, M. H., & Singh, V. (2011). Perspectives from social and affective neuroscience on the design of digital learning technologies. In *New perspectives on affect and learning technologies* (pp. 233–241). Springer. doi:10.1007/978-1-4419-9625-1_17

Inal, M., & Korkmaz, Ö. (2019). The effect of web based blended learning on students' academic achievement and attitudes towards English course. *Education and Information Technologies, 24*(4), 2603–2619. doi:10.100710639-019-09890-7

Ingersoll, J. (2019). *Academic Probation in College Why it Happens & How to Get Back on Track.* https://www.accredited-schoolsonline.org/resources/academic-probation/#:~:text=Students%20on%20academic%20probation%20may,more%20about%20this%20possibility%20below)

Insitute, S. (2003). *Global Information Assurance Certification Paper.* Academic Press.

ITU-D. (n.d.). *ICT statistics.* Available online www.itu.int/en/ITU-D/Statistics/Pages/stat/default.aspx

Iyer, M., & Sharma, R. (2020). Impact of internet overuse on psychological well-being of adolescents during covid 19 shutdown. *J Crit Rev, 7*(9), 2832–2839.

Jain, M. R., Gupta, P., & Anand, N. (2012). Impact of Social Networking Sites in the Changing Mindset of Youth on Social Issues – A Study of Delhi-NCR Youth. *Journal of Arts, Science & Commerce, 2*(2), 36–43.

Janicki, M. P., Davidson, P. W., Henderson, C. M., McCallion, P., Taets, J. D., Force, L. T., Sulkes, S. B., Frangenberg, E., & Ladrigan, P. M. (2002). Health characteristics and health services utilization in older adults with intellectual disability living in community residences. *Journal of Intellectual Disability Research, 46*(4), 287–298. doi:10.1046/j.1365-2788.2002.00385.x PMID:12000580

Jantjies, M. (2020), April 29. How South Africa can address digital inequalities in e-learning. *Conversation.* Available at: https://theconversation.com/how-south-africa-can-address-digital-inequalities-in-e-learning-137086

Jarrett, J. (2021). *Is There a Link Between Social Media and Eating Disorders?* Magnolia Creek. Retrieved 6 May 2021, from https://www.magnolia-creek.com/eating-disorder-recovery-blog/social-media-and-eating-disorders/

Jensen, K. B. (1991). Media audiences Reception analysis: Mass Communication as the Social Production of Meaning. In K. B. Jensen & N. W. Jankowski (Eds.), *Handbook of Qualitative Methodologies for Mass Communication Research.* Routledge.

Jiang, Cao, & Chen. (2013). Detecting Product Review Spammers Using Activity Model. *Proceeding of International Conference on Advanced Computer Science and Electronics Information (ICACSEI 2013)*, 650–653.

Jindal, N., Liu, B., & Lim, E. P. (2010). Finding Unusual Review Patterns Using Unexpected Rules. *International Conference on Information and Knowledge Management, Proceedings.* 10.1145/1871437.1871669

Johari & Shahrina. (2012). Ketagihan Penggunaan Internet Di Kalangan Remaja Sekolah Tingkatan 4 Di Bandaraya Johor Bahru. *Journal of Technical, Vocational & Engineering Education*, 23-43.

Johnson, J. G., Cohen, P., Smailes, E. M., Kasen, S., & Brook, J. S. (2002). Television viewing and aggressive behavior during adolescence and adulthood. *Science*, *295*(0), 2468–2471. doi:10.1126cience.1062929 PMID:11923542

Johnson, K. F., & Kalkbrenner, M. T. (2017). The Utilization of Technological Innovations to Support College Student Mental Health: Mobile Health Communication. *Journal of Technology in Human Services*, *35*(4), 314–339. doi:10.1080/15228835.2017.1368428

Jones, S. (2013). *Digital literacy, HIV/AIDS information and English language learners in Uganda*. https://www.researchgate.net/publication/292424756

Jones, E. A., Mitra, A. K., & Bhuiyan, A. R. (2021). Impact of COVID-19 on mental health in adolescents: A systematic review. *International Journal of Environmental Research and Public Health*, *18*(5), 2470.

Juberg, M., Spencer, S. D., Martin, T. J., Vibell, J., da Costa Ferro, A., Kam, B., & Masuda, A. (2019). A mindfulness-based intervention for college students, faculty, and staff: A preliminary investigation. *Clinical Case Studies*, *18*(3), 185–199. doi:10.1177/1534650119836166

Junaidi. (2015). Internet Of Things, Sejarah, Teknologi Dan Penerapannya: Review. *Jurnal Ilmiah Teknologi Informasi Terapan*, *1*(3).

Junni, P. (2007). Students seeking information for their Masters' theses: the effect of the Internet. *Information Research*, *12*(2) paper 305. Available at http://InformationR.net/ir/12-2/paper305.html

Kamaruddin, H., Rosilah, H., Norasmah, O., Wan, M. D., Zaki, A. W., & Md Sum, S. (2016). Meeting the Needs of Fourth Industrial Revolution (4IR) in Entrepreneurial Education in Malaysia: The Government's Role. *Intech, Tourism*, 13. https://www.intechopen.com/books/advanced-biometric-technologies/liveness-detection-in-biometrics

Kamas, L., Preston, A., & Baum, S. (2008). Altruism in individual and joint-giving decisions: What's gender got to do with it? *Feminist Economics*, *14*(3), 23–50. doi:10.1080/13545700801986571

Kamba, M. A. (2008). *The changing roles of researchers in Nigeria: The internet as an alternative future to modernity*. Library of Philosophy and Practice.

Kann, L., McManus, T., Harris, W. A., Shanklin, S. L., Flint, K. H., Hawkins, J., Queen, B., Lowry, R., Olsen, E. O., Chyen, D., Whittle, L., Thornton, J., Lim, C., Yamakawa, Y., Brener, N., & Zaza, S. (2016). Youth Risk Behavior Surveillance - United States, 2015. *MMWR. Surveillance Summaries*, *65*(6), 1–174. doi:10.15585/mmwr.ss6506a1 PMID:27280474

Kaufman, M. (2011). *How families facilitate the development of empathy in children: A family systems theory perspective.* https://www.semanticscholar.org/paper/How-families-facilitate-the-development-of-empathy-Kaufman/e7d280f10a3ed-2cec5005379ff10766b1f11831d

Kaur, R., & Bashir, L. (2015). Impact of social media on mental health of adolescents. *International Journal of Education*, *5*, 22–29.

Kaur, R., & Bhatt, L. (2016). Impact of stress on mental health of students: Reason and intervention. *International Journal of Education*, *5*, 30–35.

Kazemi, D. M., Borsari, B., Levine, M. J., Shehab, M., Nelson, M., Dooley, B., Stinson, B., Fang, F., & Li, S. (2019). Real-time demonstration of a mHealth app designed to reduce college students hazardous drinking. *Psychological Services*, *16*(2), 255–259. doi:10.1037er0000310 PMID:30407059

KeeleryS. (n.d.). www.statista.com/topics/2157/internet-usage-in-india/

Kellner, D. (1995). *Media Culture: Cultural Studies, Identity and Politics between the Modern and the Postmodern.* Routledge.

Kelly, C. T. (2019). *The Impact of an Autonomy-Supportive Yoga Intervention on the Self-Determination and Anxiety of Individuals with Intellectual and Developmental Disabilities: A Multi-Method Approach* (MSc Thesis). Clemson University. Retrieved on January 29, from https://tigerprints.clemson.edu/all_theses/3083

Kemp, S. (2021). *Digital 2021: Global Overview Report — DataReportal – Global Digital Insights.* Retrieved 4 May 2021, from https://datareportal.com/reports/digital-2021-global-overview-report

Keong, T. C., & Abu, C. (2013). Pengaplikasian Video YouTube: Bahan Bantu Mengajar (BBM) Dalam Proses Pengajaran dan Pembelajaran Mata Pelajaran Sains Sosial. Seminar Pendidikan Sejarah dan. Universiti Malaysia Sabah: Sabah.

Key, B. L., Rowa, K., Bieling, P., McCabe, R., & Pawluk, E. J. (2017). Mindfulness-based cognitive therapy as an augmentation treatment for obsessive–compulsive disorder. *Clinical Psychology & Psychotherapy, 24*(5), 1109–1120. doi:10.1002/cpp.2076 PMID:28194835

Kim, J. (2020). Learning and Teaching Online During Covid-19: Experiences of Student Teachers in an Early Childhood Education Practicum. *International Journal of e-Collaboration, 52*, 145–158. https://doi.org/10.1007/s13158-020-00272-6

Kim, J., & Park, N. (2020). Blockchain-based data-preserving AI learning environment model for AI cybersecurity systems in IoT service environments. *Applied Sciences (Switzerland), 10*(14), 4718. Advance online publication. doi:10.3390/app10144718

King, D. L., Delfabbro, P. H., Kaptsis, D., & Zwaans, T. (2014). Adolescent simulated gambling via digital and social media: An emerging problem. *Computers in Human Behavior, 31*, 305–313.

Kirmayer, L. J. (2015). Mindfulness in cultural context. *Transcultural Psychiatry, 52*(4), 447–469. doi:10.1177/1363461515598949 PMID:26264787

Kirschner, P. A., & Karpinski, A. C. (2010). Facebook® and academic performance. *Computers in Human Behavior, 26*(6), 1237–1245. doi:10.1016/j.chb.2010.03.024

Kolhar, M., Kazi, R. N. A., & Alameen, A. (2021). Effect of social media use on learning, social interactions, and sleep duration among university students. *Saudi Journal of Biological Sciences, 28*(4), 2216–2222. doi:10.1016/j.sjbs.2021.01.010 PMID:33911938

Komalasari, R. (2020). Manfaat Teknologi Informasi Dan Komunikasi Di Masa Pandemi Covid 19. Jurnal Teknologi Informasi Dan Komunikasi, 7(1).

Kopecký, K. (2014). Cyberbullying and Sexting between Children and Adolescents - Comparative Study. *Procedia: Social and Behavioral Sciences, 149*, 467–471. doi:10.1016/j.sbspro.2014.08.292

Krajcovic, P., & Cabyova, L. (2020). Use of Social Media for Marketing Communication of Socially Responsible Business Activities in Slovakia. In C. Karpacitis (Ed.), *Proceedings of the 7th European Conference on Social Media* (pp. 135–143). ACPI, UK.

Kramer, A., & Kramer, K. Z. (2020). *The potential impact of the Covid-19 pandemic on occupational status, work from home, and occupational mobility.* Academic Press.

Kraut, R., Patterson, M., Lundmark, V., Kiesler, S., Mukophadhyay, T., & Scherlis, W. (1998). Internet paradox: 'A social technology that reduces social involvement and psychological well-being?'. *The American Psychologist, 53*(9), 1017–1031. doi:10.1037/0003-066X.53.9.1017 PMID:9841579

Krebs, D. L., & Van Hesteren, F. (1994). The development of altruism: Toward an integrative model. *Developmental Review, 14*(2), 103–158. doi:10.1006/drev.1994.1006

Krebs, G., & Heyman, I. (2015). Obsessive-compulsive disorder in children and adolescents. *Archives of Disease in Childhood, 100*(5), 495–499. doi:10.1136/archdischild-2014-306934 PMID:25398447

Krönke, M. (2020). Africa's digital divide and the promise of e-learning. *Afrobarometer Policy Paper*, (66).

Krosnick, J. A., Anand, S. N., & Hartl, S. P. (2003). Psychosocial Predictors of Heavy Television Viewing Among Preadolescents and Adolescents. *Basic and Applied Psychology, 25*(2), 87–110. doi:10.1207/S15324834BASP2502_1

Kubey, R., Lavin, M., & Barrows, J. (2001). Article. *Journal of Communication, Internet Use and Collegiate Academic Performance Decrements: Early Findings, 51*(2), 366-382.

Kulidtod, R., & Pasagui, N. (2017). Effects of Social Networking Media to the Academic Performance of the Students. *2nd International Conference on Educational Management and Administration (CoEMA), 45*(CoEMA), 59–64. 10.2991/coema-17.2017.11

Kumar, R., & Kaur, A. (2006). Internet Use by Teachers and Students in Engineering Colleges of Punjab, Haryana, and Himachal Pradesh States of India: An Analysis. *Elect J Acad Special Libra, 7*(1). Available from URL: https://southernlibrarianship.icaap.org/con-tent/v07n01/kumar_r01.htm

Kuran, C. H. A., Morsut, C., Kruke, B. I., Krüger, M., Segnestam, L., Orru, K., Nævestad, T. O., Airola, M., Keränen, J., Gabel, F., Hansson, S., & Torpan, S. (2020). Vulnerability and vulnerable groups from an intersectionality perspective. *International Journal of Disaster Risk Reduction, 50*, 101826. doi:10.1016/j.ijdrr.2020.101826

Kushlev, K., Proulx, J. D. E., & Dunn, E. W. (2017). Digitally connected, socially disconnected: The effects of relying on technology rather than other people. In Computers in Human Behavior (Vol. 76). doi:10.1016/j.chb.2017.07.001

Labrecque, L. I., vor dem Esche, J., Mathwick, C., Novak, T. P., & Hofacker, C. F. (2013). Consumer power: Evolution in the digital age. *Journal of Interactive Marketing, 27*(4), 257–269. doi:10.1016/j.intmar.2013.09.002

Lai, J., & Widmar, N. O. (2020, October 12). (2020). Revisiting the Digital Divide in the COVID-19 Era. *Applied Economic Perspectives and Policy*. Advance online publication. doi:10.1002/aepp.13104 PMID:33230409

Lal, M. (2011). LACLO 3 - Web 3.0 in Education & Research. *International Journal of Information Technology, 3*(2), 973–5658.

Lam, J., & Duan, G. (2012). A review of the mobile learning environment in the higher education sector of Hong Kong: Technological and social perspectives. In S. K. Cheung, J. Fong, L. Kwok, K. Li, & R. Kwan (Ed.), Hong Kong (pp. 165-173). Springer. doi:10.1007/978-3-642-32018-7_16

Lamas, H. (2015). School Performance Review. *Journal of Educational Psychology, 3*(1), 313–386. http://revistas.usil.edu.pe/index.php/pyr

Lambić, D. (2016). Correlation between Facebook use for educational purposes and academic performance of students. *Computers in Human Behavior, 61*, 313–320. doi:10.1016/j.chb.2016.03.052

Lancaster, F. W., & Sandore, B. (1997). Technology and Management in Library and Information Services. Univ. of Illinois Graduate School of Library and Information Science.

LaRose, R., Eastin, M. S., & Gregg, J. (2001). Reformulating the Internet paradox: Social cognitive explanations of Internet use and depression. *Journal of Online Behavior, 1*. Retrieved January 21, 2003, from http://www.behavior.net/job/v1n2/paradox.html

Lauri, M. A., Borg, J., & Farrugia, L. (2015). *Children's Internet Use and Parents' Perceptions Of their Children's Online Experience.* Report of Malta Communications authority commissioned by University of Malta. Accessed online from www.mca.org.mt/

Lavuri, R., Navulla, D., & Yamini, P. (2019). Effect of social media networks on academic performance of Indian students. *Journal of Critical Reviews*, *6*(4), 71–78. doi:10.22159/jcr.06.04.13

Leal-Rodrígueza, A. L., & Albort-Morant, G. (2017). Promoting innovative experiential learning practices to improve academic performance: Empirical evidence from a Spanish Business School. *Journal of Innovation and Kno*, *4*, 50728.

Lee, C. J., Geiger-Brown, J., & Beck, K. H. (2016). Intentions and willingness to drive while drowsy among university students: An application of an extended theory of planned behavior model. *Accident; Analysis and Prevention*, *93*, 113–123. doi:10.1016/j.aap.2016.05.002 PMID:27180286

Lee, H. E., Mustaffa, N., & Salman, A. (2015). Faktor-Faktor yang Mempengaruhi Remaja Lembah Klang untuk Terus Membaca dan Membeli Melalui Pengiklanan Facebook. *Jurnal Komunikasi*, *30*(1), 229–250.

Lennox, N., Beange, H., Parmenter, T., Santos-Teachout, R., Evenhuis, R., Kerr, M., McElduff, A., Fraser, W., Böhmer, C. J., Davis, R., Turner, G., & Stewart, L. (2002). *Health Guidelines for Adults with an Intellectual Disability.* Retrieved on February 15 from http://www.iassid.org/pdf/healthguideline-2002.pdf

Lennox, N. G., Diggens, J. N., & Ugoni, A. M. (1997). The general practice care of people with intellectual disability: Barriers and solutions. *Journal of Intellectual Disability Research*, *41*(5), 380–390. doi:10.1111/j.1365-2788.1997.tb00725.x PMID:9373818

Levine, M. P., & Smolak, L. (1998). The mass media and disordered eating: Implications for primary prevention. In The prevention of eating disorders. Athlone.

Liberati, N. (2019). Emotions and Digital Technologies. *HUMANA MENTE Journal of Philosophical Studies*, *12*(36), 292–309.

Lin, J. W., & Mai, L. J. (2018). Impact of mindfulness meditation intervention on academic performance. *Innovations in Education and Teaching International, 55*(3), 366-375. http://dx.doi.org.tcsedsystem.idm.oclc.org/10.1080/14703297.2016.1231617

Lin, Y. (2014). Towards Online Anti-Opinion Spam: Spotting Fake Reviews from the Review Sequence. *ASONAM 2014 - Proceedings of the 2014 IEEE/ACM International Conference on Advances in Social Networks Analysis and Mining*, 261–64.

Linder, A. D., Liu, H.-Y., Woodson-Smith, A., & Jung, J. (2018). Physical activity behaviors among non-traditional and traditional college students: An application of ajzen's theory of planned behavior. *Negro Educational Review*, *69*(1-4), 33–50.

Lingren, T., Thaker, V., Brady, C., Namjou, B., Kennebeck, S., Bickel, J., Patibandla, N., Ni, Y., Van Driest, S. L., Chen, L., Roach, A., Cobb, B., Kirby, J., Denny, J., Bailey-Davis, L., Williams, M. S., Marsolo, K., Solti, I., Holm, I. A., ... Crimmins, N. (2016). Developing an Algorithm to Detect Early Childhood Obesity in Two Tertiary Pediatric Medical Centers. *Applied Clinical Informatics*, *7*(3), 693–706. doi:10.4338/ACI-2016-01-RA-0015 PMID:27452794

Li, S., Da Xu, L., & Zhao, S. (2015). The internet of things: A survey. *Information Systems Frontiers*, *17*(2), 243–259. doi:10.100710796-014-9492-7

Littlefield, A. (2020). *No, You Don't Have the Attention Span of a Goldfish* [Blog]. Retrieved from https://www.ceros.com/inspire/originals/no-dont-attention-span-goldfish/

Livingstone, S., Nandi, A., & Stoilova, M. (n.d.). *Young adolescents and digital media*. Academic Press.

Loiacono, E., Fulwiler, C. E., Cohanim, R., & Davis, M. (2018). An Investigation into College Students' Preferences for Technology Integration into Mindfulness-based Stress Reduction. *AIS Transactions on Human-Computer Interaction*, *10*(4), 188–204. doi:10.17705/1thci.00109

Ma, C. M. S. (2018). A Latent Profile Analysis of Internet use and Its Association with Psychological Well-Being Outcomes among Hong Kong Chinese Early Adolescents. *Applied Research in Quality of Life*, *13*(3), 727–743. doi:10.100711482-017-9555-2

Magis-Weinberg, L., Ballonoff Suleiman, A., & Dahl, R. E. (2021). Context, Development, and Digital Media: Implications for Very Young Adolescents in LMICs. *Frontiers in Psychology*, *12*, 632713. doi:10.3389/fpsyg.2021.632713 PMID:33967899

Magson, N. R., Freeman, J. Y. A., Rapee, R. M., Richardson, C. E., Oar, E. L., & Fardouly, J. (2021). Risk and Protective Factors for Prospective Changes in Adolescent Mental Health during the COVID-19 Pandemic. *Journal of Youth and Adolescence*, *50*(1), 44–57. doi:10.100710964-020-01332-9 PMID:33108542

Mahalakshmi, K., & Radha, R. (2020). Covid 19: A Massive Exposure Towards Web Based Learning. *Journal of Xidian University*, *14*(4). Advance online publication. doi:10.37896/jxu14.4/266

Mahaye, N. E. (2020). The impact of COVID-19 pandemic on education: navigating forward the pedagogy of blended learning. *Research Online*.

Mahfouz, J., Levitan, J., Schussler, D., Broderick, T., Dvorakova, K., Argusti, M., & Greenberg, M. (2018). Ensuring College Student Success Through Mindfulness-Based Classes: Just Breathe. *The College Student Affairs Journal*, *36*(1), 1–16. doi:10.1353/csj.2018.0000

Makridakis, S., Polemitis, A., Giaglis, G., & Louca, S. (2018). Blockchain: The next Breakthrough in the Rapid Progress of AI. In *Artificial Intelligence - Emerging Trends and Applications* (pp. 197–219). http://www.intechopen.com/books/trends-in-telecommunications-technologies/gps-total-electron-content-tec- prediction-at-ionosphere-layer-over-the-equatorial-region%0AInTec%0Ahttp://www.asociatiamhc.ro/wp-content/uploads/2013/11/Guide-to-Hydropower.pdf

Mak, W. W., Tong, A. C., Yip, S. Y., Lui, W. W., Chio, F. H., Chan, A. T., & Wong, C. C. (2018). Efficacy and Moderation of Mobile App–Based Programs for Mindfulness-Based Training, Self-Compassion Training, and Cognitive Behavioral Psychoeducation on Mental Health: Randomized Controlled Noninferiority Trial. *JMIR Mental Health*, *5*(4), e60. doi:10.2196/mental.8597 PMID:30309837

Mallet Official Page. (2010). http://mallet.cs.umass.edu/index.php/Main_Page

Manap (2020). Ancaman Tort Kecerdasan Buatan Terhadap Manusia: Retorik Atau Realiti? (The Threats Of Artificial Intelligence In Tort: Rhetoric Or Reality?). *Dewan Bahasa dan Pustaka, 32*(1).

Mangwaya, E., & Ndlovu, E. (2013). Rural Secondary School Teachers' Capacity to Respond to Hiv and Aids: The Case of Shurugwi District in Zimbabwe. *Asian Social Science*, *9*(2), 129. doi:10.5539/ass.v9n2p129

Mani, M., Kavanagh, D. J., Hides, L., & Stoyanov, S. R. (2015). Review and Evaluation of Mindfulness-Based iPhone Apps. *JMIR mHealth and uHealth*, *3*(3), e82. doi:10.2196/mhealth.4328 PMID:26290327

Mann, G., Lambert, L., Chang, Y., & Jurss, M. (2018). P135 - using the theory of planned behavior to understand college student snack choices. *Journal of Nutrition Education and Behavior: Supplement, 50*(7), 119. doi:10.1016/j.jneb.2018.04.164

Mansureh, K., Lipschuetz, A., & Santiague, L. (2017). Issues and Challenges for Teaching Successful Online Courses in Higher Education: A Literature Review. *Journal of Educational Technology Systems.*

Mares, M. L., & Woodard, E. (2005). Positive effects of television on children's social interactions: A meta-analysis. *Media Psychology, 7*(3), 301–322. doi:10.1207/S1532785XMEP0703_4

Marzano, G., & Ochoa-Siguencia, L. (2017). Challenges of Web-Based Participatory Learning. *Society. Integration. Education. Proceedings of the International Scientific Conference, 2*(May), 458. 10.17770ie2017vol2.2395

Masino, S., & Niño-Zarazúa, M. (2016). What works to improve the quality of student learning in developing countries? *International Journal of Educational Development, 48*, 53–65. doi:10.1016/j.ijedudev.2015.11.012

Masood, M., & Thigambaram, M. (2015). The Usability of Mobile Applications for Pre-schoolers. *Procedia: Social and Behavioral Sciences, 197*(February), 1818–1826. doi:10.1016/j.sbspro.2015.07.241

Maurya, P. (2020). Need for Digital Learning in Rural Areas: Special reference to Higher Education. *Research Review Journals., 5*(9), 175–177. Advance online publication. doi:10.31305/rrijm.2020.v05.i09.044

Mayo Clinic. (2020, May 15). *Telehealth: Technology meets Healthcare.* https://www.mayoclinic.org/healthy-lifestyle/consumer-health/in-depth/telehealth/art-20044878#:~:text=Telehealth%20is%20the%20use%20of,or%20support%20health%20care%20services

Mbodila, M. (2020). *Online learning – The Pandemic cannot Change Reality.* University World News: Africa Edition. Available at: https://www.universityworldnews.com/post.php?story=20200420130222745

McCollum, R., Taegtmeyer, M., Otiso, L., Tolhurst, R., Mireku, M., Martineau, T., & Theobald, S. (2019). Applying an intersectionality lens to examine health for vulnerable individuals following devolution in Kenya. *International Journal for Equity in Health, 18*(1), 1–15. doi:10.118612939-019-0917-2 PMID:30700299

McCrae, N., Getting, S., & Purssell, E. (2017). Social media and depressive symptoms in childhood and adolescence. *Systematic Reviews, 2*(4), 315–330.

McGuigan, S. M., Hollins, S., & Attard, M. (1995). Age-specific standardized mortality rates in people with learning disability. *Journal of Intellectual Disability Research, 39*(6), 527–531. doi:10.1111/j.1365-2788.1995.tb00573.x PMID:8746740

McIlwraith, A. (2006). *Information security and employee behaviour: how to reduce risk through employee education, training and awareness.* Gower Publishing, Ltd.

McLuhan, M. (2006). The Medium is the Message. In Media and Cultural Studies: Keyworks in Cultural Studies. Blackwell Publishing.

Mead, G. H. (1934). *Mind, Self, and Society.* University of Chicago Press.

Means, B., Toyama, Y., Murphy, R., Bakia, M., & Jones, K. (2009). Evaluation of Evidence-Based Practices in Online Learning. *Structure (London, England), 66.* www.ed.gov/about/offices/list/opepd/ppss/reports.html

Meates, J. (2020). Problematic Digital Technology Use of Children and Adolescents: Psychological Impact. *Teachers and Curriculum, 20*(1), 51–62.

Meeus, A., Beyens, I., Geusens, F., Sodermans, A. K., & Beullens, K. (2018). Managing positive and negative media effects among adolescents: Parental mediation matters—but not always. *Journal of Family Communication*, *18*(4), 270–285. doi:10.1080/15267431.2018.1487443

Meherali, S., Punjani, N., Louie-Poon, S., Abdul Rahim, K., Das, J. K., Salam, R. A., & Lassi, Z. S. (2021). Mental health of children and adolescents amidst CoViD-19 and past pandemics: A rapid systematic review. *International Journal of Environmental Research and Public Health*, *18*(7), 3432.

Meilman, P. W., & Weatherford, R. D. (2016). "Wait! Can You Do This Too?!?". College Counseling and Distance Education. *Journal of College Student Psychotherapy*, *30*(1), 1–3. doi:10.1080/87568225.2016.1105636

Melville, C. A., Hamilton, S., Hankey, C. R., Miller, S., & Boyle, S. (2007). The prevalence and determinants of obesity in adults with intellectual disabilities. *Obesity Reviews*, *8*(3), 223–230. doi:10.1111/j.1467-789X.2006.00296.x PMID:17444964

Meyerson, M. (2010). *Success Secrets Of The Social Media Marketing Superstars*. Entrepreneur Media Inc.

Michael, J. W. (2018). Blockchain Technology And Regulatory Investigations. Thomson Reuter. *The Journal Litigation*.

Mim, F. N., Islam, M. A., & Paul, G. K. (2018). Impact of the use of social media on students ' academic performance and behavior change. *International Journal of Statistics and Applied Mathematics*, *3*(1), 299–302.

Ministry of Health – Manatū Hauora. (2017 October). *Guidance on evaluating or developing a health app*. https://www.health.govt.nz/system/files/documents/pages/guidance-evaluating-developing-health-app-oct17-v2.pdf

Ministry of Health of the Republic of Indonesia. Health Law No. 36 of 2009 concerning health.

Mishra, L., Gupta, T., & Shree, A. (2020). Online teaching-learning in higher education during lockdown period of COVID-19 pandemic. *International Journal of Educational Research, 1*.

Mishra, A., Maheswarappa, S. S., & Colby, C. L. (2018). Technology readiness of teenagers: A consumer socialization perspective. *Journal of Services Marketing*.

Mitchell, D. E., & Nielsen, Y. S. (2012). *Internationalization and Globalization in Higher Education*. Academic Press.

Mitchell, S. L., Oakley, D. R., & Dunkle, J. H. (2019). White Paper: A Multidimensional Understanding of Effective University and College Counseling Center Organizational Structures. *Journal of College Student Psychotherapy*, *33*(2), 89–106. doi:10.1080/87568225.2019.1578941

Moghavvemi, S., Sulaiman, A., Jaafar, N. I., & Kasem, N. (2018). Social media as a complementary learning tool for teaching and learning: The case of youtube. *International Journal of Management Education*, *16*(1), 37–42. doi:10.1016/j.ijme.2017.12.001

Montag, C., & Elhai, J. D. (2020). Discussing digital technology overuse in children and adolescents during the COVID-19 pandemic and beyond: On the importance of considering Affective Neuroscience Theory. *Addictive Behaviors Reports*, *12*, 100313.

MoPSEC. (2021). *Curriculum development and technical services department*. http://mopse.co.zw/department/curriculum-development-and-technical-services-department

Morgan, B. (2017). Stress management for college students: An experiential multi-modal approach. *Journal of Creativity in Mental Health*, *12*(3), 276–288. doi:10.1080/15401383.2016.1245642

Morowatisharifabad, M. A., Karimi, M., & Ghorbanzadeh, F. (2015). Watching television by kids: How much and why? *Journal of Education and Health Promotion*, 4.

Ms, P., & Toro, U. (2016). *A Review of Literature on Knowledge Management using ICT in Higher Education*. Academic Press.

Mukherjee, A., Liu, B., & Glance, N. (2012). Spotting Fake Reviewer Groups in Consumer Reviews. *WWW'12 - Proceedings of the 21st Annual Conference on World Wide Web*, 191–200. 10.1145/2187836.2187863

Mustafa, A. G., Taha, N. R., Alshboul, O. A., Alsalem, M., & Malki, M. I. (2020). Using YouTube to Learn Anatomy: Perspectives of Jordanian Medical Students. *BioMed Research International*, *2020*, 1–8. Advance online publication. doi:10.1155/2020/6861416 PMID:32337267

Mustafaoğlu, R., Zirek, E., Yasacı, Z., & Özdinçler, A. R. (2018). The negative effects of digital technology usage on children's development and health. *Addicta: The Turkish Journal on Addictions, 5*(2), 13-21.

Mustafaoğlu, R., Zirek, E., Yasacı, Z., & Razak Özdinçler, A. (2018). The negative effects of digital technology usage on children's development and health. *Addicta: The Turkish Journal on Addictions, 5*(2), 13-21.

Mustafaoğlu, R., Zirek, E., Yasacı, Z., & Razak Özdinçler, A. (2018). The Negative Effects of Digital Technology Usage on Children's Development and Health. *Addicta: The Turkish Journal on Addictions*, *5*(2). Advance online publication. doi:10.15805/addicta.2018.5.2.0051

Mwantimwa, K. (2019). ICT usage to enhance firms' business processes in Tanzania. *Journal of Global Entrepreneurship Research*, *9*(1), 46. doi:10.118640497-019-0170-6

Nagatomo, S. (2019). *Japanese Zen Buddhist Philosophy*. https://plato.stanford.edu/entries/japanese-zen/

Nathanson, A. I. (2001). Parents versus Peers - Exploring the Significance of Peer Mediation of Antisocial Television. *Communication Research*, *28*(3), 251–274. doi:10.1177/009365001028003001

Nathanson, A. I. (2002). The Unintended Effects of Parental Mediation of Television on Adolescents. *Media Psychology*, *4*(3), 207–230. doi:10.1207/S1532785XMEP0403_01

National Alliance on Mental Health. (2019). *Mental Health by the Numbers*. https://www.nami.org/Learn-More/Mental-Health-By-the-Numbers

National Center for Education Statistics. (2005). *Educational Longitudinal Study of 2002*. https://nces.ed.gov/surveys/els2002/dec2005.asp

National Center for Education Statistics. (2009). https://nces.ed.gov/surveys/ELS2002/

National Health Policy. (2017). *Ministry of Health and Family Welfare Government of India*. Author.

National Planning Commission. (2012). National Development Plan 2030: Our future–make it work. Presidency of South Africa.

Nauert, R. (2015). Teenage Competition: Helpful or Harmful? *Psych Central*. Retrieved November 16, 2016 from https://psychcentral.com/news/2010/07/07/teenage-competition-helpful-or-harmful/15379.html

Nawi, A. (2019). Penerokaan Awal Terhadap Isu Dan Impak Penggunaan Teknologi Kecerdasan Buatan Terhadap Kehidupan Manusia. *Asian Journal of Civilizational Studies.*, *1*(4), 2682–8022.

Naz, M., Al-zahrani, F. A., Khalid, R., Javaid, N., Qamar, A. M., Afzal, M. K., & Shafiq, M. (2019). A Secure Data Sharing Platform Using Blockchain and Interplanetary File System. *Sustainability (Switzerland)*, *11*(24), 1–24. doi:10.3390u11247054

Ndimande, B. S. (2016). Pedagogy of Poverty: School Choice and Inequalities in Post-Apartheid South Africa. *Global Education Review*, *3*(2).

NewZimbabwe. (2021, February 7). Teachers Group Blames Govt For Low Grade 7 Pass Rate. *NewZimbabwe*. https://www.newzimbabwe.com/teachers-group-blames-govt-for-low-grade-7-pass-rate/

Ngalomba, S. (2020). *Using WhatsApp to Enhance Online Learning*. University World News: Africa Edition. Available at: https://www.universityworldnews.com/post.php?story=20200421102812987

Ng, E. S. W., Schweitzer, L., & Lyons, S. T. (2010). New Generation, Great Expectations: A Field Study of the Millennial Generation. *Journal of Business and Psychology*, *25*(2), 281–292. doi:10.100710869-010-9159-4

Ng, M. C. Y., & Bowden, D. W. (2013). Is genetic testing of value in predicting and treating obesity? *North Carolina Medical Journal*, *74*(6), 530–533. doi:10.18043/ncm.74.6.530 PMID:24316784

Niazi, A. K., & Niazi, S. K. (2011). Mindfulness-based stress reduction: A non-pharmacological approach for chronic illnesses. *North American Journal of Medical Sciences*, *3*(1), 20–23. doi:10.4297/najms.2011.320 PMID:22540058

Nisbet, M. (2020, May 2). *The mindfulness movement: How a Buddhist practice evolved into a scientific approach to life*. https://web.northeastern.edu/matthewnisbet/2017/05/24/the-mindfulness-movement-how-a-buddhist-practice-evolved-into-a-scientific-approach-to-life/

Nizami, G. N., Naeem, Z., Arzoo, K., & Ismail, S. (2017). Impact of Facebook Addiction on Academic Performance Among Undergraduate Students. *Pakistan Journal of Rehabilitation*, *6*(1), 45–50. doi:10.36283/pjr.zu.6.1/007

Nobleza, D., Hagenbaugh, J., Blue, S., Vergare, M., Pohl, C., & Stepchin, A. (2018). The use of telehealth by medical and other health professional students at a college counseling center. *Journal of College Student Psychotherapy*. Advance online publication. doi:10.1080/87568225.2018.1491362

Nonaka, I. (1994). A dynamic theory of organisational knowledge creation. *Organization Science*, *5*(1), 14–37. doi:10.1287/orsc.5.1.14

NORMA Oficial Mexicana NOM-008-SSA3-2017. (n.d.). *Para el tratamiento integral del sobrepeso y la obesidad*. Retrieved on February 16, 2021 from https://dof.gob.mx/nota_detalle.php?codigo=5523105&fecha=18/05/2018

Noviana, N. (2018). *Kesehatan Reproduksi untuk Mahasiswa Kebidanan (Revisi)*. Trans Info Media. https://scholar.google.co.id/citations?user=ac_7qq0AAAAJ&hl=en

Noviana, N., Suwitri, S., Supriyono, B., & Sutopo, P. J. (2019). Implementasi Kebijakan Kesehatan Dalam Pengendalian HIV/AIDS. Pena Persada. https://scholar.google.co.id/citations?user=ac_7qq0AAAAJ&hl=

Noviana, N. (2020). Development of Health Promotion System in Industrial Era 4.0 to Support E-Service. *International Joint Conference on Science And Technology*.

Novotney, A. (2017). *A Growing Wave of Online Therapy*. https://www.apa.org/monitor/2017/02/online-therapy

O'Connell, A., & Clarke, S. (2020). A School in the Grip of COVID-19: Musings from the Principal's Office. *International Studies in Educational Administration*, *48*(2), 4–11.

O'Keefe, G. S., & Clarke-Pearson, K.Council on Communications and Media. (2011). The impact of social media on children, adolescents and families. *Paediatrics*, *127*(4), 800–804. doi:10.1542/peds.2011-0054 PMID:21444588

O'Sullivan, K., Clark, S., McGrane, A., Rock, N., Burke, L., Boyle, N., Joksimovic, N., & Marshall, K. (2021). A qualitative study of child and adolescent mental health during the COVID-19 pandemic in Ireland. *International Journal of Environmental Research and Public Health*, *18*(3), 1–15. doi:10.3390/ijerph18031062 PMID:33504101

Ocampo, A. (2021). *How Colour Psychology can be Applied to Mobile Apps. Top mobile App Developers*. Retrieved 2 October 2021, from https://thisisglance.com/how-colour-psychology-can-be-applied-to-mobile-apps/

Odgers, C. (2018). Smartphones are bad for some teens, not all. *Nature*, *554*(7693), 432–434. doi:10.1038/d41586-018-02109-8

Odgers, C. L., & Jensen, M. R. (2020). Adolescent development and growing divides in the digital age. *Dialogues in Clinical Neuroscience*, *22*(2), 143.

OECD Publishing. (2019). Artificial Intelligence in Society. In *Artificial Intelligence in Society*. Author.

OECD. (2016). *Innovating Education and Educating for Innovation: The Power of Digital Technologies and Skills*. OECD Publishing. doi:10.1787/9789264265097-

Olasina, G. (2017). An evaluation of educational values of YouTube videos for academic writing. *The African Journal of Information Systems*, *9*(4), 2.

Orben, A., Dienlin, T., & Przybylski, A. K. (2019). Social media's enduring effect on adolescent life satisfaction. *Proceedings of the National Academy of Sciences of the United States of America*, *116*(21), 10226–10228. doi:10.1073/pnas.1902058116 PMID:31061122

Orben, A., & Przybylski, A. K. (2019). The association between adolescent well-being and digital technology use. *Nature Human Behaviour*, *3*(2), 173–182.

Osharive, P. (2015). *Social Media and Academic Performance of Students in University of Lagos*. doi:10.1007/978-3-319-67256-4_1

Paat, Y.-F., & Markham, C. (2021). Digital crime, trauma, and abuse: Internet safety and cyber risks for adolescents and emerging adults in the 21 st century. *Social Work in Mental Health*, *19*(1), 18–40. doi:10.1080/15332985.2020.1845281

Padilla-Walker, L. M., & Christensen, K. J. (2011). Empathy and self-regulation as mediators between parenting and adolescents' prosocial behavior toward strangers, friends, and family. *Journal of Research on Adolescence*, *21*(3), 545–551. doi:10.1111/j.1532-7795.2010.00695.x

Padilla-Walker, L., Coyne, S., & Collier, K. (2016). Longitudinal relations between parental media monitoring and adolescent aggression, prosocial behavior, and externalizing problems. *Journal of Adolescence*, *46*, 86–97. doi:10.1016/j.adolescence.2015.11.002 PMID:26641307

Palechor, F. M., & de la Hoz Manotas, A. (2019). Dataset for estimation of obesity levels based on eating habits and physical condition in individuals from Colombia, Peru and Mexico. *Data in Brief*, *25*, 104344. doi:10.1016/j.dib.2019.104344 PMID:31467953

Panagiotopoulos, P., Barnett, J., Bigdeli, A. Z., & Sams, S. (2016). Social Media in Emergency Management: Twitter as a Tool for. *Technological Forecasting and Social Change*, *111*(October), 1–27. Retrieved from https://www.sciencedirect.com/science/article/pii/S0040162516301196%0Ahttp://dx.doi.org/10.1016/j.techfore.2016.06.010%0Ahttps://linkinghub.elsevier.com/retrieve/pii/S0040162516301196

Pandey, S., Chavla, M., & Mathur, M. (2021). Good and Not so good of learning using Digital platforms: Experience sharing from PG students. *Turkish Journal of Computer and Mathematics Education*, *12*(7), 749–757.

Pandey, S., & Tiwari, S. (2014). Education and Prosperity through Technology – Case of Virtual Education Limited Nigeria. *Procedia: Social and Behavioral Sciences*, *157*, 55–62.

Papadakis, S., & Kalogiannakis, M. (2017). Mobile educational applications for children. What educators and parents need to know. *International Journal of Mobile Learning and Organisation, 11*(3), 256–277. doi:10.1504/IJMLO.2017.085338

Parekh, R. (2017). *What Is ADHD?* Retrieved 6 May 2021, from https://www.psychiatry.org/patients-families/adhd/what-is-adhd#:~:text=Attention%2Ddeficit%2Fhyperactivity%20disorder%20(,common%20mental%20disorders%20affecting%20children

Parikh & Atrey. (2018). Media-Rich Fake News Detection: A Survey. *2018 IEEE Conference on Multimedia Information Processing and Retrieval (MIPR)*, 436–41. https://ieeexplore.ieee.org/document/8397049/

Parmer, R. (2017, May 25). *The Importance of Self-Care for College Students.* https://collegepuzzle.stanford.edu/the-importance-of-self-care-for-college-students/.00000000

Parsons, C. E., Crane, C., Parsons, L. J., Fjorback, L. O., & Kuyken, W. (2017). Home practice in mindfulness-based cognitive therapy and mindfulness-based stress reduction: A systematic review and metaanalysis of participants' mindfulness practice and its association with outcomes. *Behaviour Research and Therapy, 95*, 29-41. http://dx.doi.org.tcsedsystem.idm.oclc.org/10.1016/j.brat.2017.05.004

Pasam, R. S. (2019). Advances in digital technology: Effect on child and adolescent behavior. *Archives of Mental Health, 20*(2), 37. doi:10.4103/AMH.AMH_25_19

Paschke, K., Austermann, M. I., Simon-Kutscher, K., & Thomasius, R. (2021). Adolescent gaming and social media usage before and during the COVID-19 pandemic. *Sucht: Zeitschrift für Wissenschaft und Praxis.*

Patil, & Patil, & Bhat. (2012). A Review on 5G Technology. *International Journal of Engineering and Innovative Technology, 1*(1), 2277–3754.

Patja, K., Iivanainen, M., Vesala, H., Oksanen, H., & Ruoppila, I. (2000). Life expectancy of people with intellectual disability: A 35-year follow-up study. *Journal of Intellectual Disability Research, 44*(5), 591–599. doi:10.1046/j.1365-2788.2000.00280.x PMID:11079356

Pelphrey, K. A., Foell, J., Zerwas, S., Esposito, G., Cataldo, I., Lepri, B., Jin, M., & Neoh, Y. (2021). Social Media Usage and Development of Psychiatric Disorders in Childhood and Adolescence. *RE:view.*

Petrie, Aladin, Ranjan, Javangwe, Gilliland, Tuominen, & Lasse. (2020). *Spotlight: Quality education for all during Covid-19 crisis.* Retrieved from https://hundred-cdn.s3. amazonaws.com/uploads/report/file/15/hundred_spotlight_covid-19_digital.pdf)

Pettersson, G. (2017). *Inre kraft och yttre tryck: perspektiv på specialpedagogisk verksamhet i glesbygdsskolor* [Inner power and outer pressure: perspectives on special education in rural schools] (Doctoral dissertation). Umeå Universitet.

Piff, P. K., Kraus, M. W., Côté, S., Cheng, B. H., & Keltner, D. (2010). Having less, giving more: The influence of social class on prosocial behavior. *Journal of Personality and Social Psychology, 99*(5), 771–784. doi:10.1037/a0020092 PMID:20649364

Piyadassi, T. (2013, November 13). *The Seven Factors of Enlightenment.* https://www.accesstoinsight.org/lib/authors/piyadassi/wheel001.html

Platt, J. (1998). *Sequential Minimal Optimization: A Fast Algorithm for Training Support Vector Machines.* Microsoft Research Technical Report.

Pooreh, S., & Hosseini Nodeh, Z. (2015). Impact of Education Based on Theory of planned behavior: An Investigation into Hypertension-Preventive Self-care Behaviors in Iranian Girl Adolescent. *Iranian Journal of Public Health, 44*(6), 839–847. PMID:26258097

Potter, W. J. (1990). Adolescents' perceptions of the primary values of television programming. *The Journalism Quarterly, 67*(4), 843–851. doi:10.1177/107769909006700439

Povey, J., Campbell, A. K., Willis, L.-D., Haynes, M., Western, M., Bennett, S., Antrobus, E., & Pedde, C. (2016). Engaging parents in schools and building parent-school partnerships: The role of school and parent organization leadership. *International Journal of Educational Research, 79*, 128–141. .Ijer.2016.07.005 doi:10.1016/j

Power, J. R. A. T. M., & Nichols, B. H. (2020). Teachers Bridging the Digital Divide in Rural Schools with 1: 1 Computing. *Rural Educator, 41*(1), 61–76. doi:10.35608/ruraled.v41i1.576

Powers, W. (2010). *Hamlet's Blackberry: A practical philosophy for building a good life in the digital age*. Harper.

Prasher, V. P., & Janicki, M. P. (2002). *The Physical Health of Adults with Intellectual Disabilities*. Blackwell Publishing Company. doi:10.1002/9780470776216

Pratama, Y., Hartanto, R., & Kusumawardani, S. S. (2018). Validating YouTube Factors Affecting Learning Performance. *IOP Conference Series. Materials Science and Engineering, 325*(1), 012003. Advance online publication. doi:10.1088/1757-899X/325/1/012003

Praveen, Chamola, Hassija, & Kumar. (2020). *Blockchain for 5G: A Prelude to Future Telecommunication*. IEEE Network.

Przybylski, A. K., Murayama, K., DeHaan, C. R., & Gladwell, V. (2013). Motivational, emotional, and behavioral correlates of fear of missing out. *Computers in Human Behavior, 29*(4), 1841–1848. doi:10.1016/j.chb.2013.02.014

Purdie-Vaughns, V., & Eibach, R. P. (2008). Intersectional invisibility: The distinctive advantages and disadvantages of multiple subordinate-group identities. *Sex Roles, 59*(5-6), 377–391. doi:10.100711199-008-9424-4

Putrawangsa & Hasanah. (2018). Integrasi Teknologi Digital Dalam Pembelajaran Di Era Industri 4.0. Kajian dari Perspektif Pembelajaran Matematika. *Jurnal Pemikiran dan Penelitian Pendidikan, 16*(1).

Python Official Site. (2019). https://www.python.org/

Querstret, D., Morison, L., Dickinson, S., Cropley, M., & John, M. (2020). Mindfulness-based stress reduction and mindfulness-based cognitive therapy for psychological health and well-being in nonclinical samples: A systematic review and meta-analysis. International *Journal of Stress Management*. http://dx.doi.org.tcsedsystem.idm.oclc.org/10.1037/str0000165

Qutteina, Y., Hallez, L., Mennes, N., De Backer, C., & Smits, T. (2019). What Do Adolescents See on Social Media? A Diary Study of Food Marketing Images on Social Media. *Frontiers in Psychology, 10*(November), 1–12. doi:10.3389/fpsyg.2019.02637

Ragelienė, T., & Grønhøj, A. (2020). The influence of peers′ and siblings′ on children's and adolescents′ healthy eating behavior. A systematic literature review. *Appetite, 148*(104592). Advance online publication. doi:10.1016/j.appet.2020.104592 PMID:31927070

Rahardja, U., Harahap, E. P., & Christianto, D. D. (2020). Pengaruh Teknologi Blockchain Terhadap Tingkat Keaslian Ijazah. Technomedia Journal, 4(2).

Rahimdel, T., Morowatisharifabad, M. A., Salehi-Abargouei, A., Mirzaei, M., & Fallahzadeh, H. (2019). Evaluation of an education program based on the theory of planned behavior for salt intake in individuals at risk of hypertension. *Health Education Research, 34*(3), 268–278. doi:10.1093/her/cyz007 PMID:30838389

Rahim, N. A., Affendi, N. R. N. M., & Awang, A. A. P. (2017). Dissemination of Values and Culture through the E-Folklore. *The Turkish Online Journal of Educational Technology, 16*(1), 32–36.

Rai, R., Chauhan, C. & Cheng, MI. (2020). Materialistic values, brand knowledge and the mass media: Hours spent on the internet predicts materialistic values and brand knowledge. *Current Psychology*, *39*, 2140–2148. doi:10.1007/s12144-018-9900-0

Raju, N. J., Valsaraj, B. P., & Noronha, J. (2015). Online Social Networking: Usage in Adolescents. *Journal of Education and Practice*, *6*(22), 80–84. http://search.proquest.com/docview/1773229886?accountid=14744

Ramesh Babu Durai, C., & Duraisamy, V. (2011). Content Based Image Retrieval Using Novel Gaussian Fuzzy Feed Forward-Neural Network. *Journal of Computational Science*, *7*(7), 958–961.

Ramler, T., Tennison, L., Lynch, J., & Murphy, P. (2016). Mindfulness and the college transition: The efficacy of an adapted mindfulness-based stress reduction intervention in fostering adjustment among first year students. *Mindfulness*, *7*(1), 179–188. doi:10.100712671-015-0398-3

Ramos Salazar, L., & Hayward, S. L. (2018). An examination of college students' problem-solving self-efficacy, academic self-efficacy, motivation, test performance, and expected grade in introductory-level economics courses. *Decision Sciences Journal of Innovative Education*, *16*(3), 217–240. doi:10.1111/dsji.12161

Rana, H., Rajiv, R., & Lal, M. (2014). E-learning: Issues and Challenges. *International Journal of Computers and Applications*, *97*(5), 20–24. doi:10.5120/17004-7154

Real, M. R. (1996). *Exploring Media Culture: A Guide*. Sage Publications. doi:10.4135/9781483327600

Regional Regulation No.11 of 2012 concerning HIV / AIDS Control in Banjarmasin City.

Richards, D., Caldwell, P. H. Y., & Go, H. (2015). Impact of social media on the health of children and young people. *Journal of Paediatrics and Child Health*, *51*(12), 1152–1157. doi:10.1111/jpc.13023 PMID:26607861

Rideout, V. G., Foehr, U. G., Roberts, D. F., & Brodie, M. (1999). *Kids & media @ the new millennium*. Kaiser Family Foundation.

Rimmer, J. H., & Yamaki, K. (2006). Obesity and intellectual disability. *Mental Retardation and Developmental Disabilities Research Reviews*, *12*(1), 22–27. doi:10.1002/mrdd.20091 PMID:16435329

Robin, B. (2008). The Effective Uses Of Digital Storytelling As A Teaching And Learning Tool. Handbook of Research on Teaching Literacy through the Communicative and Visual Arts, 2.

Robinson, L. (2021). *Social Media and Mental Health*. Retrieved 2 October 2021, from https://www.google.com/url?q=https://www.helpguide.org/articles/mental-health/social-media-and-mental-health.htm&usg=AOvVaw1FQOMnN_IiIaj6ss-JP7yqQ

Robinson, L., & Smith, M. (2021). *Social Media and Mental Health - HelpGuide.org*. Retrieved 6 May 2021, from https://www.helpguide.org/articles/mental-health/social-media-and-mental-health.htm#:~:text=Excessive%20social%20media%20use%20can,negative%2C%20self%2Dperpetuating%20cycle%3A&text=Using%20social%20media%20more%20often,depression%2C%20anxiety%2C%20and%20stress

Robinson, L. J., Schulz, J., Blank, G., Ragnedda, M., Ono, H., Hogan, B., Mesch, G. S., Cotten, S. R., Kretchmer, S. B., Hale, T. M., Drabowicz, T., Yan, P., Wellman, B., Harper, M.-G., Quan-Haase, A., Dunn, H. S., Casilli, A. A., Tubaro, P., Carvath, R., ... Khilnani, A. (2020). Digital inequalities 2.0: Legacy inequalities in the information age. *First Monday*, *25*(7). Advance online publication. doi:10.5210/fm.v25i7.10842

Robinson, T. N., Banda, J. A., Hale, L., Lu, A. S., Fleming-Milici, F., Calvert, S. L., & Wartella, E. (2017). Screen Media Exposure and Obesity in Children and Adolescents. *Pediatrics*, *140*(Suppl 2), S97–S101. doi:10.1542/peds.2016-1758K PMID:29093041

Rodgers, R. J., Tschöp, M. H., & Wilding, J. P. (2012). Anti-obesity drugs: Past, present and future. *Disease Models & Mechanisms, 5*(5), 621–626. doi:10.1242/dmm.009621 PMID:22915024

Rodriguez, A. (2021). *Social media fatigue: 10 facts you should know.* Retrieved 6 May 2021, from https://www.mycustomer.com/community/blogs/annarodriguez/social-media-fatigue-10-facts-you-should-know#:~:text=According%20to%20Techopedia%2C%20social%20media,spent%20online%20maintaining%20these%20connections.%E2%80%9D

Rodríguez, R. M. (2013). Implementation process of technology in Education: The case of Blackboard 9.1 in the University of Manchester. *Actualidades Investigativas en Educación, 13*(3). Advance online publication. doi:10.15517/aie.v13i3.12043

Rokeach, M. (1968). A theory of organization and change within value-attitude systems. *The Journal of Social Issues, 24*(1), 13–33. doi:10.1111/j.1540-4560.1968.tb01466.x

Rosenberg, M. (1979). "Which Significant Other?" and "Beyond Self-Esteem." In Conceiving the Self (pp. 83-98). Basic Books.

Rossow, A. (2021). Cyberbullying Taken To A Whole New Level: Enter The 'Blue Whale Challenge'. *Forbes.* Retrieved 3 October 2021, from https://www.forbes.com/sites/andrewrossow/2018/02/28/cyberbullying-taken-to-a-whole-new-level-enter-the-blue-whale-challenge/?sh=4e2037212673

Rothberg, M. B., Arora, A., Hermann, J., Kleppel, R., Marie, P. S., & Visintainer, P. (2010). Phantom vibration syndrome among medical staff: a cross sectional survey. *BMJ, 341*(2). doi:10.1136/bmj.c6914

Rouis, S., Limayem, M., & Salehi-Sangari, E. (2011). Impact of Facebook usage on students' academic achievement: Role of self-regulation and trust. *Electronic Journal of Research in Educational Psychology, 9*(3), 961–994. doi:10.25115/ejrep.v9i25.1465

Roy, N. (2018). *The Rise of Mental Health on College Campuses: Protecting the Emotional Health of Our Nation's College Students.* https://www.higheredtoday.org/2018/12/17/rise-mental-health-college-campuses-protecting-emotional-health-nations-college-students/

Royal College of General Practitioners Working Party. (1990). *Primary Care for People with a Mental Handicap.* Royal College of General Practitioners.

Rutherford, A., & Davidson, T. (2019). Intersectionality and the history of psychology. In Oxford Research Encyclopedia of Psychology. doi:10.1093/acrefore/9780190236557.013.468

Saad & Ismail. (2005). Pengajaran dan Pembelajaran Bahasa Arab Berasaskan Internet. In Z. Ismail & M. S. Othman (Eds.), *Pendidikan Bahasa Arab di Malaysia: Cabaran dan Inovasi.* Minda Imtiyaz Publications.

Saari & Mahiyaddin. (2017). Analisis linguistik terhadap penggunaan media sosial dalam kalangan pelajar di IPTS. *Advances in Language and Literary Studies, 1*(1), 151–165.

Salas-Pilco, S. Z., & Law, N. W. Y. (2018). ICT curriculum planning and development: Policy and implementation lessons from small developing states. In *ICT-Supported Innovations in Small Countries and Developing Regions* (pp. 77–98). Springer. doi:10.1007/978-3-319-67657-9_4

Saleha, N. S., & Rosli, M. S. (2020). *Tinjauan Literatur Sistematik: Interaksi Media Sosial Atas Talian dalam Kluster Pendidikan dan Pekerjaan. In Sains Humanika.* Universiti Teknologi Mara.

Salleh. (2019). Mudah Guna dan Manfaat Aplikasi WhatsApp terhadap hubungan antara Peribadi masyarakat dan Polis. *Malaysian Journal of Communication Jilid, 35*(4), 135–154.

Salman, A., & Pulungan, S. M. (2017). Pendedahan diri, motivasi dan kepuasan penggunaan Facebook dalam menjalinkan persahabatan. *Jurnal Komunikasi: Malaysian Journal of Communication, 33*(1), 438–459.

Samah, A., Marwah, A., Saira, G., & Ali, A. (2019). applied sciences Blockchain-Based Applications in Education. *Applied Sciences (Basel, Switzerland), 9*(2400), 1–18.

Sandhya Keelery (2020). Social media usage in India- statistics & facts.

Sarah, F. N. (2019). *Understanding Students' Self-Regulation in Asynchronous Online Learning*. University of Minnesota.

Sargent, J. D., Beach, M. L., Adachi-Mejia, A. M., Gibson, J. J., Titus-Ernstoff, L. T., Carusi, C. P., Swain, S. D., Heatherton, T. F., & Dalton, M. A. (2005). Exposure to movie smoking: Its relation to smoking initiation among US adolescents. *Pediatrics, 116*(5), 1183–1191. doi:10.1542/peds.2005-0714 PMID:16264007

Sarwar & Soomro. (2013). Impact of Smartphone's on Society. *European Journal of Scientific Research, 98*(2), 216-226. http://www.europeanjournalofscientificresearch.com

Sawyer, R. (2011). *The Impact of New Social Media on Intercultural Adaptation*. Senior Honors Projects: Honors Program at the University of Rhode Island. http://digitalcommons. uri.edu/cgi/viewcontent.cgi?article=1230&context=srhonorsprog

Sayed, Y., & Singh, M. (2020). Evidence and education policy making in South Africa during Covid-19: Promises, researchers and policymakers in an age of unpredictability. *SARE, 26*(1), 20–39.

Scalco, A., Noventa, S., Sartori, R., & Ceschi, A. (2017). Predicting organic food consumption: A meta-analytic structural equation model based on the theory of planned behavior. *Appetite, 112*, 235–248. doi:10.1016/j.appet.2017.02.007 PMID:28188865

Scharrer, E. (2008). Media Exposure and Sensitivity to Violence in News Reports: Evidence of Desensitization? *Journalism & Mass Communication Quarterly, 85*(2), 291–310. doi:10.1177/107769900808500205

Schneider, J. (2020). Perspective: How COVID-19 laid bare the vast inequities in U.S. public education. *Washington Post*. Retrieved from https://www.washingtonpost.com/education/2020/04/14/how-covid-19-has-laid-bare-vast-inequities-us-public-education/

Schneider, S. (2001). *Implementing AIDS Policy in Post-Apartheid South Africa*. Academic Press.

Serdyukov, P. (2017). Innovation in education: What works, what doesn't, and what to do about it? *Journal of Research in Innovative Teaching & Learning, 10*(1), 4–33. doi:10.1108/JRIT-10-2016-0007

Seyal, A. H., & Abd Rahman, M. N. (Eds.). (2017). *Theory of planned behavior: new research (Ser. Psychology of emotions, motivations and actions)*. Nova Science.

Sgantzos, K., & Grigg, I. (2019). Artificial intelligence implementations on the blockchain. Use cases and future applications. *Future Internet, 11*(8), 1–15. doi:10.3390/fi11080170

Shafer, L. (2017). *Social Media and Teen Anxiety*. Retrieved 6 May 2021, from https://www.gse.harvard.edu/news/uk/17/12/social-media-and-teen-anxiety

Shama, S., & Ikbal, J. (2020). *Education during A Pandemic: A feasibility study of online classes in Bangladesh to counteract potential study gaps caused by COVID-19 related lockdowns*. Academic Press.

Sharma, A., Bahl, S., Bagha, A. K., Javaid, M., Shukla, D. K., & Haleem, A. (2020). Blockchain technology and its applications to combat COVID-19 pandemic. *Research on Biomedical Engineering*. Advance online publication. doi:10.100742600-020-00106-3

Sharma, K., & Lin, K. I. (2013). Review Spam Detector with Rating Consistency Check. *Proceedings of the Annual Southeast Conference*, 34–39.

Shatkin, J. P., Diamond, U., Zhao, Y., DiMeglio, J., Chodaczek, M., & Bruzzese, J. (2016). Effects of a risk and resilience course on stress, coping skills, and cognitive strategies in college students. *Teaching of Psychology*, *43*(3), 204–210. doi:10.1177/0098628316649457

Shefet, O. (2018). Ultra-Brief, Immediate, and Resurgent: A College Counseling Paradigm Realignment. *Journal of College Student Psychotherapy*, *4*(32), 291–311. doi:10.1080/87568225.2017.1401790

Sheil, W. C. (2018, December 11) *Medical Definition of Stress*.https://www.medicinenet.com/script/main/art.asp?articlekey=20104

Shelesky, K., Weatherford, R. D., & Silbert, J. (2016). Responding to the Increased Needs of College Students: A Case Study of Case Management. *Journal of College Student Psychotherapy*, *30*(4), 284–299. doi:10.1080/87568225.2016.1219613

Sheng, E. L., & Basaruddin, F. (2014). Ways of improving students performances in assessments from instructors and students perspectives. *International Journal of Asian Social Science*, *4*(2), 301–306.

Shenglin, B., Simonelli, F., Zhang, R., & Li, W. (2017). *Digital infrastructure: Overcoming Digital Divide in Emerging Economies.* Available from https://economic-policy-forum.org/wp-content/uploads/20 17/03/Policy-Brief_ZJUCEPS_Digital-infrastructure_Over coming-Digital-Divide-in-Emerging-Economies_2017022 8.pdf

Shi, J., & Kim, H. K. (2020). Integrating risk perception attitude framework and the theory of planned behavior to predict mental health promotion behaviors among young adults. *Health Communication*, *35*(5), 597–606. doi:10.1080/10410236.2019.1573298 PMID:30810392

Shin, W., & Huh, J. (2011). Parental mediation of teenagers' video game playing: Antecedents and consequences. *New Media & Society*, *13*(6), 945–962. doi:10.1177/1461444810388025

Shrestha, A. K., Vassileva, J., & Deters, R. (2020). A Blockchain Platform for User Data Sharing Ensuring User Control and Incentives. *Frontiers in Blockchain*, *3*(October), 1–22. doi:10.3389/fbloc.2020.497985

Siljebo, J. (2020). Digitalization and digital transformation in schools: A challenge to educational theory? *Education in the North*, *27*(2), 24–37.

Silva, Da, Coletta, & Hruschka. (2016). A Survey and Comparative Study of Tweet Sentiment Analysis via Semi-Supervised Learning. *ACM Computing Surveys*, *49*(1).

Sima, V., Gheorghe, I. G., Subić, J., & Nancu, D. (2020). Influences of the industry 4.0 revolution on the human capital development and consumer behavior: A systematic review. *Sustainability (Switzerland)*, *12*(10), 4035. Advance online publication. doi:10.3390u12104035

Simpson, J. (2009). *Everyone belongs: a toolkit for applying intersectionality. Ottawa. South Africa: Broken and unequal education perpetuating poverty and inequality. 2020, February 11.* Amnesty International. Available at: https://www.amnesty.org/en/latest/news/2020/02/south-africa-broken-and-unequal-education-perpetuating-poverty-and-inequality/

Singh. (2020). Impact of Technology in Indian Education. *International Journal of Advance and Innovative Research*, *7*(1).

Singh, A., Kumar, H., Kumari, S., Scholar, R., & Professor, A. (2020). Impact of Social Media on Adolescent's Mental Health. *International Journal of Creative Research Thoughts*, *8*(7).

Sipe, W. E., & Eisendrath, S. J. (2012, February). Mindfulness-based cognitive therapy: Theory and practice. *Canadian Journal of Psychiatry*, *57*(2), 63–69. doi:10.1177/070674371205700202 PMID:22340145

Siponen, M. T., & Oinas-Kukkonen, H. (2007). A review of information security issues and respective researcher contributions. *ACM SIGMIS Database: the DATABASE for Advances in Information Systems*, *38*(1), 60–80. doi:10.1145/1216218.1216224

Siraj, I. & Taggart, B. (2014). *Exploring Effective Pedagogy in Primary Schools : Evidence from Research*. Academic Press.

Siu, A. M. H., Daniel, T. L., & Ben, L. (2013). Prosocial norms as a positive youth development construct: A conceptual review. *Journal of Alternative Medical Research*, *5*(1), 19–27. doi:10.1100/2012/832026 PMID:22666157

Skyfactor. (2017). *Second-Term Behavior of Students on Academic Probation*. https://www.skyfactor.com/wp-content/uploads/2017/06/2017-Fall-Probation-Students-in-Spring-Research-Note.pdf

Slater, M. D., Henry, K. L., Swaim, R. C., & Anderson, L. L. (2003). Violent Media Content and Aggressiveness in Adolescents: A Downward Spiral Model. *Communication Research*, *30*(6), 713–736. doi:10.1177/0093650203258281

Small, G. W., Lee, J., Kaufman, A., Jalil, J., Siddarth, P., Gaddipati, H., Moody, T. D., & Bookheimer, S. Y. (2020). Brain health consequences of digital technology use . *Dialogues in Clinical Neuroscience*, *22*(2), 179–187. doi:10.31887/DCNS.2020.22.2/gsmall PMID:32699518

Smith, D. H. (1981). Altruism, volunteers, and volunteerism. *Journal of Voluntary Action Research*, *10*(1), 21–36. doi:10.1177/089976408101000105

Smith, J., Hewitt, B., & Skrbiš, Z. (2015). Digital socialization: Young people's changing value orientations towards internet use between adolescence and early adulthood. *Information Communication and Society*, *18*(9), 1022–1038. doi:10.1080/1369118X.2015.1007074

Social media and celebrity culture 'harming young people'. (2021). *The Guardian*. Retrieved 2 October 2021, from https://amp-theguardian-com.cdn.ampproject.org/v/s/amp.theguardian.com/lifeandstyle/2018/jul/23/social-media-and-celebrity-culture-harming-young-people?amp_js_v=a6&amp_gsa=1&usqp=mq331AQKKAFQArABIIACAw%3D%3D#aoh=16331002636511&referrer=https%3A%2F%2Fwww.google.com&amp_tf=From%20%251%24s&ampshare=https%3A%2F%2Fwww.theguardian.com%2Flifeandstyle%2F2018%2Fjul%2F23%2Fsocial-media-and-celebrity-culture-harming-young-people

Song, Y., & Lindquist, R. (2018). Critical review of mindfulness-based stress reduction for anxiety and depression among patients with heart disease. *Asia Life Sciences. Supplement*, *15*(1), 1–8.

Sri, Sagala, & Santoso. (2009). *Implementation of HIV and AIDS Prevention Policies in Central Java (Review of Central Java Provincial Regulations Number 5 of 2009)*. Academic Press.

Steele, C. (2019). *What is the Digital Divide?* Digital Divide Council. Available at: http://www.digitaldividecouncil.com/what-is-the-digital-divide/

Stenman, S., & And Pettersson, F. (2020). Remote teaching for equal and inclusive education in rural areas? An analysis of teachers' perspectives on remote teaching. *The International Journal of Information and Learning Technology*, *37*(3), 87–89. doi:10.1108/IJILT-10-2019-0096

Stern, S. R. (2005). Self-Absorbed, Dangerous, and Disengaged: What Popular Films Tell Us About Teenagers. *Mass Communication & Society*, *8*(1), 23–38. doi:10.120715327825mcs0801_3

Stewart, D. W., & Shamdasani, P. (2017). Online focus groups. *Journal of Advertising*, *46*(1), 48–60. doi:10.1080/00913367.2016.1252288

Stone, M. (1974). Cross-Validatory Choice and Assessment of Statistical Predictions. *Journal of the Royal Statistical Society. Series B. Methodological, 36*(2), 111–133. doi:10.1111/j.2517-6161.1974.tb00994.x

Strahan, B. E., & Elder, J. H. (2015). Video Game Playing Effects on Obesity in an Adolescent with Autism Spectrum Disorder: A Case Study. *Autism Research and Treatment, 128365*, 1–7. Advance online publication. doi:10.1155/2015/128365 PMID:26783457

Strasburger, V. C., Jordan, A. B., & Donnerstein, E. (2012). Children, Adolescents, and the Media: Health Effects. *Pediatric Clinics of North America, 59*(3), 533–587. doi:10.1016/j.pcl.2012.03.025 PMID:22643165

Subrahmanyam, K., & Greenfield, P. (1998). Computer games for girls: What makes them play? In J. Cassell & H. Jenkins (Eds.), *From Barbie to Mortal Kombat: Gender and computer games* (pp. 46–71). The MIT Press.

Subrahmanyam, K., Greenfield, P., Kraut, R., & Gross, E. (2001). The Impact of Computer Use on Children's and Adolescents' Development. *Journal of Applied Developmental Psychology, 22*, 7–30. doi:10.1016/S0193-3973(00)00063-0

Subramanian, K. R. (2017). Influence of social media in interpersonal communication. International. *Journal of Scientific Progress and Research, 38*(2), 70–75.

Sun, H., Wang, X., & Wang, X. (2018). Application of blockchain technology in online education. *International Journal of Emerging Technologies in Learning, 13*(10), 252–259. doi:10.3991/ijet.v13i10.9455

Supplement, A. (2019). Vodacom has partnered with government for teacher development. *Mail & Guardian.* Available at: https://mg.co.za/article/2019-10-11-00-vodacom-has-partnered-with-government-for-teacher-development/

Sutter, N., & Paulson, S. (2016). Predicting college students' intention to graduate: A test of the theory of planned behavior. *College Student Journal, 50*(3), 409–421.

Tabassum, M., Mannan, S. E., Parvej, M. I., & Ahmed, F. (2021). Online Education during COVID-19 in Bangladesh: University Teachers' Perspective. *Aquademia, 5*(1).

Tahir. (2017). Kepentingan Penggunaan Media Sosial Teknologi Maklumat dalam pendidikan IPTA. *Journal of Social Sciences and Humanities, 12*(3).

Talaue, G., AlSaad, A., AlRushaidan, N., AlHugail, A., & AlFahhad, S. (2018). The Impact of Social Media on Academic Performance of Selected College Students. *International Journal of Advanced Information Technology, 8*(4/5), 27–35. doi:10.5121/ijait.2018.8503

Tan, A., Nelson, L., Dong, Q., & Tan, G. (1997). Value Acceptance in Adolescent Socialization" A Test of a Cognitive-Functional Theory of Television Effects. *Communication Monographs, 64*(1), 82–97. doi:10.1080/03637759709376406

Tariq Banday, M., & Mattoo, M. M. (2013). Social Media in e-Governance: A Study with Special Reference to India. *Social Networking*, 47–56.

Tausch, A. P., & Menold, N. (2016). Methodological aspects of focus groups in health research: Results of qualitative interviews with focus group moderators. *Global Qualitative Nursing Research, 3*, 2333393616630466. doi:10.1177/2333393616630466 PMID:28462326

Tayie, S. (2005). *Research methods and writing research proposals*. Pathways to Higher Education.

Teras, M., Suoranta, J., Teras, H., & Curcher, M. (2020). Post-Covid-19 Education and Education Technology 'Solutionism': A Seller's Market. *Postdigital Science and Education, 2*(3), 863–878. doi:10.100742438-020-00164-x

Terhoven, R., & Fataar, A. (2018). The role of leadership practices in establishing a curriculum policy platform at working-class schools. *South African Journal of Education, 38*(2), 1–10.

The 4 Features That Make Social Media Apps Addictive. (2021). *Medium*. Retrieved 2 October 2021, from https://bettermarketing.pub/the-4-features-that-make-social-media-apps-addictive-1ce7190f1702

The American College Health Association. (2018). *Role of ACHA*. https://www.acha.org/HealthyCampus/About/Role_of_ACHA/HealthyCampus/Role_of_ACHA.aspx?hkey=c825e8f5-670c-4b99-a12f-22b19c4ec863

The Lancet Child & Adolescent Health. (2018). Growing up in a digital world: Benefits and risks. *The Lancet Child & Adolescent Health, 2*(2), 79. doi:10.1016/S2352-4642(18)30002-6

The Nielsen Company. (2009). *How Teens Use Media*. Author.

The RP Group. (n.d.). *Who We Are*. https://rpgroup.org/About-Us

Thomson, K.-L., & Von Solms, R. (2005). Information security obedience: A definition. *Computers & Security, 24*(1), 69–75. doi:10.1016/j.cose.2004.10.005

Thomson, K.-L., Von Solms, R., & Louw, L. (2006). Cultivating an organisational information security culture. *Computer Fraud & Security, 2006*(10), 7–11. doi:10.1016/S1361-3723(06)70430-4

Timmins, K. A., Green, M. A., Radley, D., Morris, M. A., & Pearce, J. (2018). How has big data contributed to obesity research? A review of the literature. *International Journal of Obesity, 42*(12), 1951–1962. doi:10.103841366-018-0153-7 PMID:30022056

Toba, H. (2015). *Big Data: Menuju Evolusi Era Informasi Selanjutnya*. https://www.researchgate.net/publication/274712686

Torr, J. D. (Ed.). (2001). *Is Media Violence a Problem?* Greenhaven Press.

Tracey Crosbie & Jeanne Moore. (2004). *Work-Life Balance and Working from Home*. Cambridge University Press.

Tripathi, D., & Shahri, M. A. (2019). Digital Communication Controlling Youngsters in Delhi, India, and Salalah, Oman: A Case Study. *International Journal of Communication and Media Science, 6*(3), 7–14. doi:10.14445/2349641X/IJCMS-V6I3P102

Tsai, M. J., & Tsai, C. C. (2003). Information searching strategies in Web-based science learning: The role of Internet self-efficacy. *Innovations in Education and Teaching International, 40*(1), 43–50. doi:10.1080/1355800032000038822

Tsai, S., & Machado, P. (2002). E-learning, Online Learning, Web-based Learning, or Distance Learning. *Unveiling the Ambiguity in Current Terminology., 2001*, 3–5.

Twenge, J. M. (2020). Why increases in adolescent depression may be linked to the technological environment. *Current Opinion in Psychology, 32*, 89–94.

Twenge, J. M., Martin, G. N., & Spitzberg, B. H. (2019). Trends in U.S. Adolescents' media use, 1976–2016: The rise of digital media, the decline of TV, and the (near) demise of print. *Psychology of Popular Media Culture, 8*(4), 329–345. doi:10.1037/ppm0000203

UK Department for Education. (2019). *Realising the potential of technology in education: A strategy for education providers and the technology industry*. https://www.gov.uk/government/publications/realising-the-potential-of-technology-in-education

Umar, T. I., & Idris, M. (2018). Influence of Social Media on Psychosocial Behaviour and Academic Performance of Secondary School Students. *Journal of Education and Entrepreneurship, 5*(2), 36–46. doi:10.26762/jee.2018.40000013

UNAIDS. (2015). *HIV/AIDS & education*. UNAIDS.

UNESCO. (2006). *Good Policy and Practice in HIV/AIDS and Education. Overview Booklet 1*. UNESCO.

UNESCO. (2020). *COVID-19 Educational Disruption and Response*. https://en.unesco. org/news/covid-19-educational-disruption-and-response

UNICEF. (2019). *Investing in a safe, healthy and productive transition from childhood to adulthood is critical.* https://data.unicef.org/topic/adolescents/overview/#:~:text=Adolescents

UNICEF. (2021). *Harnessing the Power of Technology and Digital Innovation Fpor Children. Technology for Development Report, Information Communication and Technology Division.* www.unicef.org

Unterkalmsteiner, M. (2015). *Coordinating Requirements Engineering and Software Testing.* Blekinge Institute of Technology.

Valverde-Berrocoso, J., del Carmen Garrido-Arroyo, M., Burgos-Videla, C., & Morales-Cevallos, M. B. (2020). Trends in educational research about e-Learning: A systematic literature review (2009-2018). *Sustainability (Switzerland), 12*(12), 5153. Advance online publication. doi:10.3390u12125153

van Emmerik, A. A. P., Berings, F., & Lancee, J. (2018). Efficacy of a Mindfulness-Based Mobile Application: A Randomized Waiting-List Controlled Trial. *Mindfulness, 9*, 187–198. https://doi.org/10.1007/s12671-017-0761-7

Van Gorp, B. (2007). The constructionist approach to framing: Bringing culture back in. *Journal of Communication, 57*(1), 60–78. doi:10.1111/j.0021-9916.2007.00329.x

Vanden Abeele, M. M. P., De Wolf, R., & Ling, R. (2018). Mobile media and social space: Howanytime, anyplace connectivity structures everyday life. *Media and Communication, 6*(2), 5–14. doi:10.17645/mac.v6i2.1399

Vandevijvere, S., & Kraak, V. (2019). Future directions to prevent obesity within the context of the Global Syndemic. *Obesity Reviews, 20*(S2), 3–5. doi:10.1111/obr.12887 PMID:31743591

Vass, A.A. (1986). *AIDS: A Plague in Us: A social perspective. The Condition and its Social Consequences.* St.Ives: Venus Academica.

Vidal, C., Lhaksampa, T., Miller, L., & Platt, R. (2020). Social media use and depression in adolescents: A scoping review. *International Review of Psychiatry (Abingdon, England), 32*(3), 235–253. doi:10.1080/09540261.2020.1720623 PMID:32065542

Vidic, Z., & Cherup, N. (2019). Mindfulness in classroom: Effect of a mindfulness-based relaxation class on college students' stress, resilience, self-efficacy and perfectionism. *College Student Journal, 53*(1).

Vrasidas, C. (2015). The rhetoric of reform and teachers' use of ICT. *British Journal of Educational Technology, 46*(2), 370–380. doi:10.1111/bjet.12149

Walsh, P. N., Kerr, M., & Van Schrojenstein Lantman-de Valk, H. M. (2003). Health indicators for people with intellectual disabilities: A European perspective. *European Journal of Public Health, 13*(Supplement 1), 47–50. doi:10.1093/eurpub/13.suppl_1.47 PMID:14533748

Walters, A. (2020). Inequities in access to education: Lessons from the COVID-19 pandemic. *The Brown University Child and Adolescent Behavior Letter, 36*(8), 8–8. doi:10.1002/cbl.30483

Walters, A. (2020). Supporting youth and families during COVID-19. *The Brown University Child and Adolescent Behavior Letter, 36*(6), 8–8. doi:10.1002/cbl.30472

Wang, J., Tigelaar, D. E. H., & Admiraal, W. (2019). Connecting rural schools to quality education: Rural teachers' use of digital educational resources. *Computers in Human Behavior, 101*, 68–76. doi:10.1016/j.chb.2019.07.009

Wang, Y., Min, J., Khuri, J., Xue, H., Xie, B., Kaminsky, L. A., & Cheskin, L. J. (2020). Effectiveness of Mobile Health Interventions on Diabetes and Obesity Treatment and Management: Systematic Review of Systematic Reviews. *JMIR mHealth and uHealth*, *8*(4), e15400. doi:10.2196/15400 PMID:32343253

Wargo, E., Chellman, C., Budge, K., & Davis, C. (2020). On the digital frontier: Stakeholders in rural areas take on educational technology and schooling. *Journal of Research on Technology in Education*, 1–19. doi:10.1080/15391523.2020.1760753

Warren, R., Gerke, P., & Kelly, M. A. (2002). Is there enough Time on the Clock? Parental Involvement and Mediation of Children's Television Viewing. *Journal of Broadcasting & Electronic Media*, *46*(1), 87–111. doi:10.120715506878jobem4601_6

Warr, M. (1993). Parents, Peers and Delinquency. *Social Forces*, *72*(1), 247–265. doi:10.1093f/72.1.247

Warschauer, M., Knobel, M., & Stone, L. A. (2004). Technology and equity in schooling: Deconstructing the digital divide. *Educational Policy*, *18*(4), 562–588. doi:10.1177/0895904804266469

Wästlund, E., Norlander, T., & Archer, T. (2001). Internet blues revisited: Replication and extension of an Internet paradox study. *Cyberpsychology & Behavior*, *4*(3), 385–391. doi:10.1089/109493101300210295 PMID:11710264

Watt, E. (2018). *Digital learning service is helping a million children at school and home in Zimbabwe*. https://theirworld.org/news/digital-learning-service-helps-1m-children-in-zimbabwe

Weay, A. L., Abdullah, S. H., Masood, M., & Muniandy, B. (2015). Design issues and challenges for web-based history learning system for adolescents in Malaysia. *ACM International Conference Proceeding Series*, 50–54. 10.1145/2846439.2846443

Weinstein, E., & Selman, R. (2014). Digital stress: Adolescents' personal accounts. *New Media & Society*, *18*(3), 391–409. doi:10.1177/1461444814543989

Wen, L. M., Rissel, C., & He, G. (2017). The Effect of Early Life Factors and Early Interventions on Childhood Overweight and Obesity 2016. *Journal of Obesity*, *3642818*, 1–3. Advance online publication. doi:10.1155/2017/3642818 PMID:28261496

Wettlaufer, V. (2006). *Computer & your health*. Retrieved September 20, 2006, From www.slais.ubc.ca/courses/libr500/03-04-wt1/assigments/www/V_Wettlaufer/index.htm

Whitehead, P. M., & Wright, R. (2017). Becoming a college student: An empirical phenomenological analysis of first generation college students. *Community College Journal of Research and Practice*, *41*(10), 639–651. doi:10.1080/10668926.2016.1216474

Wiley, A., McCormac, A., & Calic, D. (2020). More than the Individual: Examining the relationship between culture and Information Security Awareness. *Computers & Security*, *88*, 101640. doi:10.1016/j.cose.2019.101640

Willgress, L. (2020). 12 best mindfulness apps to help you keep calm during a crisis. *The Independent*.

Williams, P. (2019). *To what extent is Social Media responsible for a rise in mental health issues in Students in America*. Academic Press.

Williams, D. M., Nawaz, A., & Evans, M. (2020). Drug Therapy in Obesity: A Review of Current and Emerging Treatments. *Diabetes Therapy: Research, Treatment and Education of Diabetes and Related Disorders*, *11*(6), 1199–1216. doi:10.100713300-020-00816-y PMID:32297119

Williams, R. S. (2012). *Adolescence: An ethological perspective*. Springer Science & Business Media.

Willis, J. C., & Templeton, N. R. (2018). Investigating the establishment and sustainability of professional learning communities in rural east Texas: The principals' perspectives. *Rural Educator, 38*(1), 30–37. doi:10.35608/ruraled.v38i1.233

Wilson, J. (2000). Volunteering. *Annual Review of Sociology*, 215-240. doi:10.1146/annurev.soc.26.1.215

Wilson, B. J. (2008). Media and children's aggression, fear, and altruism. *The Future of Children, 18*(1), 87–118. doi:10.1353/foc.0.0005 PMID:21338007

Winerman, L. (2017, November). *By the numbers: Stress on Campus College and University Counseling Centers are Seeing an Uptick in the Number of Students Seeking Help*. https://www.apa.org/monitor/2017/09/numbers

Witt, A., Ordóñez, A., Martin, A., Vitiello, B., & Fegert, J. M. (2020). Child and adolescent mental health service provision and research during the Covid-19 pandemic: Challenges, opportunities, and a call for submissions. *Child and Adolescent Psychiatry and Mental Health, 14*(1), 19. doi:10.1186/s13034-020-00324-8

Woodard, E. H., & Gridina, H. (2000). *Media in the Home The Fifth Annual Survey of Parents and Children*. The Annenberg Public Policy Center of The Univeristy of Pennsylvania.

World Health Organization. (1998). *Obesity: Preventing and Managing the Global Epidemic*. WHO.

Wright, J. W., & Ross, S. D. (1997). Trial by media? Media reliance, knowledge of crime and perception of criminal defendants. *Communication Law and Policy*, 2(4), 397–416.

WS, W. I., Bahar, N., & Mahadevan, R. (2018). Internet Addiction among Secondary School Students in Klang Valley, Malaysia: What is the Association with Depressive Symptoms, Anxiety Symptoms, and SelfEsteem? *IIUM Medical Journal Malaysia, 17*(2).

Xi, Y. (2012). Chinese Review Spam Classification Using Machine Learning Method. *Proceedings - 2012 International Conference on Control Engineering and Communication Technology, ICCECT 2012*, 669–72.

Xiang, B., & Zhou, L. (2014). Improving Twitter Sentiment Analysis with Topic-Based Mixture Modeling and Semi-Supervised Training. *52nd Annual Meeting of the Association for Computational Linguistics, ACL 2014 - Proceedings of the Conference*, 434–39.

Xu, X. (2013). Design and Implementation of Chinese Spam Review Detection System. Lecture Notes in Computer Science (Including Subseries Lecture Notes in Artificial Intelligence and Lecture Notes in Bioinformatics), 508–18.

Xuan, S., Zheng, L., Chung, I., Wang, W., Man, D., Du, X., Yang, W., & Guizani, M. (2020). An incentive mechanism for data sharing based on blockchain with smart contracts. *Computers & Electrical Engineering, 83*, 106587. Advance online publication. doi:10.1016/j.compeleceng.2020.106587

Yaga, Mell, Roby, & Scarfone. (2018). *Blockchain Technology Overview*. National Institute of Standards and Technology Internal Report 8202.

Yagoda, R. (2016, November 9). *College Students and Binge Drinking: When a Rite of Passage Becomes a Path to Destruction*. https://health.usnews.com/health-care/for-better/articles/2016-11-09/college-students-and-binge-drinking-when-a-rite-of-passage-becomes-a-path-to-destruction

Yahaya, A., Gooh, M.L., & Ma'alip, H. (2014). Pengaruh media berunsur agresif dan kesannya terhadap tingkah laku pelajar di sekolah menengah. *National Action Research Conference 2014, Official Conference Proceeding*.

Ye, L., & Yang, H. (2020). From digital divide to social inclusion: A tale of mobile platform empowerment in rural areas. *Sustainability, 12*(6), 2424. doi:10.3390u12062424

Yeora, K. (2003). *The Impact Of The Internet On Children's Daily Lives: Physical, Social And Psychological Well-Being*. University of Georgia.

Yli-Huumo, J., Choi, S., Park, S., & Smolander, K. (2016). Where is current research on Blockchain technology? - A systematic review. PLoS ONE, 11(10). doi:10.1371/journal.pone.0163477

Yoo, H., Feng, X., & Day, R. D. (2013). Adolescents' empathy and prosocial behavior in the family context: A longitudinal study. *Journal of Youth and Adolescence, 42*(12), 1858–1872. doi:10.100710964-012-9900-6

Youniss, J. (1987). *Adolescent Relations with Mothers, Fathers and Friends*. The University of Chicago Press.

Yudhanto. (2007). *Apa itu IOT (Internet of Things)?* Komunitas eLearning IlmuKomputer.Com.

Yusof, N. A., & Saadon, N. (2012). The Effects of Web-based Language Learning on University Students' Grammar Proficiency. *Procedia - Social and Behavioral Sciences, 67*(November), 402–408. doi:10.1016/j.sbspro.2012.11.344

Yusuf, S., Osman, M. S., Hassan, S. J., & Teimoury, M. (2014). Parents' Influence on Children's Online Usage. *Procedia - Social and Behavioral Sciences, 155*, 81-86. doi:10.1016/j.sbspro.2014.10.260

ZARNet. (2021). *eNhava - Reinventing Education*. http://www.zarnet.ac.zw/index.php/enhava/

Zhang, J., Dong, B., & Yu, P. S. (2018). FAKEDETECTOR: Effective Fake News Detection with Deep Diffusive Neural Network. *2020 IEEE 36th International Conference on Data Engineering (ICDE)*, 1826–29. https://ieeexplore.ieee.org/document/9101457/

Zheng, Z., & Ruggiero, K. (2017). Using machine learning to predict obesity in high school students. *IEEE International Conference on Bioinformatics and Biomedicine*, 2132-2138. 10.1109/BIBM.2017.8217988

Zilka, G. C. (2016). Reducing the digital divide among children who received desktop or hybrid computers for the home. *Journal of Information Technology Education, 15*, 233–251. doi:10.28945/3519

Zu, N., Chen, Z., Duan, Z., Li, L., & Meng, Z. (2017). *A review of social media in education: Effects and attitudes*. Academic Press.

Zucco, R., Lavano, F., Anfosso, R., Bianco, A., Pileggi, C., & Pavia, M. (2018). Internet and social media use for antibiotic-related information seeking: Findings from a survey among adult population in Italy. *International Journal of Medical Informatics, 111*(December), 131–139. doi:10.1016/j.ijmedinf.2017.12.005 PMID:29425624

About the Contributors

Rohit Bansal is working as Assistant Professor in Department of Management Studies in Vaish College of Engineering, Rohtak. He obtained Ph.D. in Management from Maharshi Dayanand University, Rohtak. With a rich experience of 13 years, he has achieved growth through robust and proactive academic initiatives. He has authored & edited 8 books as well as published 95 research papers in national and international journals of repute including chapters in edited books. He has also presented papers in 30 conferences and seminars. His area of interest includes marketing management, human resource management, organizational behaviour, services marketing and organizational development. He has proficiency in handling graduate and post graduate students by using interactive discussions and "hands-on" approaches to help them undergo experiential learning He is on Editorial Advisory Board as a member in 110 national and international peer reviewed journals. He is Managing Editor of International Journal of 360° Management Review & International Journal of Techno – Management Research.

Amit Kumar Tyagi (GATE, NPwD-JRF, UGC-NET, and ICAR-NET) received his Ph.D degree in 2018 from Pondicherry Central University, Puducherry, India, in area of "Vehicular Ad-hoc Networks". His research interests include Formal Language Theory, Smart and Secure Computing, Privacy (including Genomic Privacy), Machine Learning with Big data, Blockchain Technology, Cyber Physical System etc. He has completed his M.Tech degree from the Pondicherry Central University, Puducherry, India. He joined the Lord Krishna College of Engineering, Ghaziabad (LKCE) for the periods of 2009-2010, and 2012-2013. With more than 08 (Eight) years of teaching and research experience across India, currently he is working as an Assistant Professor in Vellore Institute of Technology, Chennai Campus, Chennai 600127, Tamilnadu, India. Additionally, He is also the recipient/ awarded of the GATE and NPwD-JRF fellowship in 2009, 2016 and 2013. He has been published one major book titled "Know Your Technical (IT) Skills". Also He is a member of various Computer/ Research Communities like IEEE, ISOC, CSI, ISTE, DataScience, MIRLab, etc.

* * *

Wilfreda Indira Chawarura is a lecturer at Midlands State University in the department of Banking and Finance. She holds a MSc in Finance and Investments from National University of Science and Technology (NUST), Zimbabwe. Her research interests include, developmental economics, personal financial literacy and behavioral finance and their application in various research areas.

Collector Tendeukai Chipfumbu Kangara is a lecturer at the Midlands State University, in the Department of Information Systems. She holds a PhD Information Technology from Nelson Mandela Metropolitan University, South Africa.

Ebru Efeoglu received her B.S. degree in Geophysics Engineering from Kocaeli University and Management Information Systems from Anadolu University, Turkey. She received her Ph.D. degree in Computer Engineering from Trakya University, Turkey in 2021. She is currently an Assistant Professor at Kütahya Dumlupınar University instead of İstanbul Gedik University, Department of Software Engineering instead of Management Information systems. She has authored several papers in international conference proceedings and SCI-Expanded journals. Her research interests include machine learning and data mining, and their applications in various research domains.

Adi Fahrudin, Ph.D., is a Professor in Social Work at the Ministry of Education and Culture Republic of Indonesia seconded at the Faculty of Social and Political Sciences, University of Muhammadiyah Jakarta. He was appointed as Visiting Professor at Japan College of Social Work, Tokyo Japan (2008-2024), Associate Research Fellow at the Psychology and Social Health Unit, Faculty of Psychology and Education, Universiti Malaysia Sabah (2015-2021), Associate Research Fellow at the Asian Research Institute for International Social Work (ARIISW), Shukutoku University, Chiba, Japan. (2016-2022). Previously he was also appointed as Visiting Professor of Social Work at Faculty of Applied Social Sciences, Universiti Sultan Zainal Abidin, Malaysia (2014-2015), and Visiting Associate Professor at Universiti Sains Malaysia (2007-2008). Prof Adi was early trained at Social Work Senior High School in Palembang, Indonesia (1986). He obtained Doctorandus Degree (Drs) in Social Work from the Bandung College of Social Welfare in 1991, Master of Social Science (Social Work) at the University of Science Malaysia in 1996, and then obtained a Ph.D. in Social Work from the same university in 1999, and Bachelor of Psychology (B.Psych) from Pasim University in Bandung (2010). His research focuses on the field of micro social work practice, psychosocial of disaster, aging, indigenous social work, social work education and training, child abuse and neglect, youth and family welfare, and psychosocial aspects of chronic and terminal illness (oncology, mental illness). Prof Adi currently research focus on psychosocial disaster in Indonesia and Malaysia, transnational marriage, indigenous social work, and religious social work especially the Islamic social work model. He published widely on disaster social work, child abuse, and clinical social work evidence-based-practice. Prof Adi has been active in national and international social work communities such as; Asian and Pacific for Social Work Education (APASWE) Board Member (2017-2021), Asian and Pacific Island of Social Work Education Association (APISWEA), and International Consortium on Social Development (ICSD) the Asia Pacific.

Ntombizandile Gcelu currently serves a as a lecturer at the University of Free where she has developed modules to ensure that exceptional quality of education is delivered in-line with best industry practices. Ntombizandile is heavily involved in the professional development of people through research supervision and mentoring of student. She was the Co-chair of the Ad-hoc Committee for Research Strategy for Faculty of Education in UFS. Ntombizandile is also a member of Ethics research committee and a coordinator of the Research Interest Group. She is also currently a Thuthuka Grantholder for 2019-2021 in a project where she collaborates with colleagues in the University of the Free State and Walter Sisulu University. She is World Education Research Association (WERA) International Research

Networks (IRN) convener. Recently, coordinated an international virtual symposium, 2020. She has internally/externally moderated modules and has assessed internal and external thesis and dissertation. Ntombizandile has also published in accredited journals.

Aida Shakila Ishak is a Senior Lecturer at the Faculty of Applied Sciences and Humanities, Universiti Malaysia Perlis. Her area of expertise is counseling psychology and her sub-field expertise is related to drug abuse. She has served in the field of Education since 2019 until now. After being awarded a Doctor of Philosophy Degree from Universiti Malaysia Perlis, she actively continued to contribute in terms of scholarly writing, namely papers in national and international journals.

Fauziah Ismail is a lecturer at the Centre for Liberal Sciences, Faculty of Applied Sciences and Humanities, Universiti Malaysia Perlis. She holds a Bachelor's Degree in Human Development and Management (Universiti Putra Malaysia), Master of Arts Communication (Universiti Sains Malaysia). Her area of expertise is Human Communication and Health Communication. Started working at Universiti Malaysia Perlis on 17 March 2017. Before appointed as a lecturer, she has worked as a Medical Social Worker at University Malaya Medical Center, Kuala Lumpur for 3 years. Apart from that, she has served in the Department of Statistics Malaysia, Penang, and also in an industry sector, Kasatani Advance Technology, Penang.

Fumane Khanare, Ph.D., is a Senior Lecturer at the Faculty of Education, University of the Free State, South Africa. Her research focuses on health-promoting schools, psychosocial well-being of learners in rural schools, learner support, and HIV and AIDS education. She combines asset-based approaches and arts-based research methods as pathways to advance young people's voices and agency for systematically enabling learning environments. She has been a MAC AIDS Fund Leadership Initiative Fellow at Columbia University, the University of California in Los Angeles (UCLA), and the Human Sciences Research Council (HSRC). In 2017 she was honored to be one of the Child and Youth Institute Laureates under the Council for the Development of Social Science Research in Africa (CODESRIA). Dr. Khanare is currently an executive member of the Education Association of South Africa (EASA) and the co-convener of WERA-IRN (World Education Research Association International Research Network (2019- 2022). Her scholarship has expanded to use personal stories to explore opportunities and complexities of historically marginalized girls and women in higher education from an African perspective.

Sunil Lavadiya received the M.Tech. Degree in communication engineering from the Nirma University, Ahmedabad, in 2010 and a pursuing Ph.D. in liquid antenna from the Marwadi University. He currently works as an assistant professor in the department of information and communication engineering, Marwadi University. He has 11 years of teaching experience. His current research interests include the liquid antenna design, graphene-based optical absorber.

Delin Louw has been an educator at a secondary school in the Motheo district for the past seven years. He completed his bachelor's degree at the Central University of Technology. He enrolled for a post graduate diploma at the University of the Free State and completed it with a distinction. He is currently doing his master's in education management and leadership.

Cyncia Matsika holds a Master's in information systems management. She is an experienced database administrator who lectures in the Department of Information Systems. She possesses an arsenal of extensive current Information Technology knowledge and holds 38 International Business Machines Corporation (IBM) certifications and 1 certification from Cisco Networking Academy in various fields that include Artificial Intelligence, Data Science & Analytics, Blockchain, Machine Learning, Internet of Things, Computer Vision, Cloud, and Natural Language Processing.

Nana Nadarsyah currently works at the Regional Research and Development Agency of South Kalimantan Province as a researcher with public health expertise and is a Non-Permanent Lecturer at the Midwifery Academy. Previously, the author was assigned to work in health services both at the Puskesmas and at the Government Hospital. The author completed his Masters in Health Promotion at Diponegoro University, Semarang with a specialization in Reproductive Health and HIV / AIDS in 2012. In 2019 the author has completed his doctoral education in Public Administration at Diponegoro University, Semarang, Indonesia.

Pamela Nyambuya is an MSc Information System Management Graduate Student at Midlands State University.

Amy Sarah Padayachee, Ph.D., is an educator at Stanger Secondary School, KwaZulu-Natal, South Africa. She completed her PhD in Education Management and Leadership in South Africa at the University of the Free State, South Africa. She has presented a paper at the South Africa International Conference on Education (SAICEd) and published in the conference proceedings found in the Rethinking Teaching and learning in the 21st Century Journal. She has participated in the Global education Network (GEN) and published in the conference proceedings found in the Multidisciplinary Research Journal. She has been a delegate of The South African Education Research Association (SAERA) and has presented a paper and contributed to the conference proceedings. Amy has co-authored a paper and published in the African Journal of Gender, Society and Development. She has participated in the World Education Research Association-International Research Networks (WERA-IRN) virtual symposium.

Suruchi Pandey is an associate professor at Symbiosis Institute of Management Studies. PhD in Management.

Melanie Richards has been working in the research, analytics and account planning world for close to 20 years across a wide variety of industries. She has spent the majority of her career in Fortune 500 companies and major nonprofit organizations, leading research, analytics, and planning teams. Richards enjoys helping organizations make smart and efficient decisions that focus on customer needs, preferences, and behaviors, which in turn optimize business performance. She mentors her student teams to approach business questions with the same attitude. Some of her research interests include brand strategy, public health communication and advocacy, generational communication, political communication, digital media/new media technologies, and experiential learning.

Nurul Naimah Rose graduated with a Master of Guidance and Counselling Education and a Bachelor Degree of Human Sciences from the International Islamic University Malaysia. She obtained a Ph.D. in Counselling from Universiti Sains Malaysia. She has served in the education institution since 2008 and currently a Senior Lecturer at the Faculty of Applied Sciences and Humanities, Universiti Malaysia Perlis. Her research interests are mental health counseling, human development, and social sciences. She also actively contributes to academic writing publications in national and international journals.

Trymore Ruvinga is an MSc Information System Management Graduate Student at Midlands State University.

Shaakira Sharif is a 3rd year student in the Educational Psychology and Technology doctoral program at The Chicago School of Professional Psychology. Currently, she is a licensed mental health therapist and provides therapy services for children and adults in the school and community settings.

Disha Sharma is a Finance Faculty, Presently working as an Assistant Professor at Amity Business School, Amity University Chhattisgarh. She has qualified UGC-NET & SET, did Masters in three subjects and completed her PhD. She has five years of teaching experience. She also has experience as Co-Principal Investigator in ICSSR Project and currently working on another Project as Principal Director.

Vishal Sorathiya has received a master's degree in electronics and communication from Marwadi University, Rajkot India. He is currently working on the field of the graphene-based nanoantenna for the IoT and smart sensor applications. He is serving as an Assistant professor at Marwadi University since 2013. He has published several research articles in the various reputed SCI index journal and international conferences. He has also filed several patents on different novel Ideas researched by him. He is recently received the honorable mentioned travel grant to visit the IEEE APS-URSI symposium in Atlanta, USA. He has been also a professional origami artist. His key interest in the field of the reconfigurable antenna for IoT application, image processing, origami antenna, solar cell, and photonics devices.

Aubrey Statti, Ed.D., earned a Bachelor of Arts in Political Science and Spanish from the University of Florida and a Master's of Professional Counseling and a Doctorate of Education from Liberty University. She has worked in higher education for 15 years and has taught at the undergraduate, graduate, and post-graduate levels in areas of psychology, counseling, education, and research. Additionally, she has worked in the Florida public schools as both a high school teacher and a school counselor. Dr. Statti is a faculty member in The Chicago School of Professional Psychology's (TCSPP) Educational Psychology and Technology program, primarily teaching research and educational technology courses. Her research interests in the areas of educational technology include K-12 education, online education, early childhood education, digital storytelling, mentorship, and rural education.

Nor Hafizan Habib Sultan is currently a Senior Lecturer at the Faculty of Applied and Human Sciences, Universiti Malaysia Perlis (UniMAP). She graduated with a bachelor's degree in Social Sciences (Anthropology and Sociology) from Universiti Sains Malaysia (USM), Penang Malaysia in 2009. She pursued her M.Sc. degree in the field of Anthropology and Sociology and graduated in the year 2015 from USM. She completed her Ph.D. in Anthropology and Sociology at the USM in 2020 specializing in Anthropology. Her research interests are community development, gender, and social issues.

Kelly Torres, Ph.D., is the Department Chair of the Educational Psychology and Technology program at TCSPP. Her research interests are focused on international education, teacher certification programs, innovative technologies, and online learning.

Theo Tsokota is a Senior lecturer at the Midlands State University, in the Department of Information Systems. His areas of interest include innovation, ICT4D, ecommerce and IT governance. Theo holds a Bachelor of Science(Honours) in Information Systems from the Midlands State University, a Postgraduate Diploma in Education from Zimbabwe Open University and a Master of Business Administration -Management Information System from Maastricht School of Management and PhD Information Technology from Nelson Mandela Metropolitan University, South Africa.

Ayşe Tuna is a lecturer at Trakya University, Turkey since 2005. She has authored papers in international conference proceedings, and has been actively serving as a reviewer for international conferences and journals. Her research interests are support services for the elderly and disabled, education of autistic and related communication handicapped children, human-robot interaction, human-computer interaction, and data management methodologies.

Chirag Visani has completed his masters from Marwadi Education Foundations group of institutions. He is currently working as an assistant professor at Marwadi University. He has research expertise in Machine learning, Artificial intelligence, and computer algorithms.

Munyaradzi Zhou is a Senior Lecturer in the Department of Information Systems at Midlands State University. Research areas and interests include Data Science and Analytics, Programming, Mobile Health, and Emerging Technologies. He is a holder of a Ph.D. in Information Systems from the University of South Africa. Currently, he is teaching programming and advanced databases.

Tinashe Gwendolyn Zhou is a Lecturer in the department of Information and Marketing Sciences at Midlands State University. She is a holder of an MSc Information Systems Management and currently studying for a PhD in Information Systems. A dynamic and experienced lecturer with over 13 years experience as an academic, Tinashe has authored many papers with 31 citations. She is also a holder of 4 IBM Explorer Badges in Introduction to Coding, Artificial Intelligence, IoT and Cloud Computing. In addition, Mrs Zhou is a holder of 4 IBM Innovator badges in IoT and 4 IBM Innovator badges in Artificial Intelligence. Her research interests include cloud computing, database management, knowledge management, IoT and ICT4D.

Index

T

V

W

Y

www.ingramcontent.com/pod-product-compliance
Lightning Source LLC
LaVergne TN
LVHW062143300926
849161LV00010B/2608